The Logic of Chance:

The Nature and Origin of Biological Evolution

Eugene V. Koonin

Vice President, Publisher: Tim Moore
Associate Publisher and Director of Marketing: Amy Neidlinger
Acquisitions Editor: Kirk Jensen
Editorial Assistant: Pamela Boland
Senior Marketing Manager: Julie Phifer
Assistant Marketing Manager: Megan Graue
Cover Designer: Alan Clements
Managing Editor: Kristy Hart
Project Editor: Betsy Harris
Copy Editor: Krista Hansing Editorial Services, Inc.
Proofreader: Kathy Ruiz
Indexer: Erika Millen
Senior Compositor: Gloria Schurick
Manufacturing Buyer: Dan Uhrig

© 2012 by Pearson Education, Inc.
Publishing as FT Press Science
Upper Saddle River, New Jersey 07458

First Printing September 2011 with corrections November 2012

Pearson Education LTD.
Pearson Education Australia PTY, Limited.
Pearson Education Singapore, Pte. Ltd.
Pearson Education Asia, Ltd.
Pearson Education Canada, Ltd.
Pearson Educación de Mexico, S.A. de C.V.
Pearson Education—Japan
Pearson Education Malaysia, Pte. Ltd.

ISBN-10: 0-13-338106-4
ISBN-13: 978-0-13-338106-1

Library of Congress Cataloging-in-Publication Data is on file.
This product is printed digitally on demand. This book is the paperback version of an original hardcover book.

To my parents

Contents

Preface:
Toward a postmodern synthesis of evolutionary biology

The title of this work alludes to four great books: Paul Auster's novel *The Music of Chance* (Auster, 1991); Jacques Monod's famous treatise on molecular biology, evolution, and philosophy, *Chance and Necessity (Le hazard et la necessite)* (Monod, 1972); the complementary book by Francois Jacob, *The Logic of Life* (Jacob, 1993); and, of course, Charles Darwin's *The Origin of Species* (Darwin, 1859). Each of these books, in its own way, addresses the same overarching subject: the interplay of randomness (chance) and regularity (necessity) in life and its evolution.

Only after this book was completed, at the final stage of editing, did I become aware of the fact that the phrase *Logic of Chance* has already been used in a book title by John Venn, an eminent Cambridge logician and philosopher who in 1866 published *The Logic of Chance: An Essay on the Foundations and Province of the Theory of Probability.* This work is considered to have laid the foundation of the frequency interpretation of probability, which remains the cornerstone of probability theory and statistics to this day (Venn, 1866). He is obviously famous for the invention of the ubiquitous Venn diagrams. I am somewhat embarrassed that I was unaware of John Venn's work when starting this book. On the other hand, I can hardly think of a more worthy predecessor.

My major incentive in writing this book is my belief that, 150 years after Darwin and 40 years after Monod, we now have at hand the data and the concepts to develop a deeper, more complex, and perhaps, more satisfactory understanding of this crucial relationship between chance and necessity. I make the case that variously constrained randomness is at the very heart of the entire history of life.

The inspiration for this book has been manifold. The most straightforward incentive to write about the emerging new vision of evolution is the genomic revolution that started in the last decade of the twentieth century and continues to unfold. The opportunity to compare the complete genome sequences of thousands of organisms from all walks of life has qualitatively changed the landscape of evolutionary biology. Our inferences about extinct, ancestral life forms are not anymore the wild guesses they used to be (at least, for organisms with no fossil record). On the contrary, comparing genomes reveals numerous genes that are conserved in major groups of living beings (in some cases, even in all or most of them) and thus gives us a previously unimaginable wealth of information and confidence about the ancestral forms. For example, it is not much of an exaggeration to claim that we have an excellent idea of the core genetic makeup of the last common ancestor of all bacteria that probably lived more than 3.5 billion years ago. The more ancient ancestors are much murkier, but even for those, some features seem to be decipherable. The genomic revolution did more than simply allow credible reconstruction of the gene sets of ancestral life forms. Much more dramatically, it effectively overturned the central metaphor of evolutionary biology (and, arguably, of all biology), the Tree of Life (TOL), by showing that evolutionary trajectories of individual genes are irreconcilably different. Whether the TOL can or should be salvaged—and, if so, in what form—remains a matter of intense debate that is one of the important themes of this book.

Uprooting the TOL is part of what I consider to be a "meta-revolution," a major change in the entire conceptual framework of biology. At the distinct risk of earning the ire of many for associating with a much-maligned cultural thread, I call this major change the transition to a post-modern view of life. Essentially, this signifies the plurality of pattern and process in evolution; the central role of contingency in the evolution of life forms ("evolution as tinkering"); and, more specifically, the demise of (pan)adaptationism as the paradigm of evolutionary biology. Our unfaltering admiration for Darwin notwithstanding, we must relegate the Victorian worldview (including its refurbished versions that flourished in the twentieth century) to the venerable museum halls where it belongs, and explore the consequences of the paradigm shift.

However, this overhaul of evolutionary biology has a crucial counterpoint. Comparative genomics and evolutionary systems biology (such as organism-wide comparative study of gene expression, protein abundance, and other molecular characteristics of the phenotype) have revealed several universal patterns that are conserved across the entire span of cellular life forms, from bacteria to mammals. The existence of such universal patterns suggests that relatively simple theoretical models akin to those employed in statistical physics might be able to explain important aspects of biological evolution; some models of this kind with considerable explanatory power already exist. The notorious "physics envy" that seems to afflict many biologists (myself included) might be soothed by recent and forthcoming theoretical developments. The complementary relationship between the universal trends and the contingency of the specific results of evolution appears central to biological evolution—and the current revolution in evolutionary biology—and this is another central theme of this book.

Another entry point into the sketch of a new evolutionary synthesis that I am trying to develop here is more specific and, in some ways, more personal. I earned my undergraduate and graduate degrees from Moscow State University (in what was then the USSR), in the field of molecular virology. My PhD project involved an experimental study of the replication of poliovirus and related viruses that have a tiny RNA molecule for their genome. I have never been particularly good with my hands, and the time and place were not the best for experimentation because even simple reagents and equipment were hard to obtain. So right after I completed my PhD project, a colleague, Alex Gorbalenya, and I started to veer into an alternative direction of research that, at the time, looked to many like no science at all. It was "sequence gazing"— that is, attempting to decipher the functions of proteins encoded in the genomes of small viruses (the only complete genomes available at the time) from the sequences of their building blocks, amino acids. Nowadays, anyone can rapidly perform such an analysis by using sleek software tools that are freely available on the Internet; naturally, meaningful interpretation of the results still requires thought and skill (that much does not change). Back in 1985, however, there were practically no computers and no software. Nevertheless, with our computer science colleagues, we managed to develop some rather handy programs (encoded at the time on punch cards). Much of the analysis was done by hand (and eye). Against all odds, and despite some missed opportunities and a few unfortunate errors, our efforts over the next five years were remarkably fruitful. Indeed, we managed to transform the functional maps of those small genomes from mostly unchartered territory to fairly rich "genomescapes" of functional domains. Most of these predictions have been subsequently validated by experiment, and some are still in the works (bench experimentation is much slower than computational analysis). I believe that our success was mostly due to the early realization of the strikingly simple but surprisingly powerful basic principle of evolutionary biology: When a distinct sequence motif is conserved over a long evolutionary span, it must be functionally important, and the higher the degree of conservation, the more important the function. This common-sense principle that is of course rooted in the theory of molecular evolution has served our purposes exceedingly well and, I believe, converted me into an evolutionary biologist for the rest of my days. What I mean is not so much theoretical knowledge, but rather an indelible feeling of the absolute centrality and essentiality of evolution in biology. I am inclined to reword the famous dictum of the great evolutionary geneticist Theodosius Dobzhansky ("Nothing in biology makes sense except in the light of evolution") (Dobzhansky, 1973) in an even more straightforward manner: *Biology is evolution.*

In those early days of evolutionary genomics, Alex and I often talked about the possibility that our beloved small RNA viruses could be direct descendants of some of the earliest life forms. After all, they were tiny and simple genetic systems, with only one type of nucleic acid involved, and their replication was directly linked to expression through the translation of the genomic RNA. Of course, this was late-night talk with no direct relevance to our daytime effort on mapping the functional domains of viral proteins. However, I believe that, 25 years and hundreds of diverse viral and host genomes later, the idea that viruses (or virus-like genetic elements) might have been central to the earliest stages of life's evolution has grown from a fanciful speculation to a concept that is compatible with a wealth of empirical data. In my opinion, this is the most promising line of thought and analysis in the study of the earliest stages of the evolution of life.

So these are the diverse conceptual threads that, to me, unexpectedly converge on the growing realization that our understanding of evolution—and, with it, the very nature of biology—have forever departed from the prevailing views of the twentieth century that today look both rather naïve and somewhat dogmatic. At some point, the temptation to try my hand in tying together these different threads into a semblance of a coherent picture became irresistible, hence this book.

Some of the inspiration came from outside of biology, from the recent astounding and enormously fascinating developments in physical cosmology. These developments not only put cosmology research squarely within the physical sciences, but completely overturn our ideas about the way the world is, particularly, the nature of randomness and necessity. When it comes to the boundaries of biology, as in the origin of life problem, this new worldview cannot be ignored. Increasingly, physicists and cosmologists pose the question "Why is there something in the world rather than nothing?" not as a philosophical problem, but as a physical problem, and explore possible answers in the form of concrete physical models. It is hard not to ask the same about the biological world, yet at more than one level: Why is there life at all rather than just solutions of ions and small molecules? And, closer to home, even assuming that there is life, why are there palms and butterflies, and cats and bats, instead of just bacteria? I believe that these questions can be given a straightforward, scientific slant, and plausible, even if tentative, answers seem to be emerging.

Recent advances in high-energy physics and cosmology inspired this book in more than only the direct scientific sense. Many of the leading theoretical physicists and cosmologists have turned out to be gifted writers of popular and semipopular books (one starts to wonder whether there is some intrinsic link between abstract thinking at the highest level and literary talent) that convey the excitement of their revelations about the universe with admirable clarity, elegance, and panache. The modern wave of such literature that coincides with the revolution in cosmology started with Stephen Hawking's 1988 classic *A Brief History of Time* (Hawking, 1988). Since then, dozens of fine diverse books have appeared. The one that did the most to transform my own view of the world is the wonderful and short *Many Worlds in One,* by Alex Vilenkin (Vilenkin, 2007), but equally excellent treatises by Steven Weinberg (Weinberg, 1994), Alan Guth (Guth, 1998a), Leonard Susskind (Susskind, 2006b), Sean Carroll (Carroll, 2010), and Lee Smolin (in a controversial book on "cosmic natural selection"; Smolin, 1999) were of major importance as well. These books are far more than brilliant popularizations: Each one strives to present a coherent, general vision of both the fundamental nature of the world and the state of the science that explores it. Each of these visions is unique, but in many aspects, they are congruent and complementary. Each is deeply rooted in hard science but also contains elements of extrapolation and speculation, sweeping generalizations, and, certainly, controversy. The more I read these books and pondered the implications of the emerging new worldview, the more strongly was I tempted to try something like that in my own field of evolutionary biology. At one point, while reading Vilenkin's book, it dawned on me that there might be a direct and crucial connection between the new perspective on probability and chance imposed by modern cosmology and the origin of life—or, more precisely, the origin of biological evolution. The overwhelming importance of chance in the emergence of life on Earth suggested by this line of enquiry is definitely unorthodox and is certain to make many uncomfortable, but I strongly felt that it could not be disregarded if I wanted to be serious about the origin of life.

This book certainly is a personal take on the current state of evolutionary biology as viewed from the vantage point of comparative genomics and evolutionary systems biology. As such, it necessarily blends established facts and strongly supported theoretical models with conjecture and speculation. Throughout the book, I try to distinguish between the two as best I can. I intended to write the book in the style of the aforementioned excellent popular books in physics, but the story took a life of its own and refused to be written that way. The result is a far more scientific, specialized text than originally intended, although still a largely nontechnical one, with only a few methods described in an oversimplified manner. An important disclaimer: Although the book addresses diverse aspects of evolution, it remains a collection of chapters on selected subjects and is by no account a comprehensive treatise. Many important and popular subjects, such as the origin of multicellular organisms or evolution of animal development, are completely and purposefully ignored. As best I could, I tried to stick with the leitmotif of the book, the interplay between chance and nonrandom processes. Another thorny issue has to do with citations: An attempt to be, if not comprehensive, then at least reasonably complete, would require thousands of references. I gave up on any such attempt from the start, so the reference list at the end is but a small subset of the relevant citations, and the selection is partly subjective. My sincere apologies to all colleagues whose important work is not cited.

All these caveats and disclaimers notwithstanding, it is my hope that the generalizations and ideas presented here will be of interest to many fellow scientists and students—not only biologists, but also physicists, chemists, geologists, and others interested in the evolution and origin of life.

1

The fundamentals of evolution: Darwin and Modern Synthesis

In this chapter and the next, I set out to provide a brief summary of the state of evolutionary biology before the advent of comparative genomics in 1995. Clearly, the task of distilling a century and a half of evolutionary thought and research into two brief, nearly nontechnical chapters is daunting, to put it mildly. Nevertheless, I believe that we can start by asking ourselves a straightforward question: What is the take-home message from all those decades of scholarship? We can garner a concise and sensible synopsis of the pregenomic evolutionary synthesis even while inevitably omitting most of the specifics.

I have attempted to combine history and logic in these first two chapters, but some degree of arbitrariness is unavoidable. In this chapter, I trace the conceptual development of evolutionary biology from Charles Darwin's *On the Origin of Species* to the consolidation of Modern Synthesis in the 1950s. Chapter 2 deals with the concepts and discoveries that affected the understanding of evolution between the completion of Modern Synthesis and the genomic revolution of the 1990s.

Darwin and the first evolutionary synthesis: Its grandeur, constraints, and difficulties

It is rather strange to contemplate the fact that we have just celebrated the 150th anniversary of the first publication of Darwin's *On the Origin of Species* (Darwin, 1859) and the 200th jubilee of Darwin himself. Considering the profound and indelible effect that *Origin* had on all of science, philosophy, and human thinking in general (far beyond the confines of biology), 150 years feels like a very short time.

What was so dramatic and important about the change in our worldview that Darwin prompted? Darwin did not discover evolution (as sometimes claimed overtly but much more often implied, especially in popular accounts and public debates). Many scholars before him, including luminaries of their day, believed that organisms changed over time in a nonrandom manner. Even apart from the great (somewhat legendary) Greek philosophers Empedokles, Parmenides, and Heraclites, and their Indian contemporaries who discussed eerily prescient ideas (even if, oddly for us, combined with mythology) on the processes of change in nature, Darwin had many predecessors in the eighteenth and early nineteenth centuries. In later editions of *Origin*, Darwin acknowledged their contributions with characteristic candor and generosity. Darwin's own grandfather, Erasmus, and the famous French botanist and zoologist Jean-Bapteste Lamarck (Lamarck, 1809) discussed evolution in lengthy tomes.[1] Lamarck even had a coherent concept of the mechanisms that, in his view, perpetuated these changes. Moreover, Darwin's famed hero, teacher, and friend, the great geologist Sir Charles Lyell, wrote about the "struggle for existence" in which the more fecund will always win. And, of course, it is well known that Darwin's younger contemporary, Alfred Russel Wallace, simultaneously proposed essentially the same concept of evolution and its mechanisms.

However, the achievements of all these early evolutionists notwithstanding, it was Darwin who laid the foundation of modern biology and forever changed the scientific outlook of the world in *Origin*. What made Darwin's work unique and decisive? Looking back at his feat from our 150-year distance, three breakthrough generalizations seem to stand out:

1. Darwin presented his vision of evolution within a completely naturalist and rationalist framework, without invoking any teleological forces or drives for perfection (or an outright creator) that theorists of his day commonly considered.

2. Darwin proposed a specific, straightforward, and readily understandable mechanism of evolution that is interplay

between heritable variation and natural selection, collectively described as the survival of the fittest.

3. Darwin boldly extended the notion of evolution to the entire history of life, which he believed could be adequately represented as a grand tree (the famous single illustration of *Origin*), and even postulated that all existing life forms shared a single common ancestor.

Darwin's general, powerful concept stood in stark contrast to the evolutionary ideas of his predecessors, particularly Lamarck and Lyell, who contemplated mostly, if not exclusively, evolutionary change within species. Darwin's fourth great achievement was not purely scientific, but rather presentational. Largely because of a well-justified feeling of urgency caused by competition with Wallace, Darwin presented his concept in a brief and readable (even for prepared lay readers), although meticulous and carefully argued, volume. Thanks to these breakthroughs, Darwin succeeded in changing the face of science rather than just publishing another book. Immediately after *Origin* was published, most biologists and even the general educated public recognized it as a credible naturalist account of how the observed diversity of life could have come about, and this was a dynamic foundation to build upon.[2]

Considering Darwin's work in a higher plane of abstraction that is central to this book, it is worth emphasizing that Darwin seems to have been the first to establish the crucial interaction between chance and order (necessity) in evolution. Under Darwin's concept, variation is (nearly) completely random, whereas selection introduces order and creates complexity. In this respect, Darwin is diametrically opposed to Lamarck, whose worldview essentially banished chance. We return to this key conflict of worldviews throughout the book.

Indeed, with all due credit given to his geologist and early evolutionary biologist predecessors, Darwin was arguably the first scholar to prominently bring the possibility of evolutionary change (and, by implication, origin) of the entire universe into the realm of natural phenomena that are subject to rational study. Put another way, Darwin initiated the scientific study of the *time arrow*—that is, time-asymmetrical, irreversible processes. By doing so, he prepared the ground not only for all further development of biology, but also for the advent of modern physics. I believe that the great physicist Ludwig Boltzmann, the founder of statistical thermodynamics and the author of the modern concept of entropy, had good reason to call Darwin a "great physicist," paradoxical as this might seem, given that Darwin knew precious little about actual physics and mathematics. Contemporary philosopher Daniel Dennett may have had a point when he suggested that Darwin's idea of natural selection might be the single greatest idea ever proposed (Dennett, 1996).

Certainly, Darwin's concept of evolution at the time *Origin* was published and at least through the rest of the nineteenth century faced severe problems that greatly bothered Darwin and, at times, appeared insurmountable to many scientists. The first substantial difficulty was the low estimate of the age of Earth that prevailed in Darwin's day. Apart from any creation myth, the best estimates by nineteenth-century physicists (in particular, Lord Kelvin) were close to 100 million years, a time span that was deemed insufficient for the evolution of life via the Darwinian route of gradual accumulation of small changes. Clearly, that was a correct judgment—the 100 million years time range is far too short for the modern diversity of life to evolve, although no one in the nineteenth century had a quantitative estimate of the rate of Darwinian evolution. The problem was resolved 20 years after Darwin's death. In the beginning of the twentieth century, when radioactivity was discovered, scientists calculated that cooling of the Earth from its initial hot state would take billions of years, just about the time Darwin thought would be required for the evolution of life by natural selection.

The second, more formidable problem has to do with the mechanisms of heredity and the so-called Jenkin nightmare. Because the concept of discrete hereditary determinants did not exist in Darwin's time (outside the obscure articles of Mendel), it was unclear how an emerging beneficial variation could survive through generations and get fixed in evolving populations without being diluted and perishing. Darwin apparently did not think of this problem at the time he wrote *Origin;* an unusually incisive reader, an engineer named Jenkin, informed Darwin of this challenge to his theory. In retrospect, it is difficult to understand how Darwin (or Jenkin or Huxley) did not think of a Mendelian solution. Instead, Darwin came up with a more extravagant concept of heredity, the so-called pangenesis, which even he himself did not seem to take quite seriously. This problem was resolved by the (re)birth of genetics, although the initial implications for Darwinism[3] were unexpected (see the next section).

The third problem that Darwin fully realized and brilliantly examined was the evolution of complex structures (*organs*, in Darwin's terms) that require assembly of multiple parts to perform their function. Such complex organs posed the classic puzzle of evolutionary biology that, in the twentieth century, has been evocatively branded 'irreducible complexity.'[4] Indeed, it is not immediately clear how selection could enact the evolution of such organs under the assumption that individual parts or partial assemblies are useless. Darwin tackled this problem head-on in one of the most famous passages of *Origin*, the scenario of evolution of the eye. His proposed solution was logically impeccable, plausible, and ingenious: Darwin posited that complex organs do evolve through a series of intermediate stages, each of which performs a partial function related to the ultimate function of the evolving complex organ. Thus, the evolution of the eye, according to Darwin, starts with a simple light-sensing patch and proceeds through primitive eye-like structures of incrementally increasing utility to full-fledged, complex eyes of arthropods and vertebrates. It is worth noting that primitive light-sensing structures resembling those Darwin postulated on general grounds have been subsequently discovered, at least partially validating his scenario and showing that, in this case, the irreducibility of a complex organ is illusory. However, all the brilliance of Darwin's scheme notwithstanding, it should be taken for what it is: a partially supported speculative scenario for the evolution of one particular complex organ. Darwin's account shows one possible trajectory for the evolution of complexity but does not solve this major problem in general. Evolution of complexity at different levels is central to understanding biology, so we revisit it on multiple occasions throughout this book.

The fourth area of difficulty for Darwinism is, perhaps, the deepest. This major problem has to do with the title and purported main subject of Darwin's book, the origin of species and, more generally, large-scale evolutionary events that are now collectively denoted as macroevolution. In a rather striking departure from the title of the book, all indisputable examples of evolution that Darwin presented involve the emergence of new varieties within a species, not new species let alone higher taxa. This difficulty persisted long after

Darwin's death and exists even now, although it was mitigated first by the progress of paleontology, then by developments in the theory of speciation supported by biogeographic data, and then, most convincingly, by comparative genomics (see Chapters 2 and 3). Much to his credit, and unlike detractors of evolution up to this day, Darwin firmly stood his ground in the face of all difficulties, thanks to his unflinching belief that, incomplete as his theory might be, there was no rational alternative. The only sign of Darwin's vulnerability was the inclusion of the implausible pangenesis model in later editions of *Origin*, as a stop-gap measure to stave off the Jenkin nightmare.

Genetics and the "black day" of Darwinism

An urban legend tells that Darwin had read Mendel's paper but found it uninspiring (perhaps partly because of his limited command of German). It is difficult to tell how different the history of biology would have been if Darwin had absorbed Mendel's message, which seems so elementary to us. Yet this was not to be.

Perhaps more surprisingly, Mendel himself, although obviously well familiar with the *Origin,* did not at all put his discovery into a Darwinian context. That vital connection had to await not only the rediscovery of genetics at the brink of the twentieth century, but also the advent of population genetics in the 1920s. The rediscovery of Mendelian inheritance and the birth of genetics should have been a huge boost to Darwinism because, by revealing the discreteness of the determinants of inheritance, these discoveries eliminated the Jenkin nightmare. It is therefore outright paradoxical that the original reaction of most biologists to the discovery of genes was that genetics made Darwin's concept irrelevant, even though no serious scientist would deny the reality of evolution. The main reason genetics was deemed incompatible with Darwinism was that the founders of genetics, particularly Hugo de Vries, the most productive scientist among the three rediscoverers of Mendel laws, viewed mutations of genes as abrupt, saltational hereditary changes that ran counter to Darwinian gradualism. These mutations were considered to be an inalienable feature of Darwinism, in full accord with *Origin.* Accordingly, de Vries viewed his mutational theory of evolution as "anti-Darwinian." So Darwin's centennial jubilee and the 50th anniversary of the *Origin* in 1909 were far from triumphant, even as genetic research surged and Wilhelm Johansson introduced the term *gene* that very year.

Population genetics, Fisher's theorem, fitness landscapes, drift, and draft

The foundations for the critically important synthesis of Darwinism and genetics were set in the late 1920s and early 1930s by the trio of outstanding theoretical geneticists: Ronald Fisher, Sewall Wright, and J. B. S. Haldane. They applied rigorous mathematics and statistics to develop an idealized description of the evolution of biological populations. The great statistician Fisher apparently was the first to see that, far from damning Darwinism, genetics provided a natural, solid foundation for Darwinian evolution. Fisher summarized his conclusions in the seminal 1930 book *The Genetical Theory of Natural Selection* (Fisher, 1930), a tome second perhaps only to Darwin's *Origin* in its importance for evolutionary biology.[5] This was the beginning of a spectacular revival of Darwinism that later became known as *Modern Synthesis* (a term mostly used in the United States) or *neo-Darwinism* (in the British and European traditions).

It is neither necessary nor practically feasible to present here the basics of population genetics.[6] However, several generalizations that are germane to the rest of the discussion of today's evolutionary biology can be presented succinctly. Such a summary, even if superficial, is essential here. Basically, the founders of population genetics realized the plain fact that evolution does not affect isolated organisms or abstract species, but rather affects concrete groups of interbreeding individuals, termed populations. The size and structure of the evolving population largely determines the trajectory and outcome of evolution. In particular, Fisher formulated and proved the fundamental theorem of natural selection (commonly known as Fisher's theorem), which states that the intensity of selection (and, hence, the rate of evolution due to selection) is proportional to the magnitude of the standing genetic variation in an evolving population, which, in turn, is proportional to the effective population size.

Box 1-1 gives the basic definitions and equations that determine the effects of mutation and selection on the elimination or fixation of mutant alleles, depending on the effective population size. The qualitative bottom line is that, given the same mutation rate, in a population with a large effective size, selection is intense. In this case, even mutations with a small positive selection coefficient ("slightly" beneficial mutations) quickly come to fixation. On the other hand, mutations with even a small negative selection coefficient (slightly deleterious mutations) are rapidly eliminated. This effect found its rigorous realization in Fisher's theorem.

Box 1-1: The fundamental relationships defining the roles of selection and drift in the evolution of populations

Nearly neutral evolution dominated by drift

$1/Ne >> |s|$

Evolution dominated by selection

$1/Ne << |s|$

Mixed regime, with both drift and selection important

$1/Ne \approx |s|$

Ne: effective population size (typically, substantially less than the number of individuals in a population because not all individuals produce viable offspring)

s: selection coefficient or fitness effect of mutation:

$s = F_A - F_a$

F_A, F_a: fitness values of two alleles of a gene

$s>0$: beneficial mutation

$s<0$: deleterious mutation

A corollary of Fisher's theorem is that, assuming that natural selection drives all evolution, *the mean fitness of a population cannot decrease during evolution* (if the population is to survive, that is). This is probably best envisaged using the imagery of a *fitness landscape*, which was first introduced by another founding father of population genetics, Sewall Wright. When asked by his mentor to present the results of his mathematical analysis of selection in a form accessible to biologists, Wright came up with this extremely lucky image. The appeal and simplicity of the landscape representation of fitness evolution survive to this day and have stimulated numerous subsequent studies that have yielded much more sophisticated and less intuitive theories and versions of fitness landscapes, including multidimensional ones (Gavrilets, 2004).[7] According to Fisher's theorem, a population that evolves by selection only (technically, a population of an infinite size—infinite populations certainly do not actually exist, but this is convenient abstraction routinely used in population genetics) can never move downhill on the fitness landscape (see Figure 1-1). It is easy to realize that a fitness landscape, like a real one, can have many different shapes. Under certain special circumstances, the landscape might be extremely smooth, with a single peak corresponding to the global fitness maximum (sometimes this is poetically called the Mount Fujiyama landscape; see Figure 1-1A). More realistically, however, the landscape is rugged, with multiple peaks of different heights separated by valleys (see Figure 1-1B). As formally captured in Fisher's theorem (and much in line with Darwin), a population evolving by selection can move only uphill and so can reach only the local peak, even if its height is much less than the height of the global peak (see Figure 1-1B). According to Darwin and Modern Synthesis, movement across valleys is forbidden because it would involve a downhill component. However, the development of population genetics and its implications for the evolutionary process changed this placid picture because of *genetic drift,* a key concept in evolutionary biology that Wright also introduced.

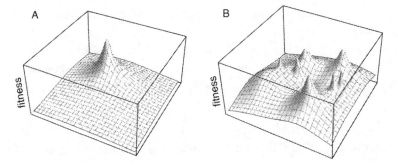

Figure 1-1   Fitness landscapes: the Mount Fujiyama landscape with a single (global) fitness peak and a rugged fitness landscape.

As emphasized earlier, Darwin recognized a crucial role of chance in evolution, but that role was limited to one part of the evolutionary process only: the emergence of changes (mutations, in the modern parlance). The rest of evolution was envisaged as a deterministic domain of necessity, with selection fixing advantageous mutations and the rest of mutations being eliminated without any long-term consequence. However, when population dynamics entered the picture, the situation changed dramatically. The founders of quantitative population genetics encapsulated in simple formulas the dependence of the intensity of selection on population size and mutation rate (see Box 1-1 and Figure 1-2). In a large population with a high mutation rate, selection is effective, and even a slightly advantageous mutation is fixed with near certainty (in an infinite population, a mutation with an infinitesimally small positive selection coefficient is fixed deterministically). Wright realized that a small population, especially one with a low mutation rate, is quite different. Here random genetic drift plays a crucial role in evolution through which neutral or even deleterious (but, of course, nonlethal) mutations are often fixed by sheer chance. Clearly, through drift, an evolving population can violate the principle of upward-only movement in the fitness landscape and might slip down (see Figure 1-2).[8] Most of the time, this results in a downward movement and subsequent extinction, but if the valley separating the local peak from another, perhaps taller one is narrow, then crossing the valley and starting a climb to a new, perhaps taller summit becomes possible (see Figure 1-2). The introduction of the notion of drift into the evolutionary narrative is central to my story. Here chance enters the picture at a new level: Although Darwin and his immediate successors saw the role of chance in the emergence of heritable change (mutations), drift introduces chance into the next phase—namely, the fixation of these changes—and takes away some of the responsibility from selection. I explore just how important the role of drift is in different situations during evolution throughout this book.

John Maynard Smith and, later, John Gillespie developed the theory and computer models to demonstrate the existence of a distinct

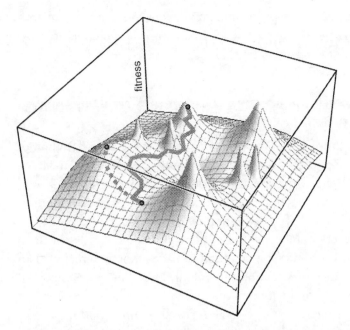

Figure 1-2   Trajectories on a rugged fitness landscape. The dotted line is an evolutionary trajectory at a high effective population size. The solid line is an evolutionary trajectory at a low effective population size.

mode of neutral evolution that is only weakly dependent on the effective population size and that is relevant even in infinite populations with strong selection. This form of neutral fixation of mutations became known as *genetic draft* and refers to situations in which one or more neutral or even moderately deleterious mutations spread in a population and are eventually fixed because of the linkage with a beneficial mutation: The neutral or deleterious alleles spread by *hitchhiking* with the linked advantageous allele (Barton, 2000). Some population-genetic data and models seem to suggest that genetic draft is even more important for the evolution in sexual populations than drift. Clearly, genetic draft is caused by combined effects of natural selection and neutral variation at different genomic sites and, unlike drift, can occur even in effectively infinite populations (Gillespie, 2000).

Genetic draft may allow even large populations to fix slightly deleterious mutations and, hence, provides them with the potential to cross valleys on the fitness landscape.

Positive and purifying (negative) selection: Classifying the forms of selection

Darwin thought of natural selection primarily in terms of fixation of beneficial changes. He realized that evolution weeded out deleterious changes, but he did not interpret this elimination on the same plane with natural selection. In the course of the evolution of Modern Synthesis, the notion of selection was expanded to include "purifying" (negative) selection; in some phases of evolution, this turns out to be more common (orders of magnitude more common, actually) than "Darwinian," positive selection. Essentially, purifying selection is the default process of elimination of the unfit. Nevertheless, defining this process as a special form of selection seems justified and important because it emphasizes the crucial role of elimination in shaping (constraining) biological diversity at all levels. Simply put, variation is permitted only if it does not confer a significant disadvantage on any surviving variant. To what extent these constraints actually limit the space available for evolution is an interesting and still open issue, and I touch on this later (see in particular Chapters 3, 8, and 9).

A subtle but substantial difference exists between purifying selection and *stabilizing selection*, which is a form of selection defined by its effect on frequency distributions of trait values. These forms include stabilizing selection that is based primarily on purifying selection, directional selection driven by positive (Darwinian) selection, and the somewhat more exotic regimes of disruptive and balancing selection that result from combinations of multiple constraints (see Figure 1-3).

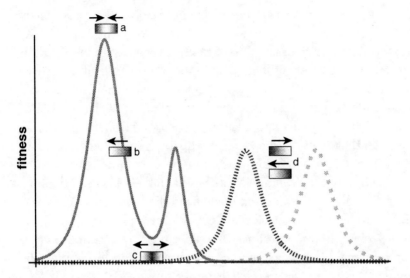

Figure 1-3 Four distinct forms of selection in an evolving population: (A) Stabilizing selection (fitness landscape represented by solid line); (B) Directional selection (fitness landscape represented by solid line); (C) Disruptive selection (fitness landscape represented by solid line); (D) Balancing selection (fitness landscape changes periodically by switching between two dotted lines).

Modern Synthesis

The unification of Darwinian evolution and genetics achieved primarily in the seminal studies of Fisher, Wright, and Haldane prepared the grounds for the Modern Synthesis of evolutionary biology. The phrase itself comes from the eponymous 1942 book by Julian Huxley (Huxley, 2010), but the conceptual framework of Modern Synthesis is considered to have solidified only in 1959, during the centennial celebration of *Origin*. The new synthesis itself was the work of many outstanding scientists. The chief architects of Modern Synthesis were arguably experimental geneticist Theodosius Dobzhansky, zoologist Ernst Mayr, and paleontologist George Gaylord Simpson. Dobzhansky's experimental and field work with the fruit fly *Drosophila melanogaster* provided the vital material support to the theory of population genetics and was the first large-scale experimental validation of the concept of natural selection. Dobzhansky's book *Genetics and the Origin of Species* (Dobzhansky, 1951) is the principal manifesto of Modern Synthesis, in which he narrowly defined evolution as "change in the frequency of an allele within a gene pool." Dobzhansky also famously declared that *nothing in biology makes sense except in the light of evolution*[9] (see more about "making sense" in Appendix A). Ernst Mayr, more than any other scientist, is to be credited with an earnest and extremely influential attempt at a theoretical framework for Darwin's quest, the origin of species. Mayr formulated the so-called biological concept of species, according to which speciation occurs when two (sexual) populations are isolated from each other for a sufficiently long time to ensure irreversible genetic incompatibility (Mayr, 1963). Simpson reconstructed the most comprehensive (in his time) picture of the evolution of life based on the fossil record (Simpson, 1983). Strikingly, Simpson recognized the prevalence of stasis in the evolution of most species and the abrupt replacement of dominant species. He also introduced the concept of quantum evolution, which presaged the punctuated equilibrium concept of Stephen Jay Gould and Niles Eldredge (see Chapter 2).

The consolidation of Modern Synthesis in the 1950s was a somewhat strange process that included remarkable "hardening" (Gould's word) of the principal ideas of Darwin (Gould, 2002). Thus, the doctrine of Modern Synthesis effectively left out Wright's concept of random genetic drift and its evolutionary importance, and remains uncompromisingly pan-adaptationist. Rather similarly, Simpson himself gave up the idea of quantum evolution, so gradualism remained one of the undisputed pillars of Modern Synthesis. This "hardening" shaped Modern Synthesis as a relatively narrow, in some ways dogmatic conceptual framework.

To proceed with the further discussion of the evolution of evolutionary biology and its transformation in the age of genomics, it seems necessary to succinctly recapitulate the fundamental principles of evolution that Darwin first formulated, the first generation of evolutionary biologists then amended, and Modern Synthesis finally codified. We return to each of these crucial points throughout the book.

1. Undirected, random variation is the main process that provides the material for evolution. Darwin was the first to allow chance as a major factor into the history of life, and this was arguably one of his greatest insights. Darwin also allowed a subsidiary role for directed, Lamarckian-type variation, and he tended to give these mechanisms more weight in later editions of *Origin*. Modern

Synthesis, however, is adamant in its insistence on random mutations being the only source of evolutionarily relevant variability.

2. Evolution proceeds by fixation of rare beneficial variations and elimination of deleterious variations: This is the process of natural selection that, along with random variation, is the principal driving force of evolution, according to Darwin and Modern Synthesis. Natural selection, which is obviously akin to and inspired by the "invisible hand" of the market that ruled economy according to Adam Smith, was the first mechanism of evolution ever proposed that was simple and plausible and that did not require any mysterious innate trends. As such, this was Darwin's second key insight. Sewall Wright emphasized that chance could play a substantial role in the fixation of changes during evolution rather than only in their emergence, via genetic drift that entails random fixation of neutral or even moderately deleterious changes. Population-genetic theory indicates that drift is particularly important in small populations that go through bottlenecks. Genetic draft (hitchhiking) is another form of stochastic fixation of nonbeneficial mutations. However, *Modern Synthesis in its "hardened" form effectively rejected the role of stochastic processes in evolution beyond the origin of variation and adhered to a purely adaptationist (pan-adaptationist) view of evolution.* This model inevitably leads to the concept of "progress," gradual improvement of "organs" during evolution. Darwin endorsed this idea as a general trend, despite his clear understanding that organisms are less than perfectly adapted, as strikingly exemplified by rudimentary organs, and despite his abhorrence of any semblance of an innate strive for perfection of the Lamarckian ilk. Modern Synthesis shuns progress as an anthropomorphic concept but nevertheless maintains that evolution, in general, proceeds from simple to complex forms.

3. The beneficial changes that are fixed by natural selection are infinitesimally small (in modern parlance, the evolutionarily relevant mutations are supposed to have infinitesimally small fitness effects), so evolution occurs via the gradual accumulation of these tiny modifications. Darwin insisted on *strict gradualism* as an essential staple of his theory: "Natural selection

can act only by the preservation and accumulation of infinitesimally small inherited modifications, each profitable to the preserved being. ...If it could be demonstrated that any complex organ existed, which could not possibly have been formed by numerous, successive, slight modifications, my theory would absolutely break down." (*Origin of Species,* Chapter 6). Even some contemporaries of Darwin believed that was an unnecessary stricture on the theory. In particular, the early objections of Thomas Huxley are well known: Even before the publication of *Origin,* Huxley wrote to Darwin, "You have loaded yourself with an unnecessary difficulty in adopting *Natura non facit saltum* so unreservedly" (http://aleph0.clarku.edu/huxley/). Disregarding these early warnings and even Simpson's concept of quantum evolution, Modern Synthesis uncompromisingly embraced gradualism.

4. An aspect of the classic evolutionary biology that is related to but also distinct from the principled gradualism is *uniformitarianism* (absorbed by Darwin from Lyell's geology). This is the belief that the evolutionary processes have remained essentially the same throughout the history of life.

5. This key principle is logically linked to gradualism and uniformitarianism: *Macroevolution* (the origin of species and higher taxa), is governed by the same mechanisms as *microevolution* (evolution within species). Dobzhansky, with his definition of evolution as the change of allele frequencies in populations, was the chief proponent of this principle. Darwin did not use the terms *microevolution* and *macroevolution;* nevertheless, the sufficiency of intraspecies processes to explain the origin of species and, more broadly, the entire evolution of life can be considered the central Darwinian axiom (or perhaps a fundamental theorem, but one for which Darwin did not have even an inkling of the proof). It seems reasonable to speak of this principle as "generalized uniformitarianism": *The processes of evolution are the same not only throughout the history of life, but also at different levels of evolutionary transformation, including major transitions.* The conundrum of microevolution versus macroevolution is, in some ways, the fulcrum of evolutionary biology, so we revisit it repeatedly throughout this book.

6. Evolution of life can be accurately represented by a "great tree," as emphasized by the only illustration in *Origin* (in Chapter 4). Darwin introduced the Tree of Life (TOL) only as a general concept and did not attempt to investigate its actual branching order. The tree was populated with actual life forms, to the best of the knowledge at the time, by the chief German follower of Darwin, Ernst Haeckel. The founders of Modern Synthesis were not particularly interested in the TOL, but they certainly embraced it as a depiction of the evolution of animals and plants that the fossil record amply supported in the twentieth century. By contrast, microbes that were increasingly recognized as major ecological agents remained effectively outside the scope of evolutionary biology.

7. A corollary of the single TOL concept deserves the status of a separate principle: All extant diversity of life forms evolved from a single common ancestor (or very few ancestral forms, under Darwin's cautious formula in Chapter 14 of *Origin*; see Darwin, 1859). Many years later, this has been dubbed the Last Universal Common (Cellular) Ancestor (LUCA). For the architects of Modern Synthesis, the existence of LUCA was hardly in doubt, but they did not seem to consider elucidation of its nature a realistic or important scientific goal.

Synopsis

In his book *On the Origin of Species,* Charles Darwin meticulously collected evidence of temporal change that permeates the world of living beings and proposed for the first time a plausible mechanism of evolution: natural selection. Evolution by natural selection certainly is one of the most consequential concepts ever developed by a scientist and even has been deemed the single most important idea in human history (Dennett, 1996). Somewhat paradoxically, it is also often branded a mere tautology, and when one thinks in terms of the survival of the fittest, there seems to be some basis for this view. However, considering the Darwinian scenario as a whole, it is easy to grasp its decidedly nontautological and nontrivial aspect. Indeed, what Darwin proposed is a mechanism for the transformation of random variation into adaptations that are not random at all, including elaborate, complex devices that perform highly specific functions and so increase the fitness of their carriers. Coached in physical terms and loosely following Erwin Schroedinger's famous treatise, Darwinian evolution is a machine for the creation of *negentropy*—in other words, order from disorder. I submit that this was the single key insight of Darwin, the realization that a simple mechanism, devoid of any teleological component, could plausibly account for the emergence, from random variation alone, of the amazing variety of life forms that appear to be so exquisitely adapted to their specific environments. Viewed from that perspective, the "invisible hand" of natural selection appears almost miraculously powerful, and one cannot help wondering whether it is actually sufficient to account for the history of life. This question has been repeatedly used as a rhetoric device by all kinds of creationists, but it also has been asked in earnest by evolutionary biologists. We shall see in the rest of this book that the answers widely differ, both between scientists and between different situations and stages in the evolution of life.

Of course, Darwinism in its original formulation faced problems more formidable and more immediate than the question of the sufficiency of natural selection: Darwin and his early followers had no sensible idea of the mechanisms of heredity and whether these mechanisms, once discovered, would be compatible with the Darwinian scenario. In that sense, the entire building of Darwin's concept was suspended in thin air. The rediscovery of genetics at the beginning of the twentieth century, followed by the development of theoretical and experimental population genetics, provided a solid foundation for Darwinian evolution. It was shown beyond reasonable doubt that populations evolved through a process in which Darwinian natural selection was a major component. The Modern Synthesis of evolutionary biology completed the work of Darwin by almost seamlessly unifying Darwinism with genetics. As it matured, Modern Synthesis notably "hardened" through indoctrinating gradualism, uniformitarianism, and, most important, the monopoly of natural selection as the only route of evolution. In Modern Synthesis, all changes that are fixed during evolution are considered adaptive, at least initially. For all its fundamental merits, Modern Synthesis is a rather dogmatic and woefully incomplete theory. To name three of the most glaring problems, Modern Synthesis makes a huge leap of faith by extending the mechanisms and patterns established for microevolution to macroevolutionary processes; it has nothing to say about evolution of microbes, which are the most abundant and diverse life forms on Earth; and it does not even attempt to address the origin of life.

Recommended further reading

Futuyma, Douglas. (2009) *Evolution*, 2d edition. Sunderland, MA: Sinauer Associates.

Probably the best available undergraduate text on evolutionary biology.

Gould, Stephen Jay. (2002) *The Structure of Evolutionary Theory*. Cambridge, MA: Harvard University Press.

The almost 1,500-page tome obviously is not for the feeble at heart, and not many will read it in its entirety. Nevertheless, at least the first part is valuable for its clear and witty presentation of the history of evolutionary biology and its pointed critique of Modern Synthesis.

Hartl, Daniel L., and Andrew G. Clark. (2006) *Principles of Population Genetics*, 4th edition. Sunderland, MA: Sinauer Associates.

An excellent, fairly advanced, but accessible textbook on population genetics.

Mayr, Ernst. (2002) *What Evolution Is*. New York: Basic Books.

A basic but clear and useful presentation of classical evolutionary biology by one of the architects of Modern Synthesis.

Schroedinger, Erwin. (1992) *What Is Life?: With "Mind and Matter" and "Autobiographical Sketches."* Cambridge, MA: Cambridge University Press.

The first edition of this wonderful book was published in 1944, on the basis of a series of lectures that Schroedinger (one of the founders of quantum mechanics) delivered in Edinburgh, where he stayed during World War II. Obviously outdated, but remarkably lucid, prescient, and still relevant in the discussion of the role of entropy and information in biology.

2

From Modern Synthesis to evolutionary genomics: Multiple processes and patterns of evolution

In this chapter, we continue our discussion of evolutionary biology in the pregenomic era. Many of these developments did not temporally succeed Modern Synthesis. Instead, they occurred in parallel with the evolution of Modern Synthesis but were shunned from the "canon" during the "hardening" of Modern Synthesis. The advances discussed here roughly span the interval of 1930 (publication of Ronald Fisher's book that started the second, mature phase in the history of evolutionary biology) to 1995 (the first comparisons of complete genomes of cellular life forms). My goal here is to briefly present the remarkably complex network of evolutionary ideas, theories, and observations that complemented the fundamentally important but rather rigid framework of Modern Synthesis and became the launching pad for the new, genome-centric study of evolution.

Replication of digital information carriers: The central principle of biology and the necessary and sufficient condition of evolution

The model of the DNA structure built by James Watson and Francis Crick (obviously, based on X-ray structures solved by Rosalind Franklin and others) certainly is one of the central discoveries in twentieth-century biology and the entire history of biology (Watson and Crick, 1953b). However, this breakthrough is not normally mentioned in the same breath with the principles of biological evolution.

Here I posit that the DNA structure and the model of replication that Watson and Crick inferred from it in the second of their classic 1953 papers (Watson and Crick, 1953a) are the most important, foundational discoveries in the study of evolution since the publication of *Origin*. In essence, Watson and Crick discovered the biological incarnation of the extremely general principle of digital storage, coding, and propagation of information that was implicit in the DNA structure. The biological information transmission system these studies revealed can be considered an extension of the Turing machine principle, first through the rules of nucleic acid base complementarity and, subsequently, through the genetic code (see Figure 2-1). In essence, even if not in actual history, these discoveries seem to supersede Darwin, in the sense that *the entire Darwinian scheme of evolution is a straightforward and necessary corollary of the replication mechanism*. In all life forms that we are aware of, biological transfer of digital information entails the following simple but fundamental principles:[1]

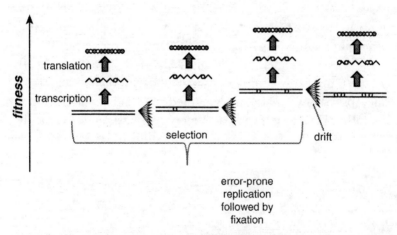

Figure 2-1   Transmission of information in biological systems and the emergence of selection and genetic drift as epiphenomena of replication. The white circles in the figure show changes in the original sequence.

1. The genetic material of any organism comprises a linear sequence of digits, the four nucleic acid bases that, directly or indirectly, encode all information required for the build-up of the organism.

2. Replication of the genetic material, which is the mechanistic basis of heredity, occurs according to one-to-one correspondence rules of complementarity between A and T(U) and G and C, which ensures replication of the genetic material. (Originally these were known as Chargaff rules, after their discoverer, Austrian and then American chemist Erwin Chargaff.)[2]

Watson and Crick stated these key features of the genetic system in their two 1953 papers. Subsequent developments added two important aspects:

1. The complementarity principle is employed not only during replication, but also during transcription of the DNA into all kinds of RNA and during translation of mRNAs into protein, via the adaptor tRNA molecules.

2. The same rules of digital replication and decoding apply to genetic systems in which the genetic material is distinct from the double-stranded (ds) DNA originally modeled by Watson and Crick (such as many viruses), and instead consists of RNA or single-stranded (ss) DNA.

Information theory is adamant in its insistence on the impossibility of error-free information transmission. The actual probability of error for any finite message can be made arbitrarily low, but any decrease in error rate of information transmission has an energy cost. This link has its roots in the second law of thermodynamics. The central principle of evolution can be formulated as follows:

> *Replication of digital information carriers is necessarily error prone and entails evolution of these replicators by natural selection and random drift, provided that the error rate of replication is below an error catastrophe threshold, a value on the order of 1 to 10 errors per genome per replication cycle.*

I refer to this generalization as the *Error-Prone Replication (EPR) principle*.[3] The EPR principle is fairly self-evident (once the existence and essentiality of replication is realized); it was first encapsulated in a straightforward mathematical theory by Manfred Eigen (Eigen, 1971), who introduced the error threshold concept (Biebricher and Eigen, 2005)—this theory and its implications are further discussed in Chapter 12. The EPR principle rests on two underlying assumptions that may be obvious but merit emphasizing:

1. Replication errors are inherited (passed through replication cycles).

2. Genotype-phenotype feedback exists: Some replication errors affect the replication efficiency (and possibly fidelity as well), either negatively or positively.

These features differentiate biological replicators with their "unlimited heredity" from replicators with "limited heredity," such as crystals or certain chemical cycles that replicate but do not pass emerging defects to the progeny (Szathmary, 2000). Less formally, the distinction is that, in nucleic acids, a substitution of one nucleotide for another affects only information, not (at least not significantly) the physical or chemical properties of the information carrier, as is the case in nonbiological systems.

In principle, a low error rate threshold for evolution should exist as well, below which evolution would be hampered by the insufficient amount of variation. Intuitively, if the expectation of error per replication cycle is close to zero, variation becomes insufficient for evolution to occur. However, it is important to note that the empirically determined error rate of nucleic acid replication in the absence of elaborate correction mechanisms (for instance, in RNA viruses) is not far below the error catastrophe threshold. Thus, the sufficiently low error rate is the critical condition of evolution. The extent to which the actual fidelity of information transmission in biological system is minimized and the extent to which it is optimized (in other words, is evolvability evolvable?) is an intriguing and much-debated issue that we discuss in Chapter 9.

Although all known naturally evolved life is based on nucleic acid replication, the EPR principle is substrate independent, as illustrated by the evolution of computer viruses and various computer models of "artificial life" evolution (Lenski, et al., 2003). However, whether digital encoding of information is necessary for evolution or whether evolution can occur in an analog system is an interesting and still open theoretical question.

Chapter 1 touched upon the quasitautological character of natural selection. In a sense, the EPR principle indeed trivializes both selection and drift by stripping these supposedly fundamental factors of evolution of their status as independent phenomena and instead presenting them as epiphenomena of error-prone replication. This does not at all belittle the historical achievements of Darwin, Wright, and other outstanding evolutionary biologists or diminish the utility of the concepts of selection and drift for high-level descriptions of evolutionary processes. Nevertheless, the discovery of replication with controlled error rate reveals more fundamental principles that underlie the classical tenets of evolutionary biology.

Molecular evolution and molecular phylogenetics

The traditional phylogeny that fleshed out Darwin's concept of the Tree of Life was based on comparisons of diagnostic features of organisms' morphology, such as the skeleton structure in animals or the flower architecture in plants (Futuyma, 2005). Evolutionary biologists did not realize that the actual molecular substrate of evolution that undergoes the changes acted upon by natural selection (the genes) could be compared for the purpose of phylogeny reconstructions, for the obvious reason that they knew almost nothing of the chemical nature of that substrate and the way it encoded the phenotype of an organism. Moreover, the pan-adaptationist paradigm of evolutionary biology seemed to imply that, regardless of their molecular nature, genes would not be significantly conserved between distant organisms, given the major phenotypic (functional) differences between them, as emphasized in particular by Ernst Mayr, one of the chief architects of Modern Synthesis.

The idea that DNA base sequence could be employed for evolutionary reconstruction seems to have been first expressed in print by Crick, even if only in passing (in the same seminal article where he formulated the adaptor hypothesis of protein synthesis—Crick, 1958). Emile Zuckerkandl and Linus Pauling developed the principles and the first actual implementation of molecular evolutionary analysis a few years later. They directly falsified Mayr's conjecture by showing that the amino acid sequences of several proteins that were available from multiple species at the time, such as cytochrome c and globins, were highly conserved even between distantly related animals (Zuckerkandl and Pauling, 1965). Zuckerkandl and Pauling also proposed the concept of the molecular clock: They predicted that the evolutionary rate of the sequence of a given protein should be constant (allowing for some fluctuations) over long time intervals, in the absence of functional change. It seems useful to note at this juncture that demonstrating that the sequences of the genes encoding "the same protein" (that is, proteins with the same activity and similar properties) in different species were highly similar—and, moreover that the degree of sequence similarity was negatively correlated with the phylogenetic distance between the respective species—may be considered the best and definitive proof of the reality of evolution.

Over the next few years, primarily through the efforts of Margaret Dayhoff and coworkers, protein sequence conservation was demonstrated to extend to the most diverse life forms, from bacteria to mammals (Dayhoff, et al., 1983). Given the discovered long-term conservation of protein sequences and the molecular clock hypothesis, it appeared natural to construct phylogenetic trees on the basis of (dis)similarity between sequences that, under the molecular clock, would reflect the relative time of divergence of the respective genes (proteins) from their common ancestors. Indeed, several distance-based methods of molecular phylogenetics have been promptly devised along with the more sophisticated maximum parsimony approach (see Box 2-1). Subsequent tests of the molecular clock hypothesis on growing sequence collections showed that, for most genes, the clock does not tick at a constant rate; instead, the clock is significantly overdispersed—that is, the variance of evolutionary rates substantially exceeds the random fluctuation predicted for a Poisson process (Bromham and Penny, 2003). The overdispersion of the molecular clock leads to a common artifact of molecular phylogeny known as long branch attraction (LBA), which seriously confounds molecular phylogenetic analysis (see Box 2-1). Molecular phylogenetics has evolved into a complex branch of applied mathematics and statistics chiefly to cope with LBA and other artifacts (Felsenstein, 2004). All the artifacts notwithstanding, molecular phylogenetics remains one of the cornerstones of modern evolutionary biology, with the contemporary methods of choice primarily centered on maximum likelihood approaches (see Box 2-1).

Box 2-1: A super-brief summary of phylogenetic methods

Sequence-based methods

All require a multiple alignment of homologous nucleotide or protein sequences.

Distance methods

These methods employ matrices of interspecies distances $<d_{ij}>$ (i,j are species) calculated from alignments, with corrections for multiple substitutions.

Ultrametric (simple hierarchical clustering) methods. Valid only under a strict molecular clock. Generally, not considered valid phylogenetic methods but can be used for purposes of classification or to generate preliminary guide trees for other methods.

Neighbor-joining (NJ). A more sophisticated bottom-up clustering method based on the minimum evolution criterion (the shortest total length of the tree branches). Sensitive to LBA and much less accurate than maximum likelihood methods, but highly computationally efficient and fast. Not used to generate definitive phylogenies, but could be the only phylogenetic method practically suitable for analyzing very large numbers of sequences.

Least squares (Fitch). A distance method based on minimization of the differences between the distances in a phylogenetic tree and the underlying distance matrix. Generally similar to NJ in terms of accuracy and efficiency.

Not considered suitable for producing definitive phylogenies, but used to generate guide trees for maximum likelihood optimization.

Maximum parsimony (MP)

Does not employ distance matrices, but instead works with character states. The character states, in particular, can be nucleotides or amino acids in individual sites of a multiple alignment. The parsimony principle, generally based on the minimum action principle in physics, postulates that the evolutionary scenario (phylogenetic tree) that includes the minimum number of events (character stated transitions) is the one that is most likely to be correct. Numerous algorithms calculate the most parsimonious trees using weighted or unweighted characters. The parsimony principle is questionable because there are numerous trees that are only slightly less parsimonious than the best one but have different topologies. Highly sensitive to LBA.

Maximum likelihood

Similar to MP, maximum likelihood (ML) methods score transitions between character states and select the tree with the highest score. Unlike MP, ML is a parametric statistical approach that employs an explicit model of character evolution to estimate the probability of the data, given a tree. The tree that has the highest probability of producing the observed data is the most likely tree. ML often yields trees similar to MP trees but is theoretically preferable because (unlike MP) it is a statistically consistent method (that is, one that is guaranteed to produce the correct tree with the maximum likelihood, given sufficient data). In practice, ML often outperforms MP. The ML methods are extremely computationally expensive and are impractical with large datasets. Therefore, ML is often used to optimize guide trees generated by fast methods such as NJ or Fitch. For phylogenetic studies in which tree accuracy is considered more important than speed, ML is the current approach of choice. Moreover, recent algorithmic developments have accelerated the construction of ML phylogenetic trees by orders of magnitude without seriously compromising accuracy (Price et al., 2010).

Bayesian inference

As with ML, this approach uses a likelihood function, but employs the Bayes theorem to connect the posterior probability of a tree with the likelihood of the data and the prior probability of a tree with the evolutionary model. Unlike MP or ML, which output the best tree or a set of trees, the Bayesian inference methods sample trees in proportion to their likelihood and yield a representative set of trees. Performs well with relatively small datasets, but impractical for large datasets.

Testing the performance of phylogenetic methods and reliability of trees

Simulated trees

Phylogenetic methods are routinely benchmarked against computationally simulated data for which the precise evolutionary history is known. Methods are compared with respect to the accuracy of the reconstruction of the true history in the resulting trees. Typically, various ML methods and Bayesian inference methods (for small data sets) outperform others. The best performers are iterative approaches that employ the initial ML tree to refine the underlying alignment and then rebuild the tree and iterate until convergence.

Bootstrap analysis

The most common test of phylogenetic tree topology reliability that samples the data (alignment columns) and estimates the tree for a large number of samples. The bootstrap support of an internal node in a tree is the percentage of samples (replications) in which the given node is recovered. The statistics of bootstrap analysis are not fully worked out, so the thresholds for "high" bootstrap support are derived by simulation or empirical analysis and can vary depending on the goals of a given study (such as greater than 90% or 70%; bootstrap values less than 50% are not normally considered reliable).

Statistical tests of phylogenetic hypotheses (tree topology)

Statistical tests based on various likelihood models have been developed to compare the likelihoods of different tree topologies for the given dataset (the best-known ones are the Kishino-Hasegawa test and the Approximately Unbiased test).

When a researcher is interested in the phylogenetic affinity of a particular taxon, the respective branch can be moved to different positions in the tree without disturbing other branches, and the statistical tests can be applied to compare the likelihood of each of the resulting trees with the likelihood of the original ML tree. Another version of the tests is used on the *constrained trees* that are employed to test phylogenetic hypotheses, such as the monophyly of a particular group (for example, archaea) in the given dataset. In this case, the likelihood of the constrained tree in which the monophyly is enforced is compared with the likelihood of the original ML tree.

The common artifacts of phylogenetic analysis

No phylogenetic method is immune from artifacts that often severely affect the tree topology. The two main classes of phylogenetic artifacts are *homoplasy* and *Long-Branch Attraction* (LBA). Homoplasy includes parallel, convergent, and reverse mutations that are phylogenetically uninformative and misleading, and are misinterpreted by phylogenetic methods. LBA refers to the extremely common case when long branches (fast-evolving lineages) in a tree cluster together only because none of them has affinity to any other groups, not because they actually form a clade. Conversely, phylogeneticicts sometimes refer to short branch attraction, artificial grouping of short branches in trees. The development of new phylogenetic methods is driven to a large extent by attempts to overcome these artifacts without making the methods computationally impractical.

Shared derived characters

An important phylogenetic approach that is complementary to traditional molecular phylogenetic is the analysis of shared derived characters (also known as synapomorphies) that can be employed to delineate monophyletic groups (clades). Synapomorphies are characters that unite all members of a monophyletic group, to the exclusion of all other species. In principle, a single valid synapomorphy can define a clade. However, this is the case only in the absence of homoplasy, which is impossible to guarantee for most characters. Potential synapomorphies are chosen to minimize the chance of homoplasy: Examples include unique inserts in conserved genes, particularly insertions of mobile elements, mutations that require multiple nucleotide substitutions, and gene fusions. In phylogenomics, there is an active search for rare genomic changes (RGC). Synapomorphies are rarely sufficient to derive definitive phylogenetic conclusions, but they provide important supporting evidence to sequence-based phylogenies.

Non-sequence-based genome trees

Phylogenetic methods can be applied not only to alignments of homologous sequences, but also to distance matrices obtained by genome-wide comparison of any number of other characters (such as shared gene content or operon organization). For example, in the case of gene content, the distance between two genomes can be calculated as $D_{ij}=n_{ij}/n_i$, where nij is the number of shared genes between the two genomes and n_i is the total number of genes in the smaller genome. The genome trees obtained by these approaches typically are not reliable phylogenies, due to the widespread homoplasy. Accordingly, these trees might be informative for comparisons of the organism lifestyles.

The neutral theory

Probably the most important conceptual breakthrough in evolutionary biology after Modern Synthesis was the neutral theory of molecular evolution. This is usually associated with Motoo Kimura (Kimura, 1983), although Jukes and King simultaneously and independently developed a similar theory. Originally, the neutral theory was derived as an extension of Wright's population-genetic ideas on the importance of genetic drift in evolution. According to the neutral theory, a substantial majority of the mutations that are fixed in the course of evolution are selectively neutral so that fixation occurs via random drift. A corollary of this theory that Kimura clearly emphasized is that gene sequences evolve in an approximately clocklike manner (in support of the original molecular clock hypothesis of Zuckerkandl and Pauling), whereas episodic beneficial mutations subject to natural selection are sufficiently rare that they can be safely disregarded for a quantitative description of the evolutionary process. Of course, the neutral theory should not be taken to mean that selection is unimportant for evolution. The theory actually maintains that the dominant mode of selection is not the Darwinian positive selection of adaptive mutations, but rather purifying selection that eliminates deleterious mutations while allowing fixation of neutral mutations by drift.

Subsequent studies have refined the theory and made it more realistic: To be fixed, a mutation needs not be literally neutral, but only needs to exert a deleterious effect that is small enough to escape efficient elimination by purifying selection—the modern "nearly neutral" theory that was developed primarily by Tomoko

Ohta (Ohta, 2002). Which mutations are "seen" by purifying selection as deleterious critically depends on the effective population size: In small populations, drift can fix even mutations with a significant deleterious effect, whereas in large population, even the slightest deleterious effect is sufficient for the elimination of a mutant allele (see Box 1-1).

The main empirical test of the (nearly) neutral theory comes from measurements of the constancy of the evolutionary rates in gene families. Although it has been repeatedly observed that the molecular clock is significantly overdispersed, such tests strongly suggest that the fraction of neutral mutations among the fixed ones is indeed substantial (Bromham and Penny, 2003; Novichkov, et al., 2004). The nearly neutral theory is a major departure from the Modern Synthesis selectionist paradigm because it explicitly posits that the majority of mutations fixed during evolution are not affected by Darwinian (positive) selection. Darwin seems to have presaged the neutralist paradigm by remarking that selectively neutral characters would serve best for classification purposes; however, he did not elaborate on this prescient idea, and it has not become part of Modern Synthesis.

Importantly, in the later elaborations of the neutral theory, Kimura, Ohta, and others realized that mutations that were nearly neutral at the time of fixation were not indifferent to evolution. On the contrary, such mutations comprised the pool of variation (a nearly neutral network of alleles) that natural selection can tap into under changed conditions, a phenomenon that could be important for both micro- and macroevolution (Kimura, 1991). This idea has become key to some of the latest advances in evolution theory, so we discuss it in more detail later in the book (see in particular Chapters 8 and 9).

Measuring selection by sequence comparison

For all its importance, Darwinian natural selection was a concept defined in qualitative terms. Within the framework of population genetics and Modern Synthesis, purifying and positive selection became concrete and mathematically tractable. Under Modern

Synthesis, selection is much like a "force" in classical physics or a "flux" in classical thermodynamics, a phenomenologically defined quantity. With the advent of sequence comparison, it became possible to define and measure selection in specific, mechanistic terms, based on counting different types of nucleotide substitutions. Two simple ideas were exploited to measure selection by sequence comparison (see Box 2-2). The approaches have much in common because they both define two classes of sites, one of which is taken as the baseline of neutral evolution. The first method involves comparing the rates of nucleotide substitutions in positions that are important for amino acid coding (known as nonsynonymous positions) and in positions that, because of the redundancy of the genetic code, are irrelevant for the sequence of the encoded protein. If the ratio of nonsynonymous to synonymous substitution rates (Ka/Ks; see Box 2-2) is significantly less than 1, the evolution of the respective gene is constrained by purifying selection which targets the encoded protein sequence, whereas $Ka/Ks > 1$ indicates evolution by positive Darwinian selection (see Box 2-2). The second, more rigorous approach employed to measure selection is known as the McDonald-Kreitman test, whereby the Ka/Ks ratio is compared for intraspecies variants (polymorphisms) and interspecies variants (fixed mutants). Because the polymorphisms that have not yet been fixed are supposed to be overwhelmingly neutral, the Ka/Ks between species should be significantly lower that the Ka/Ks for polymorphisms in the case of purifying selection, and significantly greater than the value for polymorphisms in the case of positive selection.

The advent of these quantitative approaches to the analysis of selection is notable for more than their technical utility in evolutionary studies: They are also signs of a fundamental change in the way biologists think about selection. The Darwinian qualitative idea that was embodied in an abstract mathematical quantity by Fisher and was first measured by Dobzhansky and his disciples using genetic methods now turned into a directly measurable, statistical characteristic of an ensemble of nucleotide sites. This transformation of the concept of selection is akin to the switch from abstract fluxes of classic thermodynamics to the statistical physics of Ludwig Boltzmann and Josiah Willard Gibbs (see Chapter 4).

Box 2-2: Measuring selection by sequence analysis of protein-coding genes (Hurst, 2002; Li, 1997)

Protein-coding sequences consist of two classes of sites:

- Synonymous, in which substitutions have no effect on the encoded amino acid sequence

- Nonsynonymous, in which substitutions lead to amino acid substitutions

The ratio Ka/Ks (Ka is the rate of nonsynonymous substitutions, Ks is the rate of synonymous substitutions; both are calculated with corrections for multiple substitutions) is a measure of selection that acts at the level of protein sequences.

$Ka/Ks = 1$ – neutral evolution of protein sequence (no selection on the encoded protein)

For most protein-coding genes, $Ka/Ks \ll 1$ = purifying selection

Prokaryotes: typically $Ka/Ks < 0.1$

Eukaryotes: typically $Ka/Ks \approx 0.1$-0.2

- $Ka/Ks > 1$ – positive selection; this is rare among protein-coding genes but has been detected for several categories of genes, e.g., genes involved in antiparasite defense or spermatogenesis and in viral proteins such as influenza hemagglutinin.

- Maximum likelihood methods exist to measure Ka/Ks in individual sites; many protein-coding genes contain a few sites subject to positive selection.

- Using Ka/Ks as a measure of selection assumes neutrality of synonymous sites.

- However, Ka and Ks are positively correlated, implying selection affecting synonymous sites as well.

- Noncoding sites, such as intron sequences, can be used as a proxy for neutral evolution to measure selection on synonymous sites (Ks/Ki, where Ki is the substitution n rate for intronic sites).

- The *McDonald-Kreitman test* (Aquadro, 1997; McDonald and Kreitman, 1991) is commonly used to measure selection. It compares variation within species (frequency of polymorphisms, P) with variation between species (divergence, D).
- $Dn/Ds = Pn/Ps$: Neutral evolution of the protein sequence.
- $Dn/Ds < Pn/Ps$: Purifying selection.
- $Dn/Ds > Pn/Ps$: Positive selection.

Selfish genes, junk DNA, and mobile elements

Although this was rarely stated explicitly, classic genetics certainly implies that nearly all parts of the genome (all nucleotides, in more modern, molecular terms) have a specific function. This implicit assumption is also important for Modern Synthesis, with its pan-adaptationist worldview. However, this understanding was called into doubt in the 1960s and 1970s by the accumulating data on the lack of a direct correspondence between genome size and the phenotypic complexity of organisms. Even with the crude methods available at the time, it became clear that organisms of roughly the same phenotypic complexity often have genomes that differ in size by orders of magnitude (the so-called *c*-value paradox). This paradox was conceptually resolved by two related, fundamental ideas, those of selfish genes and junk DNA.[4] The selfish gene concept was proposed by Richard Dawkins in his eponymous 1976 book (Dawkins, 2006). In a striking departure from the organism-centric paradigm of Modern Synthesis, Dawkins realized that natural selection could act not only at the level of the organism as a whole, but also at the level of an individual gene. Under a deliberately provocative formulation of this view, *genomes and the organisms are essentially vehicles for the propagation of genes*.

The selfish gene concept has many important implications, some of which we explore later in this book. The aspect that is directly relevant to the *c*-value paradox was emphasized by W. Ford Doolittle and Carmen Sapienza (Doolittle and Sapienza, 1980), and by Leslie Orgel and Francis Crick (Orgel and Crick, 1980). They proposed that much, if not most, of the genomic DNA (at least, in complex multicellular organisms) consists of various classes of repeats that originate from the amplification of selfish elements—the ultimate parasites, using the catchy language of Orgel and Crick. In other words, from the organism's standpoint, much of its genomic DNA should be considered junk. This view of the genome dramatically differs from the picture implied by the pan-selectionist paradigm intrinsic to Modern Synthesis, under which most, if not all, nucleotides in the genome would be affected by (purifying or positive) selection acting at the level of the organism.

A conceptually related major development was the discovery, first in plants by Barbara McClintock in the 1940s, and subsequently in animals, of "jumping genes" that later became known as mobile elements (that is, genetic elements that were prone to frequently change their position in the genome; McClintock, 1984). The demonstration of the ubiquity of mobile elements suggested the picture of highly dynamic, perpetually changing genomes even before the advent of modern genomics.[5]

Evolution by gene and genome duplication: Orthologs and paralogs

The central tenet of Darwin, the gradualist insistence on infinitesimal changes as the only material of evolution, was fully inherited by Modern Synthesis but was challenged by the concept of evolution by gene duplication, developed by Susumu Ohno in his classic 1970 book (Ohno, 1970). The idea that duplication of parts of chromosomes might contribute to evolution goes back to the founders of modern genetics, particularly Fisher and Haldane.[6] However, Ohno was the first to propose that gene duplication was central to the evolution of genomes and organisms, as well as the first to support this proposition with a qualitative theory. Starting from the cytogenetic evidence of a whole genome duplication (WGD) early in the evolution of chordates, Ohno hypothesized that gene duplication could be an important, if not the principal path, to the evolution of new biological functions. Under Ohno's hypothesis, duplication of a gene frees one of the copies from constraints imposed by purifying selection, so this copy would have the potential to evolve a new function (a phenomenon later denoted as neofunctionalization). Clearly, the emergence of a new gene as a result of a duplication, let alone duplication of a genomic region including multiple genes or WGD, is a far cry from Darwin's "infinitesimal" changes. If such larger events are indeed crucial for evolution, the gradualist paradigm comes into jeopardy. More recent studies on gene duplication, discussed later in the book (see Chapters 8 and 9), suggest that neofunctionalization is unlikely to be the main route of evolution of duplicated genes. However, the fact remains that duplication, as a major mechanism of evolution, flies in the face of gradualism.

In the same year Ohno's book on evolution by gene duplication appeared, Walter Fitch published a seminal paper whose true significance became clear only with the much later advances of genomics. Fitch examined the notion of gene homology (common ancestry) and distinguished between two classes of homologous genes: *orthologs* and *paralogs* (Fitch, 1970). Orthologs are genes that evolved by vertical descent from the same ancestral gene in a common ancestor of the compared organisms, whereas paralogs are genes that evolved by duplication. Obviously, the notions of orthology and paralogy are tightly linked to each other and are contingent on a particular topology of a phylogenetic tree for the given gene family, so that a duplication at a particular node of a tree gives rise to a new set of paralogs in the descendant subtree (see Chapter 3 for more details). Furthermore, the conceptually straightforward definition of orthology is complicated by lineage-specific gene loss and horizontal gene transfer (see Chapters 5 and 7). Nevertheless, as we also discuss, all these complications notwithstanding, Fitch's classification of homologs remains central to evolutionary genomics.[7]

Punctuated equilibrium and the inadequacy of gradualism

The general lack of transitional forms between species in the fossil record is a constant theme in evolutionary biology. Darwin recognized this problem and traditionally interpreted it (along with paleontologists in the Darwinian tradition) as a reflection of the dramatic incompleteness of the record. However, extensive accumulation of paleontological data in the twentieth century helped very little, if at all, so a different perspective emerged, first in the quantum evolution concept of George Gaylord Simpson and then, in a full-fledged form, in the punctuated equilibrium concept of Stephen Jay Gould and Niles Eldredge (Eldredge and Gould, 1997; Gould, 2002). Gould and Eldredge collected extensive evidence indicating that the history of the great majority of animal species, as reflected in the fossil record, represents mostly stasis—that is, virtual lack of change. Stasis is punctuated by "sudden" disappearance of a species, followed by rapid replacement by a new species. The implication of this pervasive pattern is that speciation is a rapid process, compared to the duration of stasis; that appearance of a new species in a given area typically occurs by migration from the area of speciation; and that gradualist speciation (gradual transformation of a species into a new one) is extremely rare. This punctuated equilibrium pattern seems to apply also to the evolution of higher taxa and is sometimes generalized to imply the inadequacy of gradualism in general, although the legitimacy of such generalizations has been disputed.

Spandrels, exaptation, tinkering, and the fallacy of the Panglossian paradigm of evolution

The principle of gradualism was challenged, at least implicitly, by Ohno's hypothesis on evolution by gene and genome duplication, and more explicitly by the punctuated equilibrium concept. The adaptationist program of evolutionary biology came under an unusually spirited, sweeping attack in the "Spandrels of San Marco" article of 1979 by Gould and Richard Lewontin (Gould and Lewontin, 1979), one of the most unusual and influential papers in the history of biology. Gould and Lewontin sarcastically described the adaptationist worldview as the Panglossian paradigm, after the notorious character in Voltaire's *Candide* who insisted that "everything was for the better in this best of all worlds" (even major disasters). Gould and Lewontin emphasized that, rather than hastily concoct "just so stories"[8] of plausible adaptations, evolutionary biologists should seek explanations of the observed features of biological organization with a pluralist approach that takes into account not only selection, but also intrinsic constraints, random drift, and other factors. The spandrel metaphor holds that many functionally important elements of biological organization did not evolve as specific devices to perform their current functions, but rather are products of nonadaptive architectural constraints—much like spandrels (pendentives) that appear at arches of cathedrals and other buildings solely due to constructional demands, and can be recruited for various functions such as housing key elements of the imagery adorning the cathedral (see Figure 2-2). The process of using spandrels for biological functions was given the special name *exaptation,* and Gould heralded this as an important route of evolution (Gould, 1997a). The spandrel concept is conceptually linked to the nearly neutral theory but, in a sense, goes further and closer to the core of evolutionary thinking by showing that even phenotypic features that "look like" typical adaptations might not have evolved under direct pressure of natural selection.

Spandrel (pendentive)

Figure 2-2 One of the spandrels of Basilica di San Marco in Venice. Photo by Maria Schnitzmeier, from the Wikimedia Commons, under the GNU Free Documentation License.

In an even earlier, conceptually related development, Francois Jacob (the codiscoverer of gene regulation, among other seminal discoveries in bacterial genetics; see Chapter 5) promoted the metaphor of *evolution as tinkering,* or *bricolage,* in the French original. Driving primarily from the results of comparative analysis of developmental mechanisms, Jacob posited that evolution acts not as an engineer or designer, but rather as a tinkerer that is heavily dependent on previous contingencies for solving outstanding problems:

> Natural selection has no analogy with any aspect of human behavior. However, if one wants to play with a comparison, one would have to say that natural selection does not work like an engineer works. It works like a tinkerer – a tinkerer who does not know exactly what he is going to produce but uses whatever he finds around him whether it be pieces of string, fragments of wood, or old cardboards; in short, it works like a tinkerer who uses everything at his disposal to produce some kind of workable object. (Jacob, 1977)

A key corollary of the bricolage concept is that the specific outcome of evolution is unpredictable, or at least cannot be predicted without detailed knowledge of preceding events. Put another way, in a thought experiment where the "tape of evolution is replayed" (the favorite metaphor of Gould), the resulting diversity of outcomes will be different from what we actually observe, probably beyond recognition; we return to this subject toward the end of the book (see Chapter 13).

Evolution in the world of microbes and viruses, and the three-domain Tree of Life

Perhaps, the development in biology that had the most profound effect on the changing understanding of evolution was the extension of evolutionary research into the realm of microbes, namely unicellular eukaryotes (protists), prokaryotes (bacteria and archaea), and viruses. Darwin's account of evolution and all the developments in evolutionary biology in the subsequent few decades dealt exclusively with animals and plants, with unicellular eukaryotes (Protista) and bacteria (Monera) nominally placed near the root of the Tree of Life by Ernst Haeckel and his successors. Although by the 1950s genetic analysis of bacteriophages and bacteria was well advanced, making it obvious that these life forms had evolving genomes, Modern Synthesis took no notice of these developments. That bacteria (let alone viruses) would evolve under the same principles and by the same mechanisms as animals and plants is by no means obvious, given all their striking biological differences from multicellular organisms, and specifically because they lack regular sexual reproduction and reproductive isolation that is crucial for the speciation in animals and plants.

Effectively, prokaryotes became "visible" to evolutionary biologists in 1977, through the groundbreaking work of Carl Woese and colleagues on rRNA phylogeny (Woese, 1987). Viewed in a general context, Woese's discovery is truly momentous and perhaps even merits a comparison to the discovery of DNA structure. Woese found that an actual molecular structure, the nucleotide sequence of rRNA, showed recognizable conservation throughout the entire range of cellular life forms. Furthermore, phylogenetic analysis of this universally conserved molecule proved to be informative (that is, at least roughly, rRNA evolves in a clock-like fashion) and led to another major discovery, a leading icon of evolutionary biology at the end of the twentieth century, the *three-domain Tree of Life* (see Figure 2-3; Woese, et al., 1990). The three domains are Bacteria, Archaea, and Eukaryota—the archaeal domain was discovered by George Fox and Woese through the comparative analysis of rRNA, when in the emerging tree a group of obscure "bacteria" came across as being distinct from both the rest of bacteria and the more complex eukaryotic organisms. In addition to delineating the three domains, Woese and coworkers used phylogenetic analysis of rRNA to identify multiple major lineages of archaea and bacteria (Woese, 1987). The implication was that evolution of prokaryotes was as tractable as evolution of complex eukaryotes, a concept that was alien to microbiologists before Woese's work (Stanier and Van Niel, 1962). Through the achievements of Woese, his collaborators, and his followers, a growing tendency developed to equate the phylogenetic tree of rRNA, with its three-domain structure, to the Tree of Life of Darwin and Haeckel (Pace, 2009a, 2006). Within a few years of the publication of Woese's discoveries, it became clear that the topology of the rRNA tree was (at least in its main features) congruent with the trees for some of the most conserved proteins, such as ribosomal proteins, translation factors, DNA-dependent RNA polymerase subunits, and membrane ATPases.

Two groups independently developed an ingenious idea to inject a root position into the rootless tree of the kind shown in Figure 2-3. To this end, one can use ancient paralogs that are represented in (nearly) all organisms and thus can be confidently inferred to have evolved via a duplication antedating the Last Universal Common Ancestor (LUCA). When a tree is constructed jointly for two paralogous sets of ancient orthologs, the position of the root between them is certain, so the root can be inferred for each of the orthologous sets as well (see Figure 2-4; Gogarten, et al., 1989; Iwabe, et al., 1989). The results of analysis of two pairs of primordial paralogs, translation factors, and membrane ATPase subunits were fully congruent and placed the root on the bacterial branch, thus establishing an archaeal-eukaryote clade (see Figure 2-4). However, even in the pre-genomic era, it became clear that not all trees of protein-coding genes have the same topology as the rRNA tree; the causes of the discrepancies remained murky but were thought to involve (beyond likely artifacts) Horizontal Gene Transfer (HGT; Smith, et al., 1992). These discrepancies made for just a footnote to the three-domain TOL, but things changed in the genomic era.

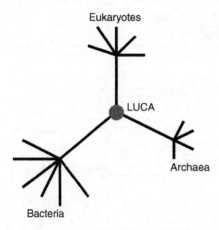

Figure 2-3   The three-domain Woeseian Tree of Life.

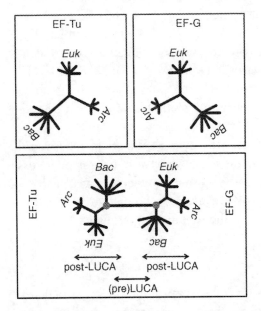

Figure 2-4   Inference of the root in the three-domain Tree of Life using ancestral paralogs. Schematics show the phylogenetic trees for two ubiquitous, paralogous translation initiation factors, EF-Tu and EF-G, reconstructed independently (top) and analyzed jointly (bottom). Black circles show the inferred positions of the roots of the two subtrees.

Viruses and the birth of evolutionary genomics

Evolutionary genomics was born more than a decade before the historic announcement of the first sequenced bacterial genome. With less fanfare (but not in obscurity, either), multiple small (roughly 4KB–100KB) genomes of diverse viruses have been sequenced, and the principles of genome comparison, along with practical computational methods, were developed. Viruses are obligate intracellular parasites, and viral genomes are both much smaller and qualitatively different from genomes of cellular life forms. Viruses typically lack certain classes of genes that are ubiquitous and essential in cellular organisms, such as genes for components of the translation system and membranes. Nevertheless, viruses follow their own "biological strategies" and encode all virion subunits, as well as at least some proteins involved in viral genome replication. (One of the central themes of this book is the key role of viruses in the evolution of the biosphere, so I return to this subject in considerable detail in Chapter 10[9])

Despite the fast sequence evolution that is characteristic of viruses (particularly those with RNA genomes), this early comparative-genomic research successfully delineated sets of genes that are conserved in large groups of viruses (Koonin and Dolja, 1993). The possibility of structural and functional mapping of an entire genome of a distinct genetic entity by means of genome comparison was realized in these studies for the first time, and this became the cornerstone of evolutionary genomics. Moreover, an unanticipated, crucial generalization emerged: Whereas some genes were conserved across an astonishing variety of viruses, genome architectures, virion structures, and biological features of viruses showed much greater plasticity (see Chapters 5 and 10 for further details and discussion).

Endosymbiosis

The hypothesis that certain organelles of eukaryotic cells, particularly the plant chloroplasts, evolved from bacteria is not that much younger than *Origin:* Several researchers proposed this in the late nineteenth century on the basis of microscopic study of plant cells that revealed conspicuous structural similarity between chloroplasts and cyanobacteria (then known as blue-green alga). The concept of symbiogenetic evolution was presented in a coherent form by Konstantin Mereschkowsky at the beginning of the twentieth century.[10] However, for the first two-thirds of the twentieth century, the endosymbiosis hypothesis remained a fringe speculation. This perception changed shortly after the appearance of the seminal 1967 article by Lynn Sagan (Margulis), who summarized the available data on the similarity between organelles and bacteria, particularly the striking discovery of organellar genomes and translation system. Sagan concluded that not only chloroplasts, but also mitochondria evolved from endosymbiotic bacteria (Sagan, 1967). Subsequent work, particularly phylogenetic analysis of both genes contained in the mitochondrial genome and genes encoding proteins that function in the mitochondria and apparently were transferred from the mitochondrial to the nuclear genome, turned the endosymbiosis hypothesis into a well-established concept, with overwhelming empirical support (Lang et al., 1999). Moreover, these phylogenetic studies convincingly demonstrated the origin of mitochondria from a particular group of bacteria, the γ-proteobacteria. The major evolutionary role assigned to unique (or extremely rare) events such as endosymbiosis is incompatible with both gradualism and uniformitarianism and is a major theme in the rest of this book, particularly in Chapters 7 and 12.

Canalization and robustness in evolution

The eminent developmental geneticist Conrad Waddington put forward the unorthodox idea of canalization of development, which is embedded within his general concept of epigenetic landscape.[11] The epigenetic landscape is a depiction of choices faced by a developing embryo, whereby development occurs by movement along valleys that encompass clusters of similar trajectories. Thus, relatively small perturbations caused by either environmental factors or mutations will not affect development—that is, biological systems are substantially robust. According to Waddington's concept, this robustness is an evolved, adaptive property of biological systems. Stress can disrupt canalization and unmask hidden variability, thus increasing the evolutionary potential (evolvability) of a population (Waddington and Robertson, 1966). In Waddington's time, these ideas were outside the mainstream of evolutionary biology, but robustness and evolvability are taking the central stage in the new vision of evolution, as discussed in Chapter 10.

Synopsis

Shortly after the completion of Modern Synthesis, evolutionary biology underwent a dramatic transformation: Evolution became traceable directly to its substrate, the evolving genome. At the deepest conceptual level, evolution by natural selection and drift is an inevitable consequence of error-prone replication of digitally encoded genetic information. Evolution is no longer a somewhat abstract process of accumulation of mutations that can be observed only indirectly through their phenotypic effect. Instead, evolution is now seen as the accumulation of concrete changes of different kinds, big and small, revealed by direct comparison of increasingly available gene and genome sequences. The existence of a gradient of sequence divergence from closely related to distant species is itself the best proof of evolution. This trend is encapsulated in the (nearly) neutral theory of molecular evolution and, on a more practical level, provides for the construction of meaningful phylogenetic trees. Molecular phylogenetics culminated in the three-domain Tree of Life that was originally discovered through the rRNA phylogeny and subsequently supported by many protein phylogenies. Analysis of ancient paralogs resulted in the placement of the root on the bacterial branch of the three-domain TOL. However, the first appearing discrepancies between individual gene tree topologies suggested that the rRNA tree might not tell the complete story of the evolution of life.

Comparison of the first available sequenced genomes, those of small viruses, marked the beginnings of evolutionary genomics. It became clear that structural and functional maps of otherwise uncharacterized genomes could be constructed through comparative analysis and that broad conservation of key genes was complemented by plasticity of genome architecture.

In parallel with the maturing of Modern Synthesis and the advent of molecular evolution and molecular phylogenies, the pregenomic evolution of evolutionary biology included several conceptual developments, such as the ideas of spandrels and canalization, which went beyond neo-Darwinism. As a result, the rapid ascent of genomics in the 1990s met with a complex, diverse landscape of evolutionary theory and methodology.

Recommended further reading

Dawkins, Richard. (1976/2006) *The Selfish Gene*. Oxford: Oxford University Press.

A book of paramount conceptual importance that introduced the seminal idea of selection at the level of individual genes for the first time.

Gould, S. J., and N. Eldredge. (1993) "Punctuated Equilibrium Comes of Age." *Nature* 366: 223–227.

A brief but compelling overview of the evidence behind the punctuated equilibrium model.

Gould, S. J., and R. C. Lewontin. (1979). "The Spandrels of San Marco and the Panglossian Paradigm: A Critique of the Adaptationist Programme." *Proceedings of the Royal Society of London B Biological Sciences* 205: 581–598.

Perhaps one of the most outstanding articles in the entire history of biology, both for the spirited advocacy of a balanced view of evolution as opposed to pan-adaptationism and for the Renaissance spirit of the entire text.

Jacob, F. (1977) "Evolution and Tinkering." *Science* 196: 1,161–1,166.

A wonderfully written article that remains important and relevant for the refutation of the view of "natural selection as a designer" and the emphasis on the overarching role of historical contingency in evolution.

Kimura, Motoo. (1983) *The Neutral Theory of Molecular Evolution*. Cambridge, MA: Cambridge University Press.

The comprehensive presentation of the neutral theory by its founder combines rigorous population genetic analysis with perfectly lucid explanations for biologists. Although the empirical aspects are outdated, the book remains unsurpassed in clarity and is remarkably readable.

Koonin, E. V., and V. V. Dolja. (1993) "Evolution and Taxonomy of Positive-Strand RNA Viruses: Implications of Comparative Analysis of Amino Acid Sequences." *Critical Reviews in Biochemistry and Molecular Biology* 28: 375–430.

A comprehensive overview of the early results of comparative and evolutionary genomics of a major class of small RNA viruses, one of the first attempts on a deep evolutionary reconstruction. The main conclusions remain valid despite the subsequent huge increase in the number and diversity of sequenced viral genomes.

Nei, Masatoshi, and Sudhir Kumar. (2000) *Molecular Evolution and Phylogenetics*. Oxford: Oxford University Press.

A technical but accessible description of the widely used phylogenetic methods, from the author of the neighbor-joining method and one of the authors of the MEGA package.

Nielsen, R. (2009) "Adaptionism—30 Years After Gould and Lewontin." *Evolution* 63: 2,487–2,490.

A perspective on the Spandrels of San Marco 30 years later.

Ohno, Susumu. (1970) *Evolution by Gene Duplication*. Vienna: Springer.

The classic book that established the concept of gene duplication as the central route of evolution.

Ohta, T., and J. H. Gillespie. (1996) "Development of Neutral and Nearly Neutral Theories." *Theoretical Population Biology* 49: 128–142.

A historical and conceptual discussion of the neutral and nearly neutral theories by two of the founding figures of molecular evolution.

Woese, C. R. (1987) "Bacterial Evolution." *Microbiology Reviews* 51: 221–271.

A magisterial narrative summarizing the early results of the 16S RNA-based phylogenetics.

Woese, C. R., and N. Goldenfeld. (2009) "How the Microbial World Saved Evolution from the Scylla of Molecular Biology and the Charybdis of the Modern Synthesis." *Microbiology and Molecular Biology Reviews* 73: 14–21.

An essay for Darwin's 200th anniversary, emphasizing the decisive importance of microbial genomics for the new understanding of evolution.

3

Comparative genomics:
Evolving genomescapes

The importance of going genomic

The fundamental principles of molecular evolution have been established, and many specific observations of major importance and impact on the fundamentals of evolutionary biology have been made in the pregenomic era (see the first two chapters). However, massive genome sequencing that started in the mid-1990s and rapidly progressed into the new millennium qualitatively changed the entire enterprise of evolutionary biology. The importance of large numbers of sequences with different degrees of divergence is obvious: This material allows researchers to investigate mechanisms and specific events of evolution with the necessary statistical rigor and to reveal even subtle evolutionary trends. However, diverse, complete genome sequences are critically important for evolutionary biology far beyond the sheer amount of sequence data. Indeed, only the complete genome sequence (as opposed to, say, 95% complete) provides the researcher with an objective, unbiased view of the gene repertoire of the given life form. That is, the researcher can determine which genes are present and, equally important, which ones are missing in the organism. Thus, comparing complete genomes is the only route to satisfactory reconstructions of evolution. The emerging picture is in many ways different from anything that would be imaginable within the framework of traditional evolutionary biology.

If we are serious in our attempts to "understand" evolution, it is crucial to sample the genome space both deeply (that is, to obtain genome sequences of multiple, closely related representatives of the same taxon) and broadly (to obtain representative sequences for as many diverse taxa as possible—and, eventually, all taxa). At the time of this writing, in the waning days of the year 2010, the collection of sequenced genomes consists of thousands of viral genomes, more than 1,000 genomes of bacteria and archaea, and around 100 eukaryotic genomes. By the time this book will be published, the genome database will almost double, and with the new generation of sequencing methods, the growth is expected to accelerate for years to come. Although not all major taxa are adequately represented, the rapidly growing collection of genomes increasingly satisfies the demands of both microevolutionary and macroevolutionary research.

Complementary to the advances of traditional genomics is the more recent and rapidly accelerating accumulation of extensive data from metagenomics—that is, exhaustive (or at least extensive) sequencing of nucleic acids from a particular environmental habitat. Although the current metagenomic approaches typically do not yield complete genomes, they provide invaluable, minimally biased information on the diversity of life in various environments.

This chapter is an overview of the diversity and major features of genomes. The following chapters explore in greater detail the implications of the comparative genomic results for the putative "postmodern synthesis" of evolutionary biology.

Evolving genomescapes

The striking diversity of genomes

Genome was the first "ome" term—and still is the most commonly used one.[1] As always in biology, defining the genome is not easy. Simply put, a genome is *the entirety of the genetic information in the given organism.* The existence of a stable core of inherited genetic information (or, more specifically, of genes) is implied by the very robustness of heredity—or, in more fundamental terms, the error-prone replication (EPR) principle (see Chapter 2). However, the relationship between "the entirety of genetic information" and "the stable core" is far from being simple. As soon as one asks a seemingly innocuous question such as "What is the genome of *Escherichia coli?*", a tangle of formidable complications becomes unavoidable. The question "What is the human genome?" triggers a somewhat different series of equally difficult questions. We postpone this discussion until later in the book (see Chapter 5); we now look into the diversity of genomes that have been sequenced over the last 15 years.

The modern era in genomics began at the end of summer 1995 when J. Craig Venter's laboratory published the genome sequence of the opportunistic pathogenic bacterium *Haemophilus influenzae* (Fleischmann, et al., 1995). In the process of obtaining the *H. influenzae* sequence, Venter, Hamilton Smith, and their colleagues perfected the whole-genome shotgun method, a brute-force approach that quickly made megabase sequencing routine. Several more bacterial genomes, the first archaeal genome (*Methanocaldococcus jannaschii*), and the first eukaryotic genome (baker's yeast, *Saccharomyces cerevisiae*) followed within a year (Koonin, et al., 1996). By 1999, the steady exponential growth of the collection of sequenced genomes had settled in (see Figure 3-1).

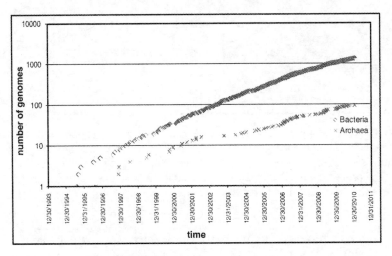

Figure 3-1   Exponential growth of the genome sequence collections. Data comes from the NCBI/Genomes website (www.ncbi.nlm.nih.gov/genome/).

From small viruses to animals, genomes span about six orders of magnitude in size, from several thousand to several billion nucleotides; for cellular life forms, excluding viruses, the range is about four orders of magnitude (see Figure 3-2). The range of gene numbers is much more narrow, only about four orders of magnitudes, from 2 to 3 genes in the simplest viruses to around 40,000 genes in some animals. Excluding viruses and parasitic (symbiotic) bacteria, the range of gene numbers becomes quite narrow, only about an order of magnitude (see Figure 3-2; Koonin, 2009a; Lynch, 2007c). It seems highly surprising that mammals or flowering plants possess only about 10-fold more (readily identifiable) genes than an average free-living bacterium and only about twice as many genes as the most complex bacterium (see Figure 3-2). Later in this book (see Chapters 5, 7, and 10), we discuss various possible explanations for these apparent constraints on the number of genes in the genomes of all life forms.

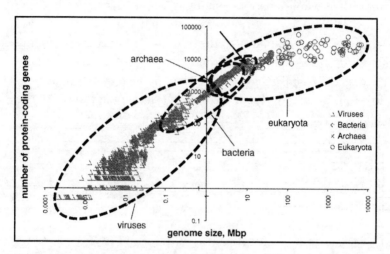

Figure 3-2   The total size of the genomes and the number of genes in viruses, bacteria, archaea, and eukaryotes. Data comes from the NCBI/Genomes website. The plot is in double logarithmic scale. Mbp stands for mega base-pairs. The arrow points to the change in the slope of curve that corresponds to the transition from "small" to "large" genomes.

Roughly, genomes can be partitioned into two distinct classes (Koonin, 2009a). The boundary between the two classes is apparent as a change in the slope of the curve in the plot in Figure 3-2.

1. Genomes with a strict proportionality between genome size and gene number. These are the small genomes of all viruses and prokaryotes, with a high gene density of approximately 0.5 to 2 genes/kilobase and, accordingly, very short intergenic regions (10% to 15% of the genome sequence or even less) that primarily consist of regulatory elements. Sometimes these genomes are described as "wall to wall" because they consist almost entirely of well-definable genes. The genomes of most unicellular eukaryotes show somewhat less tight coupling between genomes size and gene number than the genomes of viruses and prokaryotes, but they nevertheless can be included in the same class.

2. Genomes in which the gene number and the genome size are decoupled—namely, the large genomes of multicellular and some unicellular eukaryotes. Here the correlation between the total genome size and gene number is weak, at best. Accordingly, the fraction of the genome occupied by intergenic regions (and other noncoding sequences such as introns) shows enormous variance. In some of the most complex genomes, such as those of mammals, noncoding regions account for most of the genome sequence.

The variance of genome size and gene numbers is complemented by the diversity in other dimensions—for instance, the physical organization and nucleotide composition of genomes. Considering both viruses and cellular life forms, genomes come in all possible forms of nucleic acids (see Chapter 10 for details). All genomes of cellular organisms are made of dsDNA, but the number of genomic segments (chromosomes) and their relative sizes, form (circular or linear), and ploidy (copy number) broadly differ. The textbook notion is that prokaryotes possess haploid, single circular chromosomes, whereas in the genomes of eukaryotes that widely differ in ploidy, genes are distributed among multiple, linear chromosomes. Although these genomic forms might indeed be dominant, the real diversity of genomes goes far beyond such simple dichotomies. In particular, numerous prokaryotes have multiple chromosomes—in some cases, linear. Contrary to the common belief, most prokaryotic cells are not haploid—that is, they contain multiple copies of the genome.

Most of the genes in any genome are ancient and have distinct evolutionary fates

As we discussed in the preceding chapter, Ernst Mayr, one of the greatest classical evolutionary biologists of the twentieth century and a cofounder of Modern Synthesis, predicted with confidence that genes from different organisms—even rather closely related ones—would show no recognizable similarity to each other, given the major phenotypic differences between the organisms. This prediction went so spectacularly wrong that it becomes nontrivial and valuable for this dramatic failure alone. Even pregenomic sequence comparisons have revealed the high level of sequence conservation between some homologous protein and noncoding RNA molecules throughout the spectrum of life, from bacteria to mammals (see the preceding chapter). Moreover, high sequence similarity exists between ancient paralogs that apparently originate from duplications antedating the Last Universal Common Ancestor (LUCA; Gogarten, et al., 1989; Iwabe, et al., 1989). Genomics transformed this general understanding into a complete, quantitative breakdown of the genes in any genome into classes of evolutionary conservation (see Figure 3-3; Koonin and Wolf, 2008b).

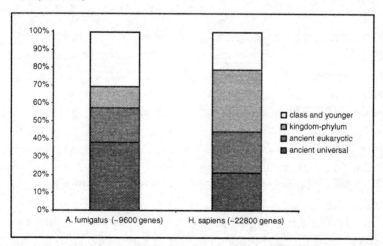

Figure 3-3   Breakdown of genes by evolutionary age. The "evolutionary age" corresponds to the deepest taxonomic node at which homologs are detectable for the protein product of the given gene. In particular, for humans, universal means "homologs detectable in prokaryotes," ancient eukaryotic means "homologs detectable in prokaryotes outside the Unikont supergroup" (see Chapter 7), kingdom-phylum means "homologs detectable in animals outside mammals," and class and younger means "no homologs reliably detected outside mammals." Data comes from Wolf, et al., 2009.

The key finding of comparative genomics is that the majority of the genes in any genome can be considered "highly conserved"—that is, they possess readily detectable homologs in organisms separated by hundreds of millions years of evolution (for example, at the level of the common ancestor of vertebrates, in the case of human genes; see Figure 3-3; Wolf, et al., 2009). This finding shows the remarkable resilience of RNA and protein sequences during evolution: *The typical time of the decay of sequence similarity between homologous genes is comparable with the time of life's existence on Earth.* Beyond its fundamental importance, this fact has huge practical consequences: This is what makes the comparative genomics enterprise highly informative and feasible in the first place.

More is involved in the structure of the evolutionary process than mere sequence conservation, though. Not only are RNA and protein sequences conserved through extremely long evolutionary spans, but the unique identity of genes tends to persist. In other words, most of the genes evolve as *orthologous lineages*, with occasional duplications (Koonin, 2005). The persistence of gene orthology becomes apparent through a simple comparative genomic procedure that can effectively identify orthologous genes sets. In this procedure, orthologs are detected as "bidirectional best hits": All protein sequences encoded in a genome are compared to all proteins encoded in another genome, and the procedure is repeated in the reverse direction (Tatusov, et al., 1997). The pairs of genes that produce the best hits (that is, the ones that show the greatest sequence similarity) in both directions are denoted as putative orthologs; it is easy to extend this procedure to multiple species by merging triangles of bidirectional best hits that share a common side (see Box 3-1). Remarkably, this straightforward approach works much of the time: Some 70% of the genes from organisms separated, say, by about 100 million years of evolution, such as human and mouse, come across as bidirectional best hits (Wolf, et al., 2009). With a simple modification to the algorithm to incorporate lineage-specific gene duplications (duplications that occurred after the divergence of the compared species), this approach allows us to identify sets of apparent orthologs (that became known as *clusters of orthologous genes*, or *COGs*) in many genomes, often as distant from each other as archaea and bacteria, the two domains of prokaryotes (see Chapter 5). More accurate and powerful methods for orthology detection involve explicit analysis of phylogenetic trees (see Box 3-1); however, the results of such analyses typically are compatible with those obtained from the simpler approaches based on sequences conservation alone (bidirectional best hits). Of course, for a fraction of genes, the history of duplications and losses is so complex that COGs are not readily detectable, so they become fuzzy clusters with an uncertain internal structure. Fortunately, these "difficult" genes are a minority in each genome.

Box 3-1: Classification of homologous relationships between genes: Orthologs, paralogs, and methods for their identification

Evolutionary relationships between genes:

- Homology: genes that share a common origin
- Orthology: homologous genes evolved by speciation at their most recent point of origin
- Paralogy: homologous genes evolved by duplication at their most recent point of origin
- Xenology: homologous genes mimicking orthologs but derived by HGT from another lineage
- In-/Out-paralogy: paralogous genes arising from lineage-specific duplication(s) after/before a given speciation event
- Co-orthology: in-paralogous genes that are collectively orthologous to genes in another lineage (due to their common origin by speciation)
- Orthologous group (COG): collection of all descendents of an ancestral gene that diverged from (after) a given speciation event

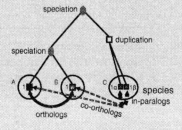

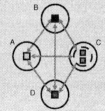

COG with (co) orthologs with 4 species (A,B,C,D) and scheme of bidirectional best hits

Originally, the acronym COGs rather awkwardly stood for Clusters of Orthologous Groups (of proteins), to account for co-orthology relationships resulting from gene duplication (see Box 3-1; Tatusov, et al., 1997). I now prefer to speak simply of Clusters of Orthologous Genes, but the acronym COG still applies and seems rather appropriate as a reference to the fundamental character of these gene clusters. This three-letter word is widely used in the literature, and I use it in this book as a shortcut to denote sets of orthologous genes.[2] Typically, COGs include more than 70% of the genes in each sequenced genome (see Figure 3-4). This seems to be an important quantity in genome evolution, to which we return more than once in this book. So a substantial majority of the genes in each genome are highly conserved—that is, are represented by orthologs in multiple distant organisms.

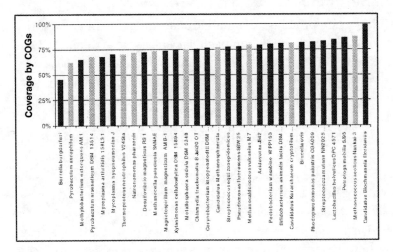

Figure 3-4 Coverage of archaeal and bacterial genomes by COGs. The complete sets of proteins encoded in 20 selected bacterial genomes (black) and 10 selected archaeal genomes (gray) were assigned to the COGs (Tatusov et al., 2003) using the automated COGNITOR method (Makarova, et al., 2007b).

Multidomain proteins and the complexity of orthologous relationships

The main thrust of the discussion in this chapter is the interplay between stability and variation in evolution. In this section, we focus on the discrete units of protein structure, domains, and the multidomain organization of many proteins (Doolittle, 1995). In doing so, we reveal the other side of gene evolution, which contrasts with and complements the stability of orthologous lineages emphasized earlier. A domain is a central concept in protein science that can be defined on at least two levels. Under the first definition, domains are compact units of protein structure, with the characteristic size of around 100 amino acid residues. Here we are concerned with the relationships between genomes, particularly orthology, so there is no need to focus on structural domains. The second definition of *domain* pertains to a distinct unit of evolution that can encompass one or more structural domains; here we are interested in these *evolutionary domains*.

Multidomain proteins are found in all forms of life but are particularly common in complex, multicellular eukaryotes (Koonin, et al., 2000a; Koonin, et al., 2000b). The domain architectures of these proteins show various degrees of evolutionary plasticity. Variability is particularly pronounced among the protein architectures that include so-called promiscuous domains—that is, domains that have a tendency to combine with different kinds of other domains (Basu, et al., 2009). Variable multidomain architectures of proteins confound the concept of orthology. Orthologous genes are assumed to maintain their identity, including a conserved biological role over a long course of evolution (to share the same evolutionary history). This is no longer the case when genes that otherwise fit the definition of orthology (see Box 3-1) evolve different domain architectures (see Figure 3-5): In these instances, only parts of the respective proteins in different organisms share the same history and perform the same function (and even the latter cannot be guaranteed because interactions between domains might well have substantial functional consequences).

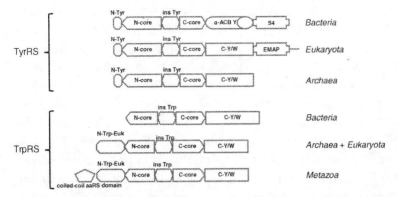

Figure 3-5   Diversity of multidomain architectures of homologous proteins. The schematic compares the domain architectures of two paralogous sets of ancient, essential orthologous proteins: tyrosine-tRNA synthetases (TyrRS) and tryptophane-tRNA synthetases (TrpRS). Each domain is shown with a unique shape. Data comes from Wolf, et al., 1999a.

Fluidity of genomes in contrast to the conservation of genes

We saw that the majority of the genes in each genome are highly conserved—that is, homologs for these genes, most often readily identifiable orthologs, are found in distant organisms. However, this striking evolutionary conservation of genes is only one side of the coin of comparative genomics. The other, contrasting face is the fluidity of the gene composition and architecture of genomes of all life forms. Genomes of prokaryotes are particularly malleable. An emblematic example is the comparison of different strains of the classic model bacteria, the laboratory K12 strain, and several pathogenic strains of the enterobacterium *Escherichia coli* (Perna, et al., 2001). The sequences of orthologous genes in these bacteria are nearly identical, but the pathogenic strains contain up to 30% more genes than the K12 strain, and the gene repertoires of the pathogenic strains differ dramatically. The inevitable conclusion is that the extra genes that form so-called pathogenicity islands have been acquired by some strains or lost by others (we return to these themes in Chapter 5).

On a more global scale, we can measure the distance between genomes, first by comparing sequences of highly conserved marker genes, such as rRNA or r-proteins, and second by examining the fraction of genes that form readily definable, one-to-one pairs of orthologs (see Box 3-1). The steep decay of the congruence between the gene repertoires is apparent, as opposed to the gradual, relatively slow decay of sequence conservation (see Figure 3-6). Note that there is no contradiction between this finding and the observation that, for the great majority of genes in the genome of any bacterium or archaeon, there are orthologs in *some* distant organisms: here *some* is the key word because many genes in any genome have different evolutionary origins and histories, so their closest relatives might be found in different taxa (see Chapter 5). The distance between genomes in terms of the fraction of shared genes can be used to depict the "genomic universe," discussed later in this chapter, and to construct a special kind of evolutionary tree (see Chapter 5).

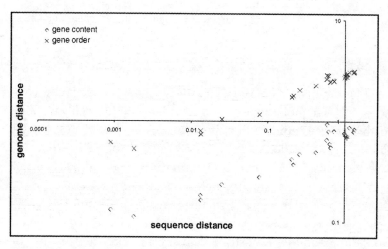

Figure 3-6 Divergence of gene repertoires and gene orders between bacteria, compared to the divergence of highly conserved protein sequences. The distances were computed from Escherichia coli K12 substr. MG1655 to 24 other diverse proteobacteria. Sequence distance: maximum likelihood distance between concatenated alignments of ribosomal proteins computed using the PROTDIST program of the Phylip software package for phylogenetic analysis (Felsenstein, 1996). Gene content distance: $-\ln(J_{COG})$, where J_{COG} is the Jaccard similarity coefficient for the COG sets in the two genomes. Gene order distance: $-\ln(J_{Pair})$, where J_{Pair} is the Jaccard similarity coefficient for the sets of unordered pairs of adjacent COGs in the two genomes. The plot is in double logarithmic coordinates.

Even to a greater extent than the gene composition of genomes, the genome architecture, that is, the arrangement of genes in a genome, shows evolutionary instability that sharply contrasts with the conservation of gene sequences (Koonin, 2009a; Novichkov, et al., 2009). With the exception of the organization of small groups of functionally linked genes in operons, relatively little conservation of gene order exists even among closely related organisms.[3] In prokaryotes, the long-range conservation of gene order disappears even in some groups of genomes that retain an almost one-to-one correspondence of orthologous genes and a greater than 99% mean sequence identity between orthologous proteins (see Figure 3-6). Eukaryotes show a somewhat greater conservation of gene order. However, even in this case, there are few shared elements of genome architecture between, for instance, different animal phyla, and there are no shared elements at all between animals and fungi, or animals and plants.

The genomescapes: Distribution of evolutionary constraints across different classes of sites in genomes

Any genome can be presented as a genomescape, a skyline-like plot in which each nucleotide site is assigned a height proportional to the strength of evolutionary constraints that affect it. In principle, constraints can be reasonably viewed as varying from 0 (unconstrained, neutrally evolving, functionally irrelevant positions) to 1 (fully constrained, essential positions in which no change is permissible; see Figure 3-7; Koonin and Wolf, 2010b). The overall distributions of constraints across genomes are dramatically different in life forms with distinct genome architectures. This is particularly true of the comparison between viruses and prokaryotes, on the one hand, with their "wall-to-wall" genomes that consist mostly of protein-coding and RNA-coding genes, and, on the other hand, multicellular eukaryotes, in whose genomes the coding nucleotides are in the minority (see Figure 3-7). On a per-site basis, the constraints on compact genomes, particularly those of prokaryotes, are orders of magnitude stronger than the constraints on the larger genomes of multicellular eukaryotes. Protein-coding sequences and sequences coding for structural RNAs are the most strongly constrained sequences in all genomes. The great majority of protein-coding genes, especially in prokaryotes, show low *Ka/Ks* values, indicating strong purifying selection affecting the encoded protein sequences (see Figure 3-8 and the preceding chapter). At the same time, in all groups of organisms, there is a significant positive correlation between *Ka* and *Ks*, indicating that even synonymous sites in protein-coding genes are constrained roughly in proportion to the constraints on nonsynonymous sites (Drummond and Wilke, 2008; also see Chapter 4). Given that prokaryotic genomes consist almost entirely of protein-coding genes, with the addition of genes for structural RNAs and short intergenic regions that are largely taken up by variously constrained regulatory regions, these compact genomes contain few unconstrained sites. A notable exception includes pseudogenes that are rare in most prokaryotes but common in some parasitic bacteria, particularly those that grow inside eukaryotic cells, such as *Rickettsia* or *Mycobacterium leprae* (Harrison and Gerstein, 2002). The genomes of most viruses are even more compact than prokaryote genomes, with nearly all of the genome sequence taken up by protein-coding genes.

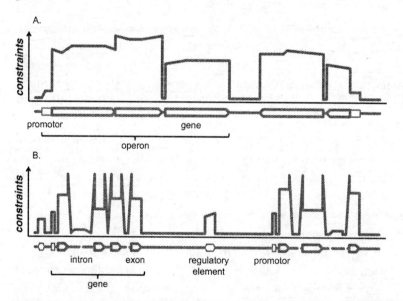

Figure 3-7   Schematic genomescapes: Distributions of evolutionary constraints across different classes of sites in genomes of prokaryotes and eukaryotes reveal distinct genome architecture principles. The top (A) shows a prokaryote genome. The bottom (B) shows a eukaryote genome.

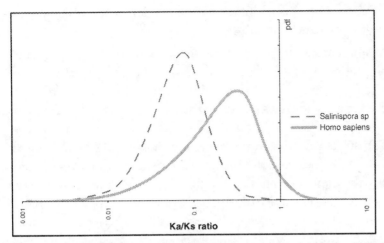

Figure 3-8   Cumulative distributions of the Ka/Ks values in a prokaryote genome and a eukaryote genome. Salinispora sp.: computed for orthologs in Salinispora arenicola CNS-205 and S. tropica CNB-440 (Actinobacteria). Homo sapiens: computed for orthologs in Homo sapiens and Macaca mulatta (Primates). The Ka and Ks values were estimated using the PAML software (Yang, 2007). The plot is in semilogarithmic coordinates; pdf stands for probability density function.

Unicellular eukaryotes that resemble prokaryotes in their overall genome architecture show a roughly similar distribution of evolutionary constraints, although the fraction of apparently unconstrained noncoding sequences in these genomes is somewhat greater. However, genomes of multicellular eukaryotes (plants and especially animals) present a stark contrast. These organisms have intron-rich genomes with long intergenic regions; a substantial, albeit variable, fraction of these noncoding sequences appears to undergo unconstrained evolution. The fractions of the nucleotides in genomes that are subject to evolutionary constraints have been estimated using methods based on the McDonald-Kreitman test (see Box 2-2). These estimated fractions substantially differ even between animals: In *Drosophila*, about 70% of the nucleotide sites in the genome, including 65% of the noncoding sites, appear to be subject to selection (including positive selection); in mammals, this fraction is estimated at 3% to 6% only (Koonin and Wolf, 2010b). Notably, however, the absolute numbers of sites subject to selection in these animal genomes of widely different size are quite close. By contrast, in *Arabidopsis*, a plant that is comparable to *Drosophila* in terms of

genome size and overall architecture, the fraction of constrained noncoding sites appears to be substantially lower.

To summarize the current understanding of the constraints
affecting different classes of sites across the known diversity of the
genomes (see Figure 3-9), some fundamental, straightforward conclusions appear indisputable. In particular, there is no doubt that
nonsynonymous sites in protein-coding sequences and sequences
encoding structural RNAs are among the most strongly constrained
parts of all genomes, and that the characteristic distributions of constraints (the genomescapes) strongly correlate with the genome
architecture (Koonin and Wolf, 2010b). However, beyond these basic
principles, and rather unexpectedly, the evolutionary regimes seem to
widely differ even for some relatively close taxa, such as arthropods
and vertebrates. Much additional research on diverse organisms is
required to develop a comprehensive picture of the constraints and
pressures that shape genome evolution. The chapters that follow
address various manifestations of selective pressures that affect different parts of genomes.

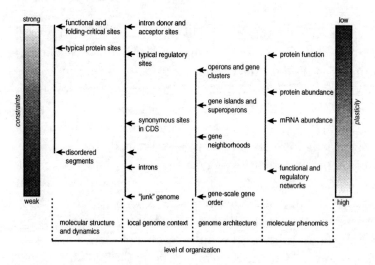

Figure 3-9   A schematic summary of evolutionary constraints that affect different classes of sites in genomes.

The gene universe

Integration of comparative genomic results allows us to start mapping the entire "gene universe." The global evolutionary resilience of genes, manifested primarily in the conservation of protein and RNA sequences, became apparent in the very first comparisons of sequenced prokaryotic and eukaryotic genomes, the bacteria *Haemophilus influenzae* and *Mycoplasma genitalium,* the archaeon *Methanocaldococcus jannaschii,* and the eukaryote yeast *Saccharomyces cerevisiae* (Tatusov, et al., 1997). A key generalization of comparative genomics is that genes are not simply conserved through varying evolutionary spans, but constitute distinct units of evolution, namely orthologous gene lineages (see Box 3-1). With the present collection of sequenced genomes, orthologs in distant taxa are found for the substantial majority of protein-coding genes in each genome. Striking examples come out of the recent genome sequencing of primitive animals: the gene repertoires of sea anemone, *Trichoplax,* and sponge show extensive conservation with mammals and birds (Putnam, et al, 2007; Srivastava, et al., 2008; Srivastava, et al., 2010). The implication is that the characteristic life span of an animal gene in these lineages covers at least hundreds millions of years. Many other animal lineages, such as insects, have lost numerous genes (Koonin, et al., 2004), so the fate of the same gene most of the time differs across lineages, resulting in patchy phyletic patterns. (As emphasized later in this chapter, the set of truly universal genes is tiny.) These lineage-specific gene fates depend on both stochastic factors and differences in selection pressure (see Chapter 9). The results of extensive comparative analysis of plant, fungal, and prokaryotic genomes are fully compatible with this conclusion. When the genes in a genome are classified by their apparent relative "age" (that is, the phylogenetic depth at which homologs are detectable), the resulting breakdown is similar for distant organisms, as illustrated in Figure 3-3 for the gene sets of humans and the fungus *Aspergillus fumigatus* (Wolf, et al., 2009). The two organisms are separated by perhaps a billion years of evolution. Nevertheless, the distributions of gene ages are strikingly similar: In each case, ancient genes with readily detectable homologs in distant taxa are significantly more numerous than "younger" genes. *Despite the widespread lineage-specific loss, genes are characterized by extreme longevity, and many might be immortal.*[4]

As discussed later in this book (see Chapters 5 and 7), the paths of genetic information transfer in prokaryotes fundamentally differ from those in eukaryotes. Nevertheless, the proportion of conserved genes is about the same. By now, this proportion is well established and quite similar in diverse bacteria and archaea, almost like a fundamental constant: For 70% to 80% of the genes, orthologs are detectable in distant organisms (Koonin and Wolf, 2008b; see Figure 3-4).

Minimal gene sets, non-orthologous gene displacement, and the elusive essential core of life

Sequencing the genomes of symbiotic and parasitic bacteria triggered the seductive idea that their gene repertoires could approximate the "minimal gene set"—that is, *the set of genes that is both necessary and sufficient to sustain a simple (prokaryotic) cell under the most favorable conditions that can be created outside other cells* (Fraser, et al., 1995; Mushegian and Koonin, 1996b). The latter qualification is critical because "minimal gene sets" are necessarily contingent on environmental conditions in which the respective organism exists (or would exist, in the case of computationally derived "conceptual" genomes). However, as soon as the first two bacterial genomes became available, the second one being the genome of *Mycoplasma genitalium*,[5] a wall-less parasitic bacterium with only about 570 genes, the obvious idea presented itself: Comparing the two differentially specialized genomes of bacterial pathogens would naturally yield the "true" minimal set (Mushegian and Koonin, 1996b). More precisely, one would expect that the orthologous genes in the two organisms would represent the set of essential biological functions that are required for the survival of a cell, regardless of the unique lifestyle of each organism.

Comparing the gene sets of *H. influenzae* and *M. genitalium* yielded 240 pairs of orthologous genes that encompassed most of the apparently essential cellular functions. However, several such functions were conspicuously missing from the conserved gene set. So far, we have discussed little "real biology"—that is, biological functions (roles) of genes. However, at this point in the narrative, we must think biologically. Defining the minimal set of essential biological functions is not trivial. It is tempting to go about this task by "reverse evolutionary engineering"—that is, from comparative genomics, to define the minimal set of essential genes as those that are conserved in all cellular life forms. However, this approach would ignore the possibility that different organisms could have evolved independent solutions for the same essential task. We shall see later in this chapter that this hypothetical possibility indeed captures a major aspect of biological reality. So to delineate the minimal set of cellular functions, we need to additionally employ the logic of biochemistry and cell biology. Knowledge in these fields is indeed sufficient to produce a reasonable catalogue of essential activities. Obviously, this knowledge is not perfect, so the actual inference of a minimal gene set requires iterative use of biological reasoning and comparative-genomic analysis. Arcady Mushegian and I surmised that the essential functions missing among the 240 *H. influenzae–M. genitalium* orthologs were probably performed by unrelated or distantly related proteins in the two bacteria. We did some guesswork to augment the putative minimal set with 16 additional genes of *M. genitalium* (see Figure 3-10). This straightforward effort in deriving a minimal gene set by combining comparative genomics and biological reasoning seems to have been reasonably successful and might approximate the functional repertoire of the simplest bacterial cell capable of independent growth under the best possible conditions. Indeed, subsequent gene knockout experiments have confirmed that most of the genes included in the minimal set are essential for bacterial survival, and the genes from the minimal set are conserved in many (though not necessarily all) newly sequenced bacterial genomes (Delaye and Moya, 2010; Koonin, 2003).

It is instructive to take a functional census of the minimal set of bacterial genes. This set is heavily dominated by genes encoding proteins involved in information transmission in the cell (that is, replication, transcription, and, above all, translation). Metabolic enzymes and transport systems are much more sparsely represented, as we might expect for an organism growing on the richest possible media. In that respect, the minimal gene set is dramatically different from the full set of COGs but resembles the set of essential bacterial genes (inactivation of the genes kills the bacterium; see Figure 3-11). This preferential evolutionary stability of the information transmission systems is one of the core generalizations of comparative genomics. We return to this subject later.

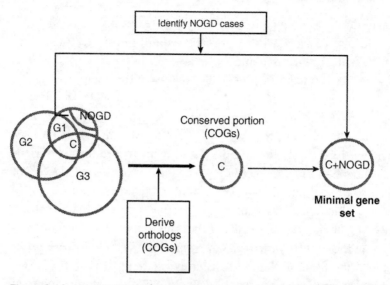

Figure 3-10   Delineation of the minimal gene set for cellular life by comparative genomics. G1, G2, G3: three compared genomes; C = set of conserved genes.

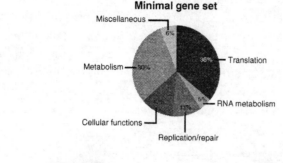

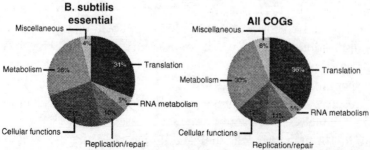

Figure 3-11   Breakdown of biological functions in the minimal gene set, the complete set of COGs, and the set of experimentally identified essential genes in the bacterium Bacillus subtilis. Data comes from Koonin, 2003.

However, it seems that the most consequential outcome of the minimal gene set exercise was the finding that several essential functions were missing from the list of readily detectable orthologs. This observation was dramatically reinforced by comparing the bacterial genomes with the first archaeal genome (*Methanocaldococcus jannaschii*), which revealed a number of additional glaring gaps in the set of conserved essential functions. These findings have been conceptualized in the notion of *non-orthologous gene displacement (NOGD)*, an evolutionary scenario under which unrelated or distantly related genes (not orthologs) become responsible for the same essential function in different organisms (Koonin, 2003). The actual evolutionary scenario is easy to imagine (see Figure 3-12): For NOGD to occur, an evolving lineage acquires an alternative, functionally redundant gene for a particular essential role and so goes through an intermediate state in which both incarnations of the function in question are present (such redundancy is often observed in organisms with more complex genomes), followed by the loss of the original version (Koonin and Mushegian, 1996). With the growth of the genome collection, it is becoming increasingly common to find organisms (typically those with larger genomes) in which both incarnations of various functions are represented, thus adding weight to the scenario of NOGD evolution shown in Figure 3-12.

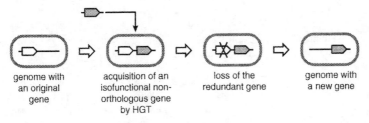

genome with an original gene   acquisition of an isofunctional non-orthologous gene by HGT   loss of the redundant gene   genome with a new gene

Figure 3-12   An evolutionary scenario for non-orthologous gene displacement.

Box 3-2 includes several examples of key biological functions for which two or more unrelated enzymes are differentially represented in partially complementary but typically overlapping sets of lineages. Even these few selected cases show that NOGD reaches across a variety of functional systems and pathways. Subsequently, with the vastly increased collection of genomes, it has become obvious that NOGD and lineage-specific gene loss are so widespread that few functions are truly monomorphic and ubiquitous (that is, represented by genes from the same orthologous lineage in all cells). Accordingly, the universal core of life has shrunk almost to the point of vanishing. All that remains ubiquitous are some 30 genes for proteins involved in translation and 3 large RNA polymerase subunits, along with an approximately equal number of structural RNA genes (rRNAs and tRNAs). Even when parasitic bacteria are disregarded, the list of universal genes does not expand much (Koonin, 2003). Thus, with the notable exception of a miniscule core of genes involved in the key steps of information transmission, *there is no universal genetic core of life, owing to the near ubiquity of NOGD and gene loss.* The concept of a relatively *small, universal set of functions* that are required to sustain a cell remains viable, but considering the combinatorics of NOGD, a vast variety of gene ensembles can fill the same minimal set of functional niches.

Box 3-2: Cases of non-orthologous gene displacement

Biological Function	Gene(s)/ Protein(s)	Phyletic Spread of Form 1	Phyletic Spread of Form 2	Phyletic Spread of Form 3
DNA replication	DNA polymerase, primase, helicase	Archaea, eukaryotes, many dsDNA viruses of prokaryotes and eukaryotes	Bacteria	—
DNA precursor biosynthesis	Thymidylate synthase	Scattered among bacteria and archaea, most eukaryotes	Scattered among bacteria and archaea, a minority of eukaryotes	—
Translation	Lysyl-tRNA synthetase	Most bacteria, a minority of archaea, most eukaryotes	Most archaea, a minority of bacteria	—
Glucose-6-phosphate isomerase	Central carbohydrate metabolism	Most bacteria, some archaea, eukaryotes	A minority of bacteria, some archaea	—

Biological Function	Gene(s)/ Protein(s)	Phyletic Spread of Form 1	Phyletic Spread of Form 2	Phyletic Spread of Form 3
Fructose bis-phosphatase	Central carbohydrate metabolism	About half bacteria, a few archaea, most eukaryotes	About half bacteria, small minority of archaea and eukaryotes	Most archaea, small minority of bacteria
Diacylglycerol kinase	Lipid metabolism	Most bacteria	Eukaryotes, small minority of bacteria	—
Carbonic anhydrase	Acid-base homeostasis, oxidative stress response	Majority of bacteria, some archaea, some eukaryotes	Minority of bacteria, most archaea, minority of eukaryotes	Minority of bacteria, majority of eukaryotes

Data comes primarily from Omelchenko, et al., 2010.

The units of evolution and the fractal structure of the gene universe

The results of comparative genomics lead to a key generalization that allows us to conduct productive evolutionary studies: The fundamental units of evolution can be fairly clearly defined and consist of sets of orthologous genes or evolutionary domains (COGs)—or, more precisely, evolving orthologous gene (domain) lineages. The histories of individual genes are often complex (extremely complex, in many cases) and include multiple gene losses, duplications, and horizontal transfers (we discuss these phenomena in greater detail later in the book; see Chapters 5 and 7). The propensities of genes for duplication, loss, and transfer differ within a broad range. However, all these complications notwithstanding, the "atomic" nature of orthologous gene sets remains solid: *The COGs are natural elements of the gene universe.*

The gene (genomic) universe (only a metaphor, but a convenient and perhaps productive one) can be represented as an evolving space-time filled with clusters consisting of genes (COGs)—or, more precisely, evolving orthologous lineages, the elementary units of evolution. Orthology is most easily traced between prokaryote genes, so here I discuss the prokaryotic domain of the genomic universe. The trends among eukaryotes are similar in principle but are more complicated because of widespread multidomain organization of proteins and extensive paralogy. In this genomic space, the COGs show a unique distribution across the genomes that can be well approximated with three exponential functions that partition the gene population into three distinct classes (see Figures 3-13A-C; Koonin and Wolf, 2008b).

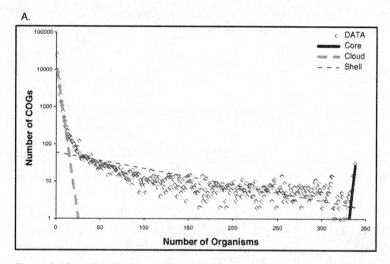

Figure 3-13A   The fractal organization of the gene universe: the gene core, shell, and cloud recapitulated at different phylogenetic depths. The deepest level: 338 prokaryotes from the EggNOG collection (Jensen, et al., 2008)

B.

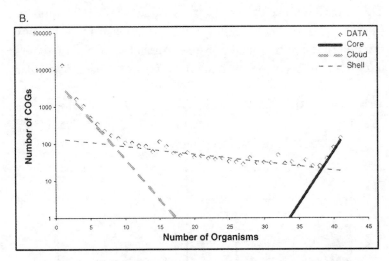

Figure 3-13B   The middle level: 41 archaea from the arCOG collection
(Makarova, et al., 2007b)

C.

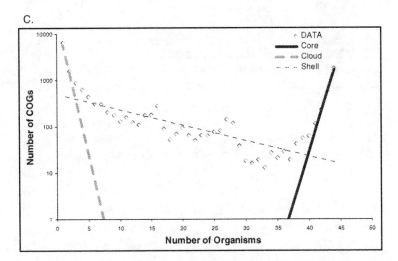

Figure 3-13C   The shallow level: 44 Escherichia, Shigella, and Salmonella
species from the COG collection (Tatusov et al., 2003). The data for all three
figures was fitted with three exponential functions (Koonin and Wolf, 2008b),
shown by broken or solid lines.

1. The (nearly) universal genes, those that are represented in (nearly) all genomes of cellular life forms make up but a tiny fraction of the entire gene universe: Altogether, this *core of cellular life* consists of, at most, about 70 genes. In each particular genome, the fraction of these core genes is no greater than 10%, even in the smallest of the genomes of cellular life forms (parasitic bacteria such as *M. genitalium*), but typically is closer to 1% of the genes or less (see Figure 3-14).

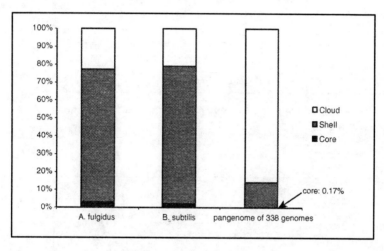

Figure 3-14   The contributions of the core, shell, and cloud to the composition of individual genomes and the entire gene universe. The calculation was done using the data from the EggNOG collection (Jensen, et al., 2008) (A. fulgidus, the archaeon Archaeoglobus fulgidus; B. subtilis, the bacterium Bacillus subtilis).

2. The moderately conserved gene *"shell"* consists of COGs represented in a broad variety but not overwhelming majority of genomes. Recent analysis of the available prokaryotic genomes puts the number of shell COGs at about 5,000. The shell genes comprise the bulk of the gene complement in any genome (see Figure 3-14).

3. The poorly conserved *"cloud"* consists of COGs that are limited to narrow groups of organisms, along with "ORFans" (genes so far identified in one genome only, but for which homologs are usually detected when additional related genome sequences

become available). The cloud genes account for a variable fraction of the genomes, usually between 10% and 30% of the genes (see Figure 3-14).

Remarkably, this structure is self-similar, or *fractal:* The same three components—the tiny core, the larger shell, and the comparatively huge cloud (let us denote this partitioning the CSC structure)—appear at any level where the gene space-time is dissected, from the entire prokaryote world to narrow groups of bacteria (see Figure 3-14). We return to the implications of this fractality of the prokaryote gene space-time in Chapter 5. An evolutionary model to explain the observed fractal pattern remains to be developed, though.

There is an apparent paradox in the distribution of the COGs in the gene space. Although in each individual genome the majority of the genes belong to the shell that is shared with distantly related organisms, when the entire gene universe is considered, the core and shell genes (or more precisely, COGs) are but a small minority (see Figure 3-14). Obviously enough, this difference arises because the shell COGs are represented in many genomes, whereas the cloud COGs and ORFans are rare or unique. Given this distinctive structure of the gene universe, evolutionary reconstructions inevitably yield a highly dynamic picture of genome evolution, with numerous genes (primarily from the cloud and to a lesser extent from the shell) lost and many others gained via HGT (mostly, in prokaryotes), and extensive gene duplication, primarily in eukaryotes (see later in this chapter).

The elementary events of genome evolution

Now that we have defined the units of genome evolution and developed an idea of the organization of the gene universe, it makes sense to complement these concepts with a list of the basic operations, the elementary events of genome evolution that we can compare with the elementary events of evolution of individual genes. The alphabets of elementary events are rather short and, actually, similar (isomorphous) at their respective levels (see Box 3-3). However, the relative contributions and frequencies of different types of events are dramatically different. A substantial difference between the evolution of individual genes and genome evolution lies in the special importance and high frequency of gene duplication, as opposed to the more limited impact of intragenic duplications. Furthermore, intragenic recombination is rarely fixed in evolution except for closely related genomes, whereas the crucial mechanisms of genome rearrangement, such as inversions and translocations, play no significant role in the evolution of individual genes. Taken together, the differences in the relative contributions of various elementary mechanisms (see Box 3-2) underlie the *substantially more dynamic character of genome evolution compared to the evolution of individual genes.*

Box 3-3: Elementary events of gene and genome evolution: A comparison

Type of Evolutionary Event	Genes	Genomes
Substitution	Nucleotide/amino acid substitutions, one of the key processes	Substitution of genes by nonorthologous or xenologous versions; important but relatively infrequent
Deletion/loss	Small deletions nearly as common as substitutions; larger deletions progressively less frequent	Lineage-specific gene loss via deletion/ inactivation common and extensive in some lineages
Insertion	Small inserts common, although typically less frequent than deletions	Acquisition of genes via HGT, a major route of genome evolution; other routes less common
Recombination/HGT	Intragenic recombination relatively rare, except between closely related genes via homologous recombination	A major route of genome evolution; dominant in prokaryotes
Duplication	Duplication of small regions common; larger intragenic duplications progressively less frequent	A major route of genome evolution; dominant in eukaryotes

Synopsis and perspective

Comparative genomics reveals a remarkable contrast between the relative evolutionary stability of individual genes, many of which retain their identity over hundreds of millions or even billions of years of evolution, and the malleability of composition and architecture of genomes that change orders of magnitude faster. Hence, the distinct organization of the gene universe emerges in which the few dense clusters consist of core genes represented in most genomes, whereas the huge number of increasingly sparse "nebulae" consisting of rare genes occupies most of the space-time. Strikingly, this organization of the gene universe is distinctly fractal—that is, it appears at all scales of evolutionary distances.

The "atomic" nature of genes (or, more precisely, COGs, orthologous evolutionary lineages) underlies the feasibility of the entire comparative genomic enterprise: Genome comparisons are highly informative, although the emerging relationships are far from simple, given the fluidity of genome architectures.

The genomescapes of different life forms, which can be defined as distributions of constraints across genomic sites, are diverse and intricate. The compact genomes of viruses, prokaryotes, and, to a lesser extent, unicellular eukaryotes mostly occupy "high altitudes," with nearly all sites subject to substantial selective constraints. The genomescapes of multicellular eukaryotes consist mostly of "valleys," with weak or effectively nonexistent selective constraints separated by rare "ridges" of strong selection. These differences reflect distinct evolutionary regimes, which we discuss in Chapter 8. Paradoxically, only the "inefficient" regime of evolution characteristic of multicellular eukaryotes allows for the emergence of organizational complexity. This paradox should give pause to anyone who is partial to the idea of "progressive evolution." We return to this issue at length in Chapters 8 and 13.

Recommended further reading

Ellegren, H. (2008) "Comparative Genomics and the Study of Evolution by Natural Selection," *Molecular Ecology* 17: 4,586–4,596.

An overview of selective factors that appear important for the evolution of different classes of genomic sequences.

Koonin, E. V. (2005) "Orthologs, Paralogs, and Evolutionary Genomics." *Annual Review of Genetics* 39: 309–338.

A detailed discussion of the concepts of orthology and paralogy, and of specific categories of evolutionary relationships between genes within these broad classes.

Koonin, E. V. (2009) "Evolution of Genome Architecture." *International Journal of Biochemistry and Cell Biology* 41: 298–306.

An overview of the diversity and evolutionary trends of genome architectures in different forms of cellular life.

Koonin, E. V. (2003) "Comparative Genomics, Minimal Gene-Sets, and the Last Universal Common Ancestor." *Nature Reviews Microbiology* 1: 127–136.

A critical discussion of the minimal gene set concept as applied to organisms with different lifestyles and a comparison of minimal gene sets with reconstructions of ancestral genomes.

Koonin, E. V., and Y. I. Wolf. (2010) "Constraints and Plasticity in Genome and Molecular-Phenome Evolution." *Nature Reviews Genetics* 11: 487–498.

An attempt on a comprehensive genome-wide census on constraints that affect different classes of sequences and sites in genomes.

Koonin, E. V., and Y. I. Wolf. (2008) "Genomics of Bacteria and Archaea: The Emerging Dynamic View of the Prokaryotic World." *Nucleic Acids Research* 36: 6,688–6,719.

A comprehensive overview of the prokaryote genomics, with a special emphasis on genome dynamics, including HGT.

Levitt, M. "Nature of the Protein Universe." (2009) *Proceedings of the National Academy of Sciences USA* 106: 11,079–11,084.

A detailed study of the novelty generation in protein evolution. The conclusion is that emergence of new multidomain architectures is a key mechanism of novelty creation.

Lynch, Michael. (2007) *The Origins of Genome Architecture* Sunderland, MA: Sinauer Associates.

The seminal book on the nonadaptive theory of genomic complexity evolution and its various implications (for a detailed discussion, see Chapter 8).

Wilkins, A. S. (1997) "Canalization: A Molecular Genetic Perspective." *Bioessays* 19: 257–262.

A reappraisal of Waddington's concept of canalization in the context of modern evolutionary biology.

4

Genomics, systems biology, and universals of evolution: Genome evolution as a phenomenon of statistical physics

In the preceding chapter, I emphasized the relative evolutionary stability of individual genes, which contrasts with the dynamic character of genome evolution. If genes or evolutionary domains can be reasonably construed as "atomic units" of genome evolution, then genomes may be viewed as statistical ensembles of these units. We can extend this oversimplified but apparently sensible and potentially productive physical analogy. The genomes can be viewed as more closely resembling gases, or perhaps liquids, where interactions between molecules are variable and important, but weak compared to the intramolecular interactions that underlie the stability of molecules—in contrast to solid states, in which intermolecular interactions are strong and defining.

Textbook knowledge is that, to a good approximation, the behavior of ensembles of weakly interacting particles (molecules) follows simple and universal statistical regularities, such as the Boltzmann distribution of particle velocities. The analogy between ensembles of genes (genomes) and ensembles of molecules (gases and liquids) prompts the search for universal statistical patterns in genome function and evolution. Moreover, this line of thinking makes one predict with some confidence that these statistical patterns should come in the form of mathematically simple, universal distributions of the values of certain variables that describe the process of evolution. We will see in this chapter that the quest for evolutionary universals is far from futile.

Before we discuss the statistical properties of gene ensembles, it is necessary to introduce another hallmark of biological research in the first decade of the third millennium, the new field that is most often referred to, perhaps somewhat disingenuously, as systems biology. The lofty ultimate goal of systems biology is to model and "understand" the functioning of biological systems in their full complexity. The reality of the current, early stage in the progression of this research field is that efforts in systems biology largely amount to collection of extensive "omic" data, including transcriptomes (the complete set of transcripts for a given cell, tissue, or organism), proteomes (the complete set of expressed proteins), and metabolomes (the complete set of metabolites) and other "omes" (Bruggeman and Westerhoff, 2007; Koonin and Wolf, 2008a). All these "omes" are characterized by systems biology in quantitative terms such as protein abundances and metabolite concentrations.

Much like with genomics at its dawn, many biologists initially considered systems biology to be boring "big science" and "busy work." (I suspect this attitude still persists.) As with genomics, this turned out to be a myopic view, to put it mildly. The high-quality, genome-wide data on gene expression, genetic and protein-protein interactions, protein localization within cells, and other types of system-level data has opened up new dimensions of evolutionary analysis (sometimes denoted evolutionary systems biology) that blends with evolutionary genomics in a new synergy. Research in systems biology has already yielded unexpected insights into the genome-wide connections of sequence evolution, gene expression, protein structure, and other characteristics of genes and proteins. These findings seem to be generally compatible with the view of genomes as statistical ensembles of genes and illuminate in new ways the selective and neutral components of the evolution of genome structure and function.

Correlations between evolutionary and phenomic variables; universals of gene, protein, and genome evolution; and physical models of the evolutionary process

As the preceding chapter illustrated, protein-coding genes (at least, the nonsynonymous positions that determine the encoded amino acid sequence) are among the most strongly constrained sequences in all genomes. However, already in the early days of molecular evolution studies, it was realized that evolutionary rates of protein-coding genes differ within a broad range (Wilson, et al., 1977). These broad distributions were generally attributed to the wide spectrum of protein functions that differentially constrain the evolution of the respective genes. Indeed, it stands to reason that, say, the function of a DNA polymerase, a sophisticated enzyme that catalyzes the template-dependent incorporation of nucleotides into the growing DNA chain, would constrain the gene sequence evolution to a much greater extent than, for instance, the function of a structural protein whose only role is to maintain the integrity of the nuclear matrix. However, the prescient idea that evolution of protein-coding genes might not completely boil down to unique molecular details of the protein function emerged in these early days. Allan Wilson and colleagues hypothesized in their seminal 1977 review article that the evolution rate of a gene sequence depended both on the unique function of the encoded protein and on the importance of that protein for the survival of the organism (Wilson, et al., 1977). However, no direct methods were available at the time to study evolutionary constraints directly, so these ideas, however intriguing, belonged in the realm of speculation.

At the beginning of the third millennium, genomics and systems biology have completely transformed the scene of evolution research. With multiple genome sequences available, it has become possible to analyze and compare the distributions of evolutionary rates across complete sets of orthologous genes in different taxa, and also to examine the correlations between evolutionary rates of orthologs in different lineages. The distribution of the rates of evolution among nonsynonymous sites in orthologous genes in any pair of compared genomes spans three to four orders of magnitude and is much broader than the distribution of the rates for synonymous sites (see Figure 4-1). Remarkably, the shapes of the rate distributions for orthologous proteins are extremely similar, to the point of being virtually indistinguishable in all studied cellular life forms, from bacteria to archaea, to mammals (see Figure 4-2; Grishin, et al., 2000; Wolf, et al., 2009). All these distributions have the so called log-normal shape—that is, the logarithm of the evolutionary rate is distributed approximately normally (a Gaussian-like, bell-shape curve). In the theory of random processes, this shape of a distribution typically appears as a result of multiplication of many independent random variables. Given the dramatic differences in the functional organization and the actual number of genes between organisms, this universality of the evolutionary rate distribution is quite unexpected and hints at the existence of fundamental, simple explanations that we discuss later in this chapter.

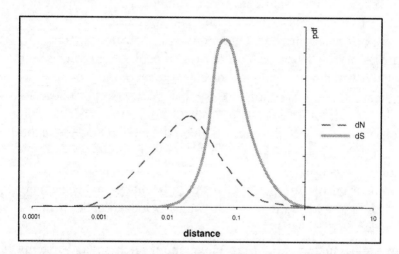

Figure 4-1   Distributions of synonymous and nonsynonymous rates of evolution in orthologous genes from human and mouse. dN = rate of evolution of nonsynonymous sites; dS = rate of evolution of synonymous sites; pdf = probability density function. Data comes from Wolf, et al., 2009; the PAML software (Yang, 2007) was used to compute the evolution rates.

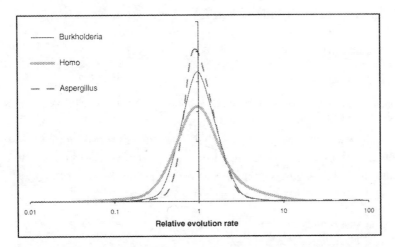

Figure 4-2 Distributions of the evolution rates for orthologous protein sets from bacteria, archaea, and eukaryotes. Burkholderia = computed for orthologs in Burkholderia cenocepacia and Burkholderia vietnamiensis (Proteobacteria). Homo = computed for orthologs in Homo sapiens and Macaca mulatta (Primates). Aspergillus = computed for orthologs in Aspergillus fumigatus and Neosartorya fischeri (Ascomycota). Data comes from Lobkovsky, et al., 2010; evolution rates were computed using the PAML software (Yang, 2007).

The progress of systems biology created the appealing opportunity to measure correlations between evolutionary rates and all kinds of "molecular phenomic" variables, such as expression level, protein abundance, protein–protein interactions, the actual phenotypic effects of gene mutation, and more (Koonin and Wolf, 2006). Such "correlomics" studies have become almost a separate research area, although the ultimate goal, of course, is not to simply describe the correlations or even their fine structure, but to develop physically sound and transparent models of genome and phenome evolution. Many significant correlations have been detected and patterns have emerged, notwithstanding the noisy molecular-phenomic data (especially in the early years of systems biology). Figure 4-3 presents a straightforward (even if inevitably oversimplified) summary (Wolf, et al., 2006). The emerging picture boils down to the existence of two broad classes of variables:

1. *Intensive, evolutionary variables*—Various rates of genome change, including sequence evolution, gene loss, genome rearrangement, and other kinds of evolutionary processes

   2. *Extensive, phenomic variables*—Expression rate, translation
   rate, protein abundance, interactivity and position in other networks, and so on

   Correlations within each of the two classes of variables are typically positive, whereas between-class correlations are negative (see
Figure 4-3). This pattern translates into a "gene status" model in
which high-status genes evolve slowly by any measure used and, conversely, are highly expressed and interact with many other genes. By
contrast, low-status genes change fast while being expressed at low
levels and interact with few partners (see Figure 4-4).

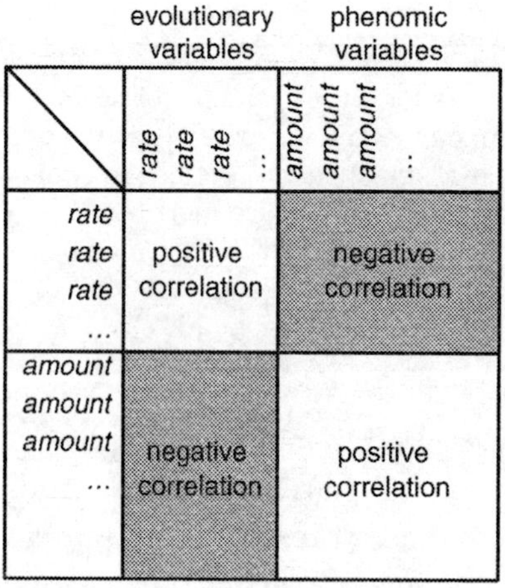

	evolutionary variables				phenomic variables			
	rate	rate	rate	⋮	amount	amount	amount	⋮
rate rate rate ...	positive correlation				negative correlation			
amount amount amount ...	negative correlation				positive correlation			

Figure 4-3   A schematic summary of correlations between evolutionary and
molecular phenomic variables.

   The strongest, universal link between evolutionary and molecular–
phenomic variables is the negative correlation between the rate of evolution of protein-coding genes and their expression levels: *Highly
expressed genes evolve slowly.* This dependence was invariably observed
in all model organisms for which expression data is available (Drummond, et al., 2006; Drummond and Wilke, 2008; Pal, et al., 2001). Given
the aforementioned positive correlation between *Ka* and *Ks*, it is not
surprising that both rates show qualitatively the same dependence.
More unexpectedly, this anticorrelation of evolutionary rate with expression has been detected also for 3'-untranslated regions (UTRs) although
not for 5'UTRs (Jordan, et al., 2004). This universal anticorrelation
comes across even stronger when the evolutionary rates are compared

directly with the experimentally measured protein abundances (Schrimpf, et al., 2009).

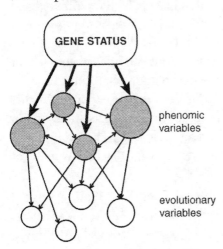

Figure 4-4   The model of "gene status."

The universal link between gene expression and evolution prompted a bold attempt on theoretical reinterpretation of protein evolution under which the primary causes of protein evolution have more to do with fundamental principles of protein structure and folding that are common across all life than with unique biological functions. It has been proposed, primarily in the work of Allan Drummond and Claus Wilke, that the principal selective factor underlying the evolution of proteins is robustness to misfolding. Under this hypothesis the primary fitness cost of mutations, both genomic and phenotypic (translation errors), comes from the deleterious effect of misfolded proteins that, in addition to the expenditure of energy, could be toxic to the cell (Drummond, et al., 2005; Drummond and Wilke, 2008). Without going into details here, this is an intuitively attractive model that quite naturally explains the anticorrelation between expression and sequence evolution: Obviously, the deleterious effect of misfolding is expected to be stronger for abundant proteins than for those produced in small quantities. Put another way, expression (protein production) is a "lens" that amplifies any deleterious effect associated with a given protein sequence, and the fitness cost of misfolding is the primary manifestation of such deleterious effects. Hence, the genes for abundant proteins would be subject to strong constraints, resulting in slow evolution. This model is compatible with the well-established preferential use of optimal codons (strong codon bias) in highly expressed and highly conserved protein-coding genes, and with the aforementioned positive correlation between Ka and Ks. Under the hypothesis of misfolding-driven protein evolution, synonymous sites are constrained, at least in part, by the same factors as the evolution of proteins because of the pressure for the preferential use of optimal codons in highly expressed proteins (hence, fast and more accurate translation) and in specific sites that are important for protein folding. The evolution of the 3'UTRs could follow the same trend, given the involvement of these regions in regulating translation.

In a study with Alexander Lobkovsky and Yuri Wolf, we asked whether two big birds could be killed with one stone—that is, whether it might be possible to explain both the universal distribution of the evolution rates of protein-coding genes and the equally ubiquitous anticorrelation between evolution rates and expression within one simple model (Lobkovsky, et al., 2010). An analysis of misfolding-dominated protein evolution that employed a simple so-called off-lattice[1] folding model yielded estimates of evolutionary rates under the assumption that misfolding was the only source of fitness cost. The results reproduced, with considerable accuracy, the universal distribution of protein evolutionary rates, as well as the dependence between evolutionary rate and expression. These findings suggest that *the universal rate distribution indeed is a consequence of the fundamental physics of protein folding.*

The *absence or weakness* of certain intuitively expected correlations between evolutionary and phenomic variables is no less striking than the correlations that actually have been detected. Indeed, as pointed out at the beginning of this section, probably the strongest biological intuition in this whole area is that the more "biologically important" a gene is, the slower it changes during evolution and the less likely it is to be lost (Wilson, et al., 1977). The general concept of biological importance can be made concrete by measuring the phenotypic effects of the knockout or other mutations in multiple genes—preferably, all genes in many organisms. One would expect that the greater the effect of knockout, the slower a gene would evolve—in particular, essential genes (those in which knockout is lethal) would evolve much more slowly than nonessential genes. By now, the comparison between the phenotypic effects of knockout and the rates of evolution has been done for a variety of model organisms, and the results are both unequivocal and almost shockingly counterintuitive: *The connection between the experimentally measured biological importance of a gene and its rate of evolution is very weak, if it exists at all* (Hurst and Smith, 1999; Jordan, et al., 2002; Krylov, et al., 2003; Wang and Zhang, 2009). Even more surprising is the lack of a strong correlation between the rate at which genes are lost during evolution (one could think of this rate as a long-term measure of gene essentiality) and the experimentally determined fitness effect: Only the set of genes that are never lost during very long spans of evolution (such as throughout the evolution of eukaryotes) is enriched for essential genes (Krylov, et al., 2003; Wang and Zhang, 2009). The first studies that demonstrated this (near) lack of connection between the rate of evolution and biological importance dealt only with the all-or-nothing effect of gene knockout (essential vs. nonessential genes); therefore, they could be questioned on the grounds that such measurements were too crude and not a good proxy for evolutionarily relevant importance. However, in the latest work of Jianzhi Zhang's laboratory, the near lack of correlation with the evolutionary rate has been demonstrated for precise fitness measurements in yeast *S. cerevisiae* under numerous conditions (Wang and Zhang, 2009). It is very difficult to make the case that all these measures are irrelevant.

What could be the underlying causes of the unexpectedly weak connection between evolution and function? With respect to sequence evolution, one could argue that the evolutionary rate depends more on intrinsic features of a gene (in particular, the structure of the encoded protein) than on its biological importance. However, this argument does not hold for the rate of gene loss. The only sensible—and, again, counterintuitive—explanation seems to be that *the phenotypic effect of a gene knockout (and, accordingly, the set of essential genes) is not at all an evolutionarily conserved feature and changes rapidly (on the evolutionary scale), conceivably due to the fast evolution of gene interaction networks.* Clearly, this is an experimentally testable prediction, even if the experiments required are laborious.

Nearly neutral networks and protein evolution

In general, the rate of evolution of a gene is determined by the size of the (nearly) neutral network—that is, the network of sequences connected by effectively single-step mutation distances (although not necessarily by single replacements) that have approximately the same fitness as the most fit sequence (Wagner, 2008a; Wolf, et al., 2010). The bigger the neutral network, the weaker the constraints on the given gene, and so the faster it can (and does) evolve (see Figure 4-5).

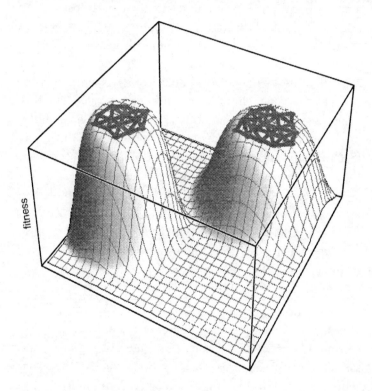

Figure 4-5 The nearly neutral networks and protein evolution. Two nearly neutral networks are schematically shown as being located on broad tops of fitness peaks for two hypothetical proteins.

Coming back to protein evolution, if the fitness of a particular sequence mostly depends on its robustness to misfolding and expression level, the size of the nearly neutral network depends on the height and shape of the peak occupied by that sequence and its neighbors in the robustness landscape (see Figure 4-6). Under this model, highly expressed proteins whose native sequences are highly robust to misfolding would occupy tall, steep peaks, with small areas of high fitness (small nearly neutral networks)—hence, strong purifying selection and slow evolution. In contrast, proteins with lower robustness occupy lower and wider peaks, with larger areas of high fitness at their typical expression levels, thus entailing weaker selection and allowing faster evolution (see Figure 4-6; Wolf, et al., 2010).

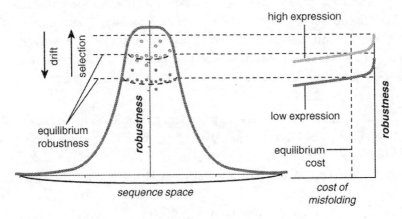

Figure 4-6   The model of misfolding-driven protein evolution.

Genome evolution by gene duplication, gene birth and death models, and the universal distribution of paralogous family sizes

We have already touched on gene duplication in the preceding chapter, particularly when compiling the list of the principal mechanisms of genome evolution. However, there are at least two good reasons for revisiting this mechanism of evolution and discussing it in greater detail. First, duplication is indeed a major route of genome evolution in all walks of life, apparently the principal one in eukaryotes (see Chapter 8). Second, evolution by gene duplication is formally a simple process that can be readily encapsulated in straightforward physical (or mathematical) models, which are the subject of this chapter.

The idea of duplication as a facile "method" of genome evolution is at the heart of our evolutionary thinking. Deliberately trivializing the matter, it seems obvious that making new functional devices (proteins and RNAs) from pre-existing evolved entities by tinkering with them (recall Jacob's key metaphor; Jacob, 1977) is so much easier than creating such devices *de novo*, from scratch (the history of this idea is outlined in Chapter 2). As ever, genomics puts the concept of

evolution by gene duplication on a firm quantitative basis by showing
that the majority of the genes in any genome belong to families of
paralogs (with the exception of the smallest genomes, like those of
Mycoplasma and other parasitic bacteria, in which the fraction of
"singletons" is greater; Jordan, et al., 2001). More detailed evolutionary reconstructions show that duplications occur, with varying intensity, at all stages of evolution, so any genome is a compendium of
duplications of all ages. However we define a lineage—say, animals,
chordates, mammals, primates, and so on—we can find in a genome
(for example, our own) all classes of lineage-specific duplications: ani-
mal-specific, chordate-specific, mammalian-specific, and so on
(Lespinet, et al., 2002).

The distribution of the size of paralogous families in any genome
is another universal statistical pattern uncovered by comparative
genomics (see Figure 4-7). The distributions for all genomes approximately follow a power law with a negative exponent: $y=ax^{-\gamma}$ (γ is a positive number, a is a coefficient; Koonin, et al., 2002; Luscombe, et al.,
2002). These distributions that conveniently become straight lines in
double-logarithmic coordinates show that the majority of the families
are small (including singletons), whereas a small fraction of families
include numerous paralogs.

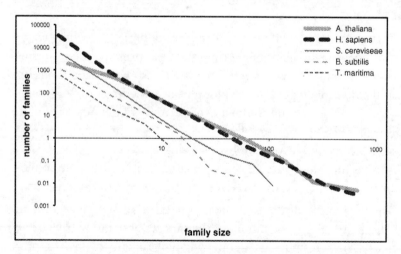

Figure 4-7   Distribution of the paralogous gene family size in diverse
genomes. The distributions are shown for the green plant Arabidopsis thaliana,
Homo sapiens, the yeast Saccharomyces cerevisiae, and the bacteria Bacillus
subtilis and Thermotoga maritima. The data comes from the EggNog collection
(Jensen, et al., 2008).

The emergence of the universal power law–like distribution of paralogous family size is described with remarkable accuracy by a simple mathematical model of the evolutionary process (see Figure 4-8). The model comes from the mathematical theory of the so-called birth and death processes and, in the case of evolution by gene duplication, is best described as a Birth, Death, and Innovation Model (BDIM; Karev, et al., 2002). In the BDIM framework, birth is a gene duplication that yields a new member of a paralogous family; death is gene loss; and innovation is the birth of a new family either via duplication followed by rapid evolution so that the "memory" of the old family is obliterated, or by HGT. The most striking result in the modeling of evolution by gene duplication was that a BDIM had to meet a set of fine requirements to reproduce the observed distributions of paralogous gene family size. The rates of gene birth and death have to be (almost) equal but depend, in a specific manner, on the size of the family—that is, large families are more dynamic than smaller ones.

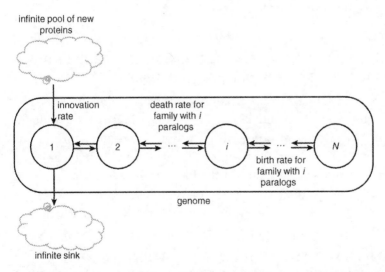

Figure 4-8   The Birth, Death, and Innovation Model of gene family evolution. In the model, "birth" corresponds to a gene duplication or possibly acquisition of a pseudo-paralog via HGT, resulting in an expansion of a paralogous family; "death" corresponds to gene elimination (regardless of the specific route); and "innovation" corresponds to acquisition of a new gene that becomes the founder of a new family (Karev, et al., 2002).

I want to emphasize that the dynamics of gene family evolution are described by a purely stochastic model of the exact kind used in statistical physics. However, for the model to be compatible with the data, a precise balance has to exist among the rates of domain birth, death, and innovation; natural selection will likely maintain this balance. Remarkably, BDIM and similar models describe the evolution of prokaryotic and eukaryotic genomes with the same precision even though the processes that lead to the formation of paralogous gene families in prokaryotes and eukaryotes appear to be markedly different. In eukaryotes, the primary, if not the only, process behind the evolution of these families is bona fide gene duplication, whereas in prokaryotes, horizontal gene transfer is likely to be quantitatively more important (hence the gene families are "pseudo-paralogous"; see Chapters 5 and 7). It speaks to the universality of the physical models of genome evolution, and at the same time to their limitations when it comes to understanding specific biology, that these models are capable of describing equally well biologically distinct processes leading to similar outcomes.

Structure and evolution of networks: The ubiquity of power laws and the underlying processes

Network is the loudest buzzword of systems biology, which permeates today's culture far beyond biology or science in general.[2] Indeed, there hardly can be a more natural way to represent connections among multiple objects than a network (which mathematically is a directed or undirected graph). In the biological context, the nodes of a network usually represent genes or proteins, whereas the edges (connections between nodes) represent their interactions, which may be physical, genetic, or regulatory (Barabasi and Oltvai, 2004).

A variety of methods have been developed to describe and compare the structures (topologies) of networks (see Box 4-1). The most popular and informative characteristic seems to be the node degree distribution—that is, the distribution of the number of edges in a given node. By this measure, all biological networks—as well as many networks of nonbiological nature, including the Internet—dramatically differ from random graphs: The latter show a bell-shaped Poisson distribution of the node degree, whereas, in the former, the distribution follows the already familiar power law (see Box 4-1). The networks with the power law distribution are often called *scale-free* because their node degree distribution looks the same on different scales (consider the straight line in double-logarithmic coordinates in Box 4-1). These networks are also "hubby": They always include a small fraction of hubs, the highly connected nodes, and many more sparsely connected nodes.

Box 4-1: Random and scale-free networks

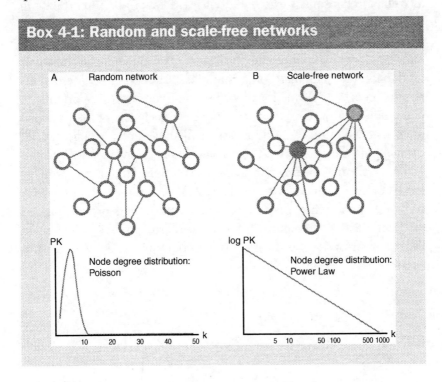

Strikingly, the power law node degree distribution seems to be an intrinsic property of *evolved* networks (including the Internet) and is not necessarily of biological origin. All kinds of biological networks, both those that describe physical interactions between proteins and those that reflect gene coregulation, are undoubtedly products of evolution and follow this type of distribution (in other words, they possess scale-free properties). To explain the emergence of the universal power law distribution, Barabasi and others proposed the mechanism of *preferential attachment* of new nodes; in the simpler, "cynical" lay parlance, this means that, in the course of a network evolution, "the rich get richer" (Barabasi, 2002). Preferential attachment is a purely stochastic, non-adaptive process. Indeed, when you create a new web site and randomly connect it with sites of the pre-existing Internet, you are more likely to connect to a hub than to an isolated node, simply because many different paths lead to any hub (see Box 4-1). This mode of evolution is inherently conservative—the network structure tends to be preserved as the network grows. Is preferential attachment the main mechanism of evolution of biological networks as well? No consensus exists yet, but it is likely to be important—in this case, preferential attachment could be caused by factors specific to biology (see Figure 4-9). Such factors might involve intrinsic "high interactivity" of the hubs, such as "stickiness" of some proteins that tend to interact with many other proteins, although these interactions are not necessarily functionally relevant. Even more importantly, a major contributor to the evolution of networks is one of the principal motors of evolution, gene duplication. When a gene is duplicated, all connections that it has with other genes are duplicated with it and gradually diverge during subsequent evolution. Under the simplest model of evolution (such as a balanced BDIM, discussed in the preceding section), when the rate of gene duplication is proportional to the family size, the structure of the network (the power law–like node degree distribution) is preserved without any selective pressure (Koonin, et al., 2002; Lynch, 2007a).

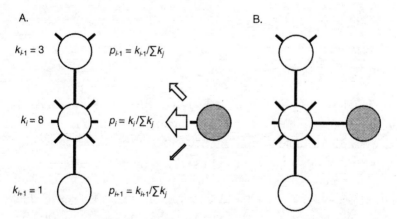

A.

$k_{i-1} = 3$ $p_{i-1} = k_{i-1}/\sum k_j$

$k_i = 8$ $p_i = k_i/\sum k_j$

$k_{i+1} = 1$ $p_{i+1} = k_{i+1}/\sum k_j$

B.

Figure 4-9 Preferential attachment in the evolution of biological networks. (A) A fragment of a network and a new element (node) to be added. (B) The result of attaching the new node. k_i = node degree, the number of nodes to which the node i is connected. p_i = the probability of attaching the new node to the node i. (See also Box 4-1.)

Genome-wide breakdown of biological functions:
Universal scaling laws

So far in our discussion of universal quantitative patterns in genome evolution, we have completely (and deliberately) stayed away from any reference to biological function. Obviously, this is an abstraction: Genomes by no means are conglomerates of faceless "molecules", but rather are ensembles of genes, each of which encodes a distinct biological function. Still, unexpected as this might seem at first, thinking borrowed from statistical physics extends even into the realm of biological functions. To this end, genes have to be classified into broad functional classes. Then we can think of these categories as distinct types of "molecules" that remain amenable to statistical analysis as long as they consist of sufficiently large numbers of genes.

As demonstrated in a series of thorough studies of Eric Van Nimwegen,[3] different functional classes of genes scale differently with the total number of genes in a genome (Molina and van Nimwegen, 2009; van Nimwegen, 2003). Some variation notwithstanding, in prokaryotes, three fundamental exponents seemingly characterize these dependencies: 0, 1, and 2. Genes for proteins involved in information processing (translation, transcription, and replication) scale with a 0 exponent—the number of these genes reaches a plateau already in the smallest genomes and effectively does not depend on the overall genomic complexity. Metabolic enzymes and transport proteins scale roughly proportionally to the total number of genes (exponent of 1). Regulators and signal transduction system components scale approximately quadratically (exponent of 2; see Figure 4-10). The characteristic exponents of the three broad functional classes of genes show little variation across prokaryotic lineages, suggesting that the differential evolutionary dynamics of genes with different functions reflect fundamental "laws" of evolution of cellular organization—or, in other words, distinct, strong constraints on the functional composition of genomes. Eukaryotic genes show similar, if less pronounced, patterns of power law gene scaling, with the exponent for the regulatory genes being substantially greater than 1 (although less than 2). All things considered, these distinct scaling laws represent another set of universals of genome evolution that is especially interesting, given the direct connection with the functional layout of the cell.

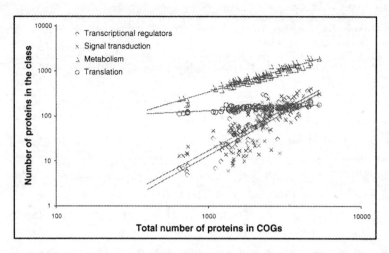

Figure 4-10   Distinct scaling of different functional classes of genes with the genome size (total number of protein coding genes) in prokaryotes. Data comes from the COG database (Tatusov et al., 2003). The plot is in double logarithmic scale.

The fundamental causes of the distinct exponents of different functional classes of genes remain to be discovered. The appealingly simple "toolbox" model of evolution of prokaryotic metabolic networks that Sergei Maslov and colleagues proposed might go a long way toward explaining the quadratic scaling of regulators (Maslov, et al., 2009). Under this model, enzymes required to utilize new metabolites are added, together with their dedicated regulators (primarily via horizontal gene transfer—see Chapter 5), to a progressively versatile reaction network. Because of the growing complexity of the preexisting network that increasingly provides enzymes for intermediate reactions, the ratio of regulators to regulated genes steadily grows. Regardless of the exact underlying mechanisms, it is notable that the superlinear scaling of the regulators could put the upper limit on the gene number in a genome. At some point (which is not easy to identify precisely), the cost of adding extra regulators ("inflating bureaucracy") will inevitably become unsustainable, curbing the growth of genomic complexity.

The "bureaucracy ceiling hypothesis" on the upper limit of genomic complexity seems particularly plausible in view of the surprising lack of dramatic gene number expansion in vertebrates, especially mammals (as in our own genomes), where the coupling between the gene number and genome size is obviously broken (see Chapters 3 and 8). In principle, genome size could be directly limited by the cost of DNA replication; however, in vertebrates, with their huge genomes, this can be ruled out as the major factor determining the upper limit. Thus, the cost of regulation, possibly along with the cost of expression, is the most likely candidate for the role of the principal constraint on the number of genes. It is not by chance, then, that vertebrates (and, to a lesser extent, other multicellular eukaryotes) have evolved other, elaborate means of increasing the proteomic complexity, such as the pervasive alternative splicing and alternative transcription, and regulatory complexity (the expansive, still under-appreciated regulatory RNome). These forms of complexity do not involve inflating the number of protein-coding genes and so lower at least some of the costs, particularly that of translation (see Chapter 8).

The universal scaling of functional classes of genes is inversely linked to the power law distribution of gene family membership, described earlier in this chapter. The greater the positive exponent of the dependency on the genome size for a functional class of genes (see Figure 4-10), the smaller the negative exponent of the family size distribution (see Figure 4-7). This connection is intuitively plausible because it should be expected that functional classes with a steep scaling on genome size will be enriched for large gene families. The inverse relationship between these two genomic universals has been derived from a simple evolutionary model that postulates a proportional recipe for the functional composition of genomes, for example, "add two regulators for each metabolic enzyme"; Grilli, et al., 2011). The predictions of this model are compatible with observations on numerous genomes of bacteria and archaea.

Stochasticity, neutrality, and selection in evolution

In the preceding sections of this chapter, we encountered numerous quantitative universals that pertain to central aspects of genome evolution and functioning. These universals include the ubiquitous power law–like distributions that describe both the structure of all kinds of biological networks and the paralogous gene families in diverse genomes, the approximately log-normal distributions of evolutionary rates of genes, and the universal correlations such as the negative correlation between gene expression and evolution rate. What is the nature of these universals? Do they reflect profound "laws" of evolution or just statistical effects that do not really help us understand biology? I argue here and toward the end of the book (see Chapter 13) that these universals signify nontrivial, relevant, and biologically important trends, even though they represent only one of at least two (and, conceivably, quite a few more) complementary (*sensu* Bohr) perspectives on life's evolution.

First, as already noted and now fully apparent, all these universals are conditioned on the behavior of genes, which are the fundamental units of evolution, as statistical ensembles. Thus, the universal dependencies and distributions are *emergent properties* of biological systems that appear because these systems consist of numerous (sufficiently numerous for the manifestation of robust statistical regularities) elements (genes or proteins, depending on the context) that only weakly interact with each other, compared to the strong interactions that maintain the integrity of each element.

Second, as we have seen, today's evolutionary analysis does not stop at demonstrating the existence of universals. Instead, at least some of the key universals, such as the distribution of evolutionary rates, the anticorrelation between evolutionary rate and expression, and the distribution of paralogous family size, are accurately reproduced in simple but well-specified, formal models of evolution. This ability of simple models that incorporate, as the elementary events, the most common evolutionary processes (such as gene duplication and loss) to explain the genome-wide universals strongly suggests that these universals reflect salient features of evolution.

Third, and perhaps most important, in terms of a new understanding of evolution, which I am attempting to sketch here, is the fact that the generative models for the genome-wide universals either do not include selection at all or more commonly incorporate only purifying selection directed at the preservation of the *status quo*, as embodied in the folds of protein molecules, the distribution of the gene family size, or the universal scaling of functional classes of genes (Koonin and Wolf, 2010b).

The analogies between the evolutionary process and statistical physics are not limited to the existence of universal dependencies and distributions, some of which can be derived from simple models. It is actually possible to draw *a detailed correspondence between the key variables in the two areas* (Barton and Coe, 2009; Sella and Hirsh, 2005). The state variables (degrees of freedom) in statistical physics such as positions and velocities of particles in a gas are analogous either to the states of sites in a nucleotide or protein sequence, or to the gene states in a genome, depending on the level of evolutionary modeling. The characteristic evolutionary rate of a site or a gene naturally corresponds to a particle velocity. Furthermore, effective population size plays a role in evolution that is clearly analogous to the role of temperature in statistical physics, and fitness is a natural counterpart to free energy.

Synopsis and perspective: The nature of the evolutionary process

The combination of the comparative-genomic and systems biology results discussed in this chapter leads to a key generalization:

> Many, if not most, gross patterns of genome and molecular phenome evolution are shaped by stochastic processes that are underpinned by the Error-Prone Replication principle and constrained by purifying selection that maintains the existing overall (but not specific) genome architecture and cellular organization.

This generalization should not be over(mis)interpreted to imply that adaptation is unimportant during evolution. Certainly, adaptations are common and indispensible for the evolution of all life. However, it is becoming increasingly clear that the overall quantifiable characteristics of genome architecture, functioning, and evolution are primarily determined by non-adaptive, stochastic processes. Adaptations only modulate these patterns. It is tempting to draw a rather obvious parallel with Kimura's neutral theory. Through the analysis of higher-level genomic and molecular phenotypic variables, we can start to discern contours of "neo-neutralism" (see also Chapter 8).

The analogy between evolution and stochastic physical processes by no means defies Jacob's tinkering metaphor. Rather, the new findings of evolutionary genomics seem to mesh well with the tinkering idea: Natural selection (the adaptive component of evolution) is thought to work like a tinkerer, and a tinkerer usually does not rebuild the house—he or she just changes and adds a few things here and there. Thus, the primary form of selection is purifying selection maintaining the *status quo*. This generalization has a striking but inevitable implication: Most of the evolution that matters—in a sense, the most interesting events in the entire history of life—occurred during the first several hundred million years of life's existence on Earth, before the emergence of modern type cells. That period in the history of life must have been qualitatively different from the rest of evolution; one is tempted to submit that *the real feat of evolution is the origin of the cell—the rest is history.* We discuss precellular evolution from this angle in Chapters 11 and 12 and return to the general nature of evolution in Chapter 13.

The parallels between evolutionary biology and statistical physics appear to be both detailed and fundamental to the degree that the conclusion seems to be justified that this is not an analogy, but rather a manifestation of the *general statistical principles (it is tempting to call them "laws") of the behavior of large ensembles of weakly interacting entities.* In both physics and evolutionary biology, such ensembles (for example, the ideal gas model in physics and the "genome as a bag of genes" model in biology) certainly are idealizations. In the real world, the deviations from the behavior that simple statistical models describe are inevitable and important. In evolutionary biology, such deviations are caused, above all, by various kinds of interaction between genes that might have unexpected effects, such as the lack of a strong correlation between the biological importance of a gene and its rate of evolution. Nevertheless, the remarkable heuristic power of the straightforward statistical approach in describing at least some of the fundamental features of both physical and biological processes is undeniable.

Recommended further reading

Barabási, A. L., and Z. N. Oltvai. (2004) "Network Biology: Understanding the Cell's Functional Organization." *Nature Reviews Genetics* 5: 101–113.

An overview of the properties of biological networks, with an emphasis on scale-free properties.

Barton, N. H., and J. B. Coe. (2009) "On the Application of Statistical Physics to Evolutionary Biology." *Journal of Theoretical Biology* 259: 317–324.

A technical but important elaboration of the thermodynamic approach in evolutionary biology.

Drummond, D. A., and C. O. Wilke. (2009) "The Evolutionary Consequences of Erroneous Protein Synthesis." *Nature Reviews Genetics* 10: 715–724.

A critical review of the concept of mistranslation and misfolding-driven evolution of proteins.

Lobkovsky, A. E., Y. I. Wolf, and E. V. Koonin. (2010) "Universal Distribution of Protein Evolution Rates As a Consequence of Protein Folding Physics." *Proceedings of the National Academy of Sciences USA* 107: 2,983–2,988.

Here evolutionary dynamics is derived from a simple model of protein folding, and the universal distribution of evolutionary rates is reproduced with a surprising accuracy.

Koonin, E. V., and Y. I. Wolf. (2006) "Evolutionary Systems Biology: Links Between Gene Evolution and Function." *Current Opinion in Biotechnology* 17: 481–487.

An overview of the correlations between evolutionary and molecular–phenomic variables.

Koonin, E. V., Y. I. Wolf, and G. P. Karev. (2002) "The Structure of the Protein Universe and Genome Evolution." *Nature* 420: 218–223.

A discussion of universal distributions and dependences in protein evolution, with an emphasis on the role of stochastic processes and preferential attachment.

Molina, N., and E. van Nimwegen. (2009) "Scaling Laws in Func-
    tional Genome Content Across Prokaryotic Clades and
    Lifestyles." *Trends in Genetics* 25: 243–247.

   An update on the universal scaling laws for different functional
   classes of genes.

Sella, G., and A. E. Hirsh. (2005) "The Application of Statistical
    Physics to Evolutionary Biology." *Proceedings of the National
    Academy of Sciences USA* 102: 9,541–9,546.

   Seminal article describing the detailed formal analogy between
   statistical physics and evolutionary dynamics.

Schroedinger, Erwin. (1944/1992) *What Is Life?: With "Mind and
    Matter" and "Autobiographical Sketches."* Cambridge, MA:
    Cambridge University Press.

   This short classic book can be recommended once again, as a
   physicist's perspective on biology that probably has not much
   changed since Shroedinger's days.

5

The web genomics of the prokaryotic world: Vertical and horizontal flows of genes, the mobilome, and the dynamic pangenomes

When Darwin wrote about evolution, he meant animals and plants—at least, he used these advanced multicellular organisms for all his concrete examples. Unicellular organisms hardly figure in *The Origin of Species* or any other work of Darwin. Nevertheless, given that Darwin seriously discussed the origin of all extant species from one or a few ancestral forms (see Chapters 2 and 11), he should have had an idea that these ancestors were unicellular.[1] Ernst Haeckel, the prolific German apostle of Darwin, placed Protista (unicellular eukaryotes often defined by the same term even now) and Monera (now known as prokaryotes—bacteria and archaea) in the foundation of his monumental Tree of Life, the first such tree that was actually populated with real life forms. Obviously, animals dominate Haeckel's tree, with Protista and Monera occupying uncertain positions near the root.

The ubiquity and importance of bacteria in the biosphere gradually became apparent in parallel with the development of evolutionary biology, first through the dramatic exploration of bacterial pathogens, and later through the advances of environmental microbiology. Early enough, microbiologists have shown that bacteria are, in a very meaningful sense, the principal agents in the biosphere: The vast majority of living cells on the planet are bacterial, bacteria display by far the greatest biochemical diversity of all life forms, and bacteria are the main geochemical force. However, despite their biological importance, their fascinating biochemical and ecological versatility, and the enormous progress of microbiology in the middle of the twentieth century (two examples include the discovery of antibiotics and the demonstration of the chemical nature of the genetic material in bacteria), microbiology contributed nothing to Modern Synthesis and was not an evolutionary discipline throughout most of that century. Not that microbiologists did not think about evolution; however, all their attempts to decipher evolutionary relationships between bacteria using cell morphology, metabolic characteristics, and other features of the phenotype, and to employ these characters to build phylogenetic taxonomy, produced inconsistent and unconvincing results. Rather ironically, approximately at the time of the consolidation of Modern Synthesis of evolutionary biology, the leading microbiologists of that era, including Roger Stanier and Cornelis Van Niel, concluded that Darwinian evolution did not apply to the microbial world and that any evolution that did occur was effectively intractable and useless for microbial taxonomy and microbiology in general (Stanier and Van Niel, 1962; Van Niel, 1955).

As Chapter 3 pointed out, everything changed abruptly in 1977, when Carl Woese and coworkers introduced phylogenetic analysis of rRNA as the method of choice for studying the evolution of microbes and constructing microbial taxonomy (Woese, 1987). The new methodology was spectacularly illustrated by the discovery of Archaea, conceivably the first major discovery in biology that was made solely by sequence analysis (Woese, et al., 1990). This breakthrough was followed by a *"sturm und drang"* period in the 1980s and early 1990s, when rRNA phylogeny was successfully applied to resolve the relationships among many groups of prokaryotes. The prevailing wisdom among the molecular evolutionists of the time was that, at least in principle, these methods yielded accurate reconstructions of microbial evolution.

However, the brave new world of microbial evolution was short-lived: Evolutionary genomics has confounded the picture again in the most dramatic manner. The first bacterial genome was sequenced in 1995, and the first archaeal genome in 1996.[2] Shortly after these breakthroughs, an exponential rate of genome sequencing was established, with a doubling time of about 20 months for bacteria and about 34 months for archaea (see Figure 3-1). Comparative analysis of the hundreds of sequenced bacterial and dozens of archaeal genomes leads to a crucial realization: *Microbes certainly evolve, but their evolution is quite different from the narrative of Modern Synthesis* (Doolittle, 1999b; Woese and Goldenfeld, 2009). The key insight is that prokaryote genomes do not behave as if they were coherent, faithfully inherited repositories of the genetic information of an organism (species). On the contrary, microbial genomes are extremely dynamic, heterogeneous entities that are relatively stable over only short time intervals, that have their characteristic rates of decay, and that persist in a dynamic equilibrium between diverse forms of life with their distinct genome organizations. Within the "prokaryotic world," these interconnected and incessantly interacting life forms include not only bacteria and archaea, but also diverse plasmids, viruses and other mobile elements. Under this new, dynamic paradigm of prokaryote evolution, the traditional concept of a species with a distinct, stable genome loses much, if not most, of its relevance (Doolittle and Zhaxybayeva, 2009). It becomes more useful to speak of a series of *"pangenomes"* at all levels, from the pangenome of, say, *Escherichia coli* or any other bacterial or archaeal "species," to the entire prokaryotic pangenome (Lapierre and Gogarten, 2009; Mira, et al., 2010).

In Chapter 3, we already discussed important aspects of the structure of the prokaryotic gene universe; it was treated mostly as a static, if complex, object—that is, in terms of distributions of various relevant quantities. Here we consider more distributions, but we primarily try to take the dynamic view and explore the prokaryotic world in terms of gene flows and interactions between replicons.

Size and overall organization of bacterial and archaeal genomes

Despite the tremendous variety of lifestyles, as well as metabolic and genomic complexity, bacterial and archaeal genomes show easily discernible, common architectural principles (see Chapter 3 for a preview). The sequenced bacterial and archaeal genomes span two orders of magnitude in size, from about 144 Kb in the intracellular symbiont *Hodgkinia cicadicola* to around 13 Mb in the soil bacterium

Sorangium cellulosum (Koonin and Wolf, 2008b). Remarkably, bacteria show a bimodal distribution of genome sizes,[3] with the peak at about 5 Mb and an additional plateau around 2 Mb (see Figure 5-1). Although there are many genomes of intermediate size, this distribution suggests the existence of two fairly distinct classes of bacteria: those with "small" genomes and those with "large" genomes. Some caution is required with these observations, because there could be a bias in genome sequencing toward smaller genomes (above all, bacterial pathogens), but with the growth of the genome collection, this explanation is becoming less plausible.

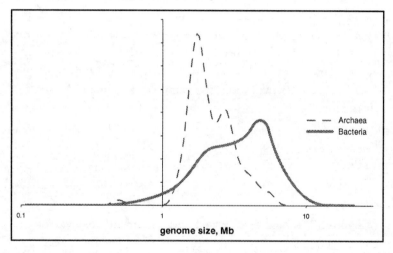

Figure 5-1   Distribution of genome sizes among bacteria and archaea. Mb = megabase.

Archaea show a narrower but also complex genome size distribution, from about 0.5 Mb in the parasite/symbiont *Nanoarchaeum equitans* to about 5.5 Mb in *Methanosarcina barkeri*, with a sharp peak at around 2 Mb that almost precisely coincides with the position of the plateau of small bacterial genomes: a second, small peak at about 3 Mb; and a heavy tail corresponding to larger genomes (see Figure 5-1). Biases in the databases might be relevant once again because there are currently about an order of magnitude fewer sequenced archaeal genomes than there are bacterial genomes, so there might not be enough data to reveal the true shape of the archaeal genome size distribution. More likely, however, archaea are indeed a less diverse group, as discussed further in this chapter.

All very small (less than 1 Mb) genomes of bacteria and archaea belong to bacterial parasites and intracellular symbionts of eukaryotes, and the only known archaeal parasite/symbiont, *Nanoarchaeum equitans*, that lives off another archaeon, *Ignicoccus hospitalis*. So it seems increasingly likely that the minimal genome size of a free-living prokaryote—at least, an autotroph that does not depend on other life forms for nutrients—is slightly greater than 1 Mb. The current record of genomic reduction among free-living cells, at about 1.3 Mb, belongs to the photosynthetic marine α-proteobacterium *Pelagibacter ubique* (SAR11), which also happens to be the most abundant known cellular life form on Earth (Giovannoni, et al., 2005). (The connection between population size and genome size is potentially important; we return to this issue in Chapter 8.)

As already discussed in Chapter 3, bacterial and archaeal genomes are characterized by a "wall-to-wall" organization, where protein-coding genes account for most of the sequence. Bacterial and archaeal genomes show unimodal and rather narrow distributions of protein-coding gene densities: The great majority of prokaryote genomes encompass between about 0.8 and 1.2 genes per kilobase of genomic DNA, so the rule of thumb is as simple as it gets: 1 gene per 1,000 base pairs. The archaeal distribution is shifted toward higher densities compared to the bacterial distribution, so, on average, archaeal genomes are even more compact than bacterial ones. It seems that both proteins and intergenic regions are slightly shorter in archaea than they are in bacteria.

Thus, archaea and bacteria are quite similar in terms of their characteristic genome sizes and overall genome architecture, but are sharply distinct from eukaryotes that span a much larger range of genome sizes, possess protein-coding genes that are typically interrupted by introns, and have longer intergenic regions (see Chapter 8). These features support the concept of a "prokaryote principle of genome organization" (see more later).

The prokaryote space-time and its evolution

The fractal genome space-time, pangenomes, and clustering of prokaryotes

In Chapter 3, we focused on the tripartite core-shell-cloud (CSC) structure of the prokaryotic genome space and showed that this structure is *fractal:* The same three components—namely, the tiny core, the larger shell, and the comparatively huge cloud—appear at any level where the gene space is dissected, from the entire prokaryote world to rather narrow groups of bacteria (see Figure 3-14). The immediate implication of the fractal CSC structure is the importance of "pangenomes"—the totality of genes represented in the genomes that belong to a "cluster" of archaea or bacteria at the given level. One would (and should) immediately ask what defines the clusters and where the levels come from. For now, let us assume that the Woesian rRNA tree (see Figure 2-3) is a reasonable depiction of the organization of the space-time in the prokaryote world and at least one of the sources of clustering. In Chapter 6, we discuss the validity and meaning of the concept of a Tree of Life in depth and show that the rRNA tree, although by no means a complete representation of the history of prokaryote evolution, is not irrelevant.

A huge number of archaeal and bacterial genes encode proteins without detectable similarity to any other available protein sequences; accordingly, these genes are often denoted ORFans (Daubin and Ochman, 2004). Typically, ORFans comprise 10%–15% of the predicted genes in archaeal and bacterial genomes. Many ORFans are very short genes, and these have received the unflattering name Evil Little Fellows (ELFs)[4] because some of them might not even be real genes, but rather false predictions made in the course of genome analysis (Ochman, 2002). Furthermore, it has been proposed that most of the ORFans that are bona fide genes have been derived from bacteriophages and, accordingly, are characterized by high horizontal mobility, although occasionally they can be recruited for a cellular function and, accordingly, fixed in a bacterial or archaeal lineage. Recent estimates from metagenomic surveys of bacteriophages suggest that the diversity of phage sequences is vast and remains largely unexplored (Edwards and Rohwer, 2005). Therefore, it seems plausible that a major fraction of bacterial and archaeal ORFans derive from this huge gene reservoir. In the tripartite CSC structure of the prokaryotic gene universe with which we are by now familiar, the ORFans naturally merge into the "cloud" of rare genes that quantitatively dominate the gene space—but not the individual genomes, as discussed in Chapter 3.

How big is the entire genome space of prokaryotes? How many genes does it contain altogether? Detailed extrapolation of the expansion of the genome space with further bacterial and archaeal genome sequencing and a reliable estimate of the actual size of this space is a difficult exercise. Nevertheless, considering the vast diversity of the microbial viromes, which are the main gene reservoirs and gene transfer vehicles (see also Chapter 10), it appears most likely that the number of elements of the prokaryotic genome space will increase by orders of magnitude—mostly, if not exclusively, through the expansion of the "cloud" (Koonin and Wolf, 2008b; Lapierre and Gogarten, 2009).

The dynamic evolution of the genome architecture in prokaryotes: Operons, überoperons, and gene neighborhood networks

As already pointed out in Chapter 3, almost immediately following the release of the first complete genome sequences, it became apparent that the gene order in bacterial and archaeal genomes is relatively poorly conserved, dramatically less so than genes themselves (see Figure 3-6). To analyze gene order evolution, one needs to obtain a robust set of orthologous genes between the compared genomes (see Box 3-1). Once such a set of orthologous genes is defined, it becomes straightforward to assess the gene order conservation—for example, using a dot-plot (one of the earliest representations of nucleotide and protein sequence similarity) in which each point corresponds to a pair of orthologs. Examination of these plots reveals rapid divergence of gene order in prokaryotes so that, even between closely related organisms, the chromosomal colinearity is broken at several points (see Figure 5-2A) and moderately diverged organisms show only a few extended collinear regions (see Figure 5-2B and 5-2C); for any pair of relatively distant organisms, the plot looks like the map of the night sky (see Figure 5-2D). Disruption of synteny during the evolution of bacterial and archaeal genomes typically shows a clear and striking pattern, with an X-shape seen in the dot-plots (see Figure 5-2B and 5-2C). It has been proposed that the X-pattern is generated by symmetric chromosomal inversions around the origin of replication (Eisen, et al., 2000). The underlying cause of these inversions could be the high frequency of recombination in replication forks that, in the circular chromosomes of bacteria and archaea, are normally located on both sides of and at the same distance from the origin site.

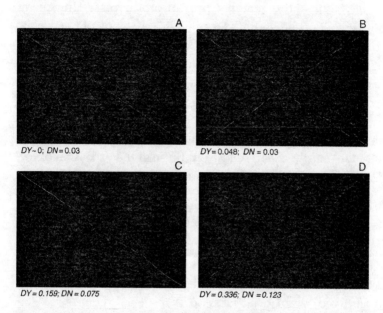

Figure 5-2   The divergence of gene orders between bacterial genomes:(A) Borrelia afzelii PKo vs. Borrelia burgdorferi B31; (B) Shewanella oneidensis MR-1 vs. Shewanella sp. ANA-3; (C) Pseudomonas fluorescens PfO-1 vs. Pseudomonas fluorescens Pf-5; (D) Pseudomonas fluorescens Pf-5 vs. Pseudomonas syringae pv. tomato str. DC3000. Each dot represents a pair of orthologous genes identified using the bidirectional best hit approach (see Box 3-1). The bright dots show pairs of orthologs that belong to conserved gene arrays; faint dots show isolated orthologs. DY is the gene order distance between genomes computed, as described in Novichkov, et al., 2009. DN is the median distance between nonsynonymous sites in protein-coding genes.

One of the earliest and central concepts of bacterial genetics is the *operon*, a group of cotranscribed and coregulated genes (Jacob and Monod, 1961). The operon hypothesis is the great conceptual breakthrough of Francois Jacob and Jacques Monod. Although an enormous amount of variation on the simple theme of regulation by the Lac repressor developed by Jacob and Monod has been discovered during the 50 years since their first publication, the operon has stood the test of comparative genomics as the major organizational principle in bacterial and archaeal genomes. Operons are much more strongly conserved during evolution than large-scale synteny. Even so, comparative analysis of gene order in bacteria and archaea reveals relatively few operons that are shared by a broad range of organisms. As noticed early on, these highly conserved operons typically encode physically interacting proteins, a trend that is readily interpretable in terms of selection against the deleterious effects of imbalance between protein complex subunits. The most striking illustration of this trend is the ribosomal superoperon that includes more than 50 genes of ribosomal proteins that are found in different combinations and arrangements in all sequenced archaeal and bacterial genomes. Analysis of the ribosomal superoperon and other, smaller groups of partially conserved operons led to the concept of an *überoperon* (Lathe, et al., 2000) or a *conserved gene neighborhood* (Rogozin, et al., 2002), an array of overlapping, partially conserved gene strings (known or predicted operons; see Figure 5-3). In addition to the ribosomal superoperon, striking examples of conserved neighborhoods are the group of predicted overlapping operons that encode subunits of the archaeal exosomal complex and the *cas* genes that comprise an antivirus defense system (see also Chapters 9 and 10).

Most of the genes in each of the conserved neighborhoods encode proteins involved in the same process and/or complex, but highly conserved arrangements that include genes with seemingly unrelated functions exist as well: A striking example is the common occurrence of the enolase gene in ribosomal neighborhoods or genes for proteasome subunits in the archaeal exosome neighborhood. The presence of these seemingly irrelevant genes in conserved gene neighborhoods might be due to hidden functional connections, "gene sharing" (multiple functionalities of the respective proteins), or "genomic hitchhiking," in which an operon combines genes without specific functional links, but with similar requirements for expression (Rogozin, et al., 2002).

The gene neighborhoods embody the paradigm of prokaryote genome evolution, if not of genome evolution in general, as they show the quintessential *interplay between partial conservation of core elements and extensive diversification at the periphery* (see Figure 5-3A). As with so many other objects and relationships in biology, these partially conserved neighborhoods can be naturally represented by networks in which genes are nodes, neighbors are connected by edges, and the weights of the edges are proportional to the frequency with which the given connection occurs in genomes (see Figure 5-3B).

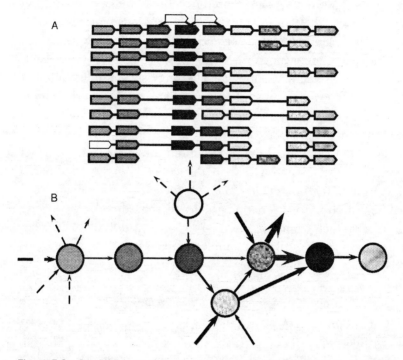

Figure 5-3   Partially conserved gene neighborhoods in genomes of prokaryotes. (A) Overlapping, partially conserved gene arrays. Genes are shown by arrow shapes with unique shading or texture. Thick connecting lines show short intergenic regions, and thin lines show long regions, separating the respective genes. (These contain additional genes and are not to scale.) In the genomes where operons are not connected, they may be located in different parts of the genome. The figure depicts actual gene arrays, but the genome and gene names are not indicated, to emphasize the generic character of this type of arrangement. The data comes from Rogozin, et al., 2002. (B) The network representation of a gene neighborhood. Filled circles show genes that belong to a neighborhood automatically delineated using the algorithm described in (Rogozin, et al., 2002); only part of the neighborhood is shown. The open circle denotes a gene that is connected to one of the genes from the neighborhood but was not included by the automatic procedure. The arrows show connections between genes in operons (solid arrows within the neighborhoods and broken arrows outside); the thickness of the arrows is roughly proportional to the number of genomes in which the given gene pair is represented.

The majority of operons do not belong to complex, interconnected neighborhoods, but instead are simple strings of two to four genes, with variations in their arrangement. Identical or similar, in terms of gene organization, operons are often found in highly diverse organisms and in different functional systems. A notable case in point are numerous metabolite transport operons that consist of similarly arranged genes encoding the transmembrane permease, ATPase, and periplasmic subunits of the so-called ABC transporters. The persistence of such common operons in diverse bacteria and archaea has been interpreted within the framework of the *selfish operon* hypothesis (Lawrence, 1999), which posits that operons are maintained not so much because of the functional importance of coregulation of the constituent genes, but due to the selfish character of these compact genetic units that are prone to horizontal spread among prokaryotes. (We return to this concept in the discussion of horizontal gene transfer later in this chapter.)

A systematic comparison of the arrangements of orthologous genes in archaeal and bacterial genomes reveals a relatively small fraction of conserved (predicted) operons and a much greater abundance of unique directons (strings of genes transcribed in the same direction and separated by short intergenic sequences; Wolf, et al., 2001). Perhaps surprisingly, directons have been shown to be quite accurate predictors of operons: Most directons actually seem to be operons (Salgado, et al., 2000). Thus, archaeal and bacterial genomes appear to be shaped by the operon organization, with a small number of highly conserved operons and a much larger number of rare or unique operons. In this respect, the pattern of operon conservation is, at least qualitatively, reminiscent of the distribution of clusters of orthologs, with its tripartite CSC pattern (see earlier): *Rare genes and rare operons are more numerous than nearly ubiquitous genes or operons by a wide margin.*

The degree of genome "operonization" widely differs among bacteria and archaea; some genomes, such as that of the hyperthermophilic bacterium *Thermotoga maritima*, are almost fully covered by (predicted) operons, whereas others, such as those of most Cyanobacteria, seem to contain only a few operons. What determines the extent of operonization in an organism remains unclear, although it stands to reason that this degree depends on the balance between the intensity of recombination, the horizontal gene flux, and selective factors that oppose disruption of operons.

Expression regulation and signal transduction in bacteria and archaea: From the basic operon scheme to überoperons, regulons, and entangled networks

Bacteria and archaea possess elaborate, elegant systems for gene expression regulation; comparative genomics dramatically changed the existing views of the organizational principles, distribution in nature, and evolution of these regulatory mechanisms. The operon concept of Jacob and Monod, which was introduced in the preceding section as the governing principle of the local architecture of bacterial and archaeal genome, is also the paradigm of gene expression regulation and signal transduction in prokaryotes. Under the Jacob-Monod model, the regulator (the *lac*-repressor, in the original study) is a sensor of extracellular or intracellular cues (in this case, the concentration of lactose) that affect the conformation of the regulator protein and, indirectly, the expression state of the operon (in the case of the *lac*-operon, the repressor binds lactose, dissociates from the operator, and allows transcription). Over the half-century that elapsed since the Jacob-Monod breakthrough, numerous variations on this subject have been discovered, including regulators that symmetrically affect transcription of adjacent divergent genes, and global regulators that control the expression of numerous, dispersed genes and operons, as opposed to the repressor of a single operon in the Jacob-Monod model. The most prominent global regulators are the catabolite repressor protein (CRP) and the stress response (SOS) regulator LexA. Considering the discovery of these and other global regulators, the operon concept was amended with the notion of a *regulon,* a set of genes that share the same *cis* regulatory signal (operator) and are regulated by the same regulator protein. Comparative-genomic analysis of regulons revealed their extreme evolutionary plasticity, with substantial differences found between regulons even in closely related organisms (Lozada-Chavez, et al., 2006). A global transcription regulator, such as LexA, can be widespread and highly conserved in diverse bacteria, but the gene composition of the LexA regulon is highly variable. The plasticity of regulons parallels the variability of genome architectures (see earlier), in accord with the idea that regulation of gene expression and genome architecture are tightly linked in the evolution of archaea and bacteria. In a striking contrast to the variability and plasticity of regulons, there is a remarkable unity in the architecture and structure of bacterial and archaeal transcription regulators. Typically, these regulators consist of a small-molecule-binding sensor domain and a DNA-binding domain. The overwhelming majority of the DNA-binding domains are variations on the same structural theme, helix-turn-helix; less common but also abundant DNA-binding domains include ribbon-helix-helix and Zn-ribbon (Aravind, et al., 2005; Aravind and Koonin, 1999).

A more complex scheme of signal transduction and expression regulation that is dedicated to sensing extracellular cues is embodied in the so-called two-component systems (Casino, et al., 2010). The two-component systems consist of a membrane histidine kinase and a soluble response regulator between which the signal is transmitted via a phosphotransfer relay. Notably, the classical transcriptional regulators and histidine kinases share many of the same sensor (input) domain, a kinship that prompts one to consider the transcriptional regulators (one-component systems) and the two-component systems within the same, integrated framework of signal transduction and expression regulation. The one-component systems that are nearly ubiquitous and typically numerically dominant in bacteria and archaea are thought to be the ancestral signal transduction devices, whereas the two-component systems are likely to be a derivative, more elaborate form of signal transduction that evolved as an adaptation for environmental signaling (Ulrich, et al., 2005).

Comparative genomics of bacteria and archaea has been instrumental in the discovery of novel, previously unsuspected but actually common systems of signal transduction. It has been known for years that a widespread form of global regulation in bacteria is mediated by cAMP, with the participation of diverse adenylate cyclases (a striking case of NOGD); numerous proteins containing cAMP sensors, such as the GAF domain; and the CRP, FNR, and other transcription regulators, also containing cAMP-binding domains. Comparative-genomic analyses revealed numerous uncharacterized proteins that contain many of the same sensor domains that are characteristic of cAMP-dependent regulators and two-component systems combined with one or two novel domains, GGDEF and EAL (so denoted after the respective conserved amino acid signatures). The genomic context of these domains and the demonstration that the GGDEF domain is a distant homolog of one of the classes of adenylate cyclases led to the hypothesis that these proteins were components of a novel signal transduction system(s). Subsequently, this predicted system has been discovered through the demonstration that the GGDEF domain possessed the activity of a di-GMP cyclase, whereas EAL is a cyclic di-GMP phosphodiesterase. The c-di-GMP-dependent signal transduction, the existence of which was not even suspected in the pregenomic era, is emerging as a major regulatory system in bacteria and archaea (Seshasayee, et al., 2010).

Another emerging theme is the abundant representation in prokaryotes of various modules of complex signal transduction systems that were previously conceived eukaryotic staples. In particular, comparative-genomic analysis convincingly showed that serine-threonine protein kinases and the corresponding phosphatases are common and diverse among archaea and bacteria, and appear to be yet another major component of the increasingly versatile prokaryotic signal transduction network. Analysis of the larger bacterial genomes unexpectedly revealed homologs of proteins previously thought to be limited in their spread to eukaryotes and involved in known eukaryotic signal transduction pathways such as programmed cell death (PCD). These proteins include proteases of the caspase superfamily, AP-ATPase family ATPases, and NACHT family GTPases, all of which are involved in various forms of plant and animal PCD (Koonin and Aravind, 2002; Leipe, et al., 2004). Typically, these proteins possess complex multidomain, modular architecture, with diverse domains mediating protein-protein interactions appended to the respective catalytic domains. These predicted signaling molecules are most common in bacteria with complex developmental phases, such as cyanobacteria, actinobacteria, and myxobacteria, and are present also in *Methanosarcinales,* so far the only identified group of archaea with relatively large genomes and complex morphology. A detailed investigation of the functions of these proteins remains to be performed, but preliminary indications show that they might be involved in PCD in some bacteria (Bidle and Falkowski, 2004). These findings indicate that at least some of the complex signaling networks of eukaryotes have their counterparts and putative evolutionary predecessors in bacteria. We return to these connections when discussing eukaryogenesis in Chapter 7.

Along with the aforementioned, roughly quadratic dependence on genome size, comparative-genomic analysis reveals great variation in the complexity of the signal transduction systems among bacteria and archaea. This variability seems to reflect the diversity of the lifestyles among the respective organisms. This variation in the fraction of the genes dedicated to signal transduction was quantitatively captured in the "bacterial IQ," a quotient that is proportional to the square root of the number of signal transduction proteins (considering the quadratic scaling) and inversely proportional to the total number of genes (Galperin, 2005). The IQ reflects the ability of bacteria and archaea to respond to diverse environmental stimuli. Accordingly, the IQ values are the lowest in intracellular symbionts (parasites); are only slightly higher in organisms with compact genomes that inhabit stable environments, such as marine cyanobacteria; and are much greater in organisms from complex and changing environments, even those with relatively small genomes.

Horizontal gene transfer: The defining process in the evolution of prokaryotes

The ubiquity of HGT in the prokaryote world

The ubiquity and major importance of horizontal gene transfer (HGT) in the evolution of archaea and bacteria can be considered the biggest novelty revealed by comparative genomics of prokaryotes. No other discovery has caused so much controversy and (sometimes acrimonious) debate, during which opposite views of HGT have been expounded, from assertions of its rampant occurrence and overarching role in the evolution of bacteria and archaea to the denial of any substantial contribution of HGT (Gogarten and Townsend, 2005; Kurland, et al., 2003; O'Malley and Boucher, 2005). The existence of HGT—*the transfer of genes between distinct organisms by means other than vertical transmission of replicated chromosomes during cell division*—was recognized long before the first genomes were sequenced (Syvanen, 1994). Moreover, it was realized that, at least under selective pressure, such as in the case of the spread of antibiotic resistance in a population of pathogenic bacteria, HGT can be rapid and extensive. However, until comparison of multiple, complete genome sequences became feasible, HGT was, by default, viewed as a marginal phenomenon, perhaps important under specific circumstances, such as evolution of resistance, but one that could be largely disregarded in the study of the evolution of organisms. One must remember that the very relevance of the question of the role of HGT in evolution stems from another revolution: Woese's demonstration that phylogenetic analysis of prokaryotic rRNA was feasible and could potentially be a reasonable depiction of evolution of bacteria and the newly discovered archaea. To most biologists, *the three-domain rRNA tree derived by Woese became synonymous with the hypothetical Tree of Life (TOL)* originally postulated by Darwin and now seemingly attained, and ready to be used as a scaffold for mapping all kinds of evolutionary events (Pace, 2006). Such was the paradigm when the HGT revolution was instigated by the advent of comparative genomics.

Historically and methodologically, the problem of HGT identification and the impact of HGT on the evolution of bacteria and archaea is sharply divided along the lines of (relatively) recent and ancient transfers on one hand and transfers between closely related and distant organisms on the other hand (Koonin, et al., 2001a). The recent HGT, especially between closely related organisms, is common, often easily detected, and noncontroversial. Indeed, genomic comparisons of bacterial strains provide clear-cut evidence of extensive HGT. Perhaps the most striking case in point is the discovery of the so-called pathogenicity islands—gene clusters that carry pathogenicity determinants, such as genes encoding various toxins, components of type III secretion systems, and others, in parasitic bacteria—and similar "symbiosis islands" in symbiotic bacteria. Pathogenicity islands are large genomic regions, up to 100 Kb in length, that are typically located near tRNA genes and contain multiple prophages, suggesting that the insertion of these islands is mediated by bacteriophages (Juhas, et al., 2009). The now classic comparative-genomic analysis of the enterohemorrhagic O157:H7 strain and the laboratory K12 strain of *E. coli* has shown that the pathogenic strain contained 1,387 extra genes distributed among several strain-specific clusters (pathogenicity islands) of widely different sizes. Thus, up to 30% of the genes in the pathogenic strain seem to have been acquired via relatively recent HGT (Perna, et al., 2001). A further, detailed analysis of individual lineages of *E. coli* O157:H7 has demonstrated continuous HGT, apparently contributing to the differential virulence of these isolates (Zhang, et al., 2007). The impact of recent

HGT certainly is not limited to the effects on pathogenicity. Most of the recent (estimated to occur within the last 100 million years) additions to the metabolic network of *E. coli* clearly were due to HGT that often involved operons encoding two or more enzymes or transporters of the same pathway; the contribution of gene duplication to the metabolic innovation appears to be quantitatively much less important (Pal, et al., 2005).

Numerous studies have revealed the pivotal contribution of HGT to the evolution of individual functional systems of prokaryotes. Perhaps the most spectacular results have been obtained with photosynthetic gene clusters of cyanobacteria and other photosynthesizing bacteria (Raymond, et al., 2002). Phylogenetic analyses strongly suggest that these clusters are complex mosaics of genes assembled via multiple HGT events; simply put, oxygenic photosynthesis that shaped the Earth's current atmosphere apparently evolved via HGT (Mulkidjanian, et al., 2006). Furthermore, the majority of cyanophages carry one or more photosynthetic genes, presumably utilizing them to augment the photosynthetic machinery of infected cells. Thus, these bacteriophages are *de facto* vehicles for the HGT of photosynthetic genes (Lindell, et al., 2005).

The discovery of the Gene Transfer Agents (GTAs) in several groups of bacteria and archaea seems to be of particular importance. The GTAs are defective derivatives of tailed bacteriophages that appear to have evolved as generalized transducing agents that package and transfer random chromosome fragments (not the prophage genes that encode the capsid and the packaging apparatus) between prokaryotes (Paul, 2008). In direct experiments with marine microbial communities, the GTAs have been shown to transfer genes with an extraordinary efficiency and without much specificity with respect to the recipient (McDaniel, et al., 2010). Thus, startling as this might be, it seems appropriate to view the GTAs *as dedicated vehicles for HGT that probably make major contributions to the gene flows in the prokaryote world*. We return to the role of viruses and GTAs in HGT and evolution of genomes in general in Chapter 10.

Apart from direct experimental demonstration and compelling genome comparisons, recent HGT is detectable through analysis of nucleotide composition, oligonucleotide frequencies, codon usage, and other "linguistic" features of nucleotide sequences that reveal horizontally acquired genes as compositionally anomalous for a given genome. However, horizontally transferred sequences are ameliorated at a relatively high rate as the acquired genes are "domesticated" during evolution—that is, the transferred genes soon become "linguistically" indistinguishable from the recipient genome (Ragan, 2001). Importantly, the molecular mechanisms of HGT between closely related organisms are well understood (even if not completely understood) and include conjugation, bacteriophage-mediated transduction, and transformation (Bushman, 2001).

In contrast to the well-established recent HGT, especially within tight groups of related organisms, the extent of HGT across long evolutionary distances, especially in the remote past, and its mechanisms and impact on the evolution of archaea and bacteria remain a matter of intense debate (Gogarten and Townsend, 2005; Kurland, et al., 2003). Comparative genomics has provided ample indications of likely HGT, including that between very distant organisms, particularly archaea and bacteria. The first clear-cut evidence of massive archaeal-bacterial HGT was obtained by showing that hyperthermophilic bacteria (namely, *Aquifex aeolicus* and *Thermotoga maritima*) contained many more homologs of characteristic archaeal proteins than mesophilic bacteria, as well as proteins with homologs in both archaea and bacteria, but with much higher sequence similarity to the archaeal counterparts than to bacterial ones (see Figure 5-4)[5]. Comparisons with mesophilic bacteria showed that the fraction of "archaeal" proteins in bacterial hyperthermophiles was much greater (with a high statistical significance) than in mesophiles (Aravind, et al., 1998; Nelson, et al., 1999). Subsequently, it has been shown that mesophilic archaea with relatively large genomes, *Methanosarcina* and halobacteria, possess many more "bacterial" genes than thermophilic archaea with smaller genomes (see Figure 5-4; Deppenmeier, et al., 2002). These admittedly crude estimates suggest that at least 20% of the genes in an organism could have been acquired through archaea-bacterial HGT, provided shared habitats.

Despite these rather striking observations, HGT between distant prokaryotes is intensely disputed, and all presented evidence is often (sometimes severely) criticized (Kurland, 2005; Kurland, et al., 2003). The taxonomic breakdown of the results of genome-wide sequence comparisons is strongly suggestive of HGT, especially as widely different results are seen for prokaryotes with different lifestyles (see Figure 5-4). However, this evidence does not "prove" HGT, and alternative explanations (even if not necessarily credible ones) have been duly proposed, such as convergence of protein sequences in distant organisms that share similar habitats (for example, archaeal and bacterial hyperthermophiles). Nevertheless, over the first decade of the third millennium, numerous phylogenomic studies—analysis of the phylogenetic trees for all or nearly all genes of prokaryotes that are sufficiently conserved and so retain enough phylogenetic information for robust conclusions—clearly revealed extensive transfer of genes between well-established groups of archaea and bacteria, including interkingdom transfers (Beiko and Hamilton, 2006; Puigbo, et al., 2009; Sicheritz-Ponten and Andersson, 2001). Moreover, these studies demonstrate beyond doubt the existence of HGT "highways"— that is, preferential routes of gene flow (Beiko, et al., 2005); major highways connect, in particular, different thermophilic organisms (see also Chapter 6).

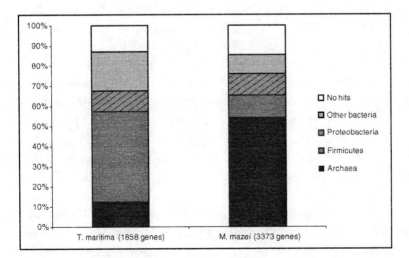

Figure 5-4   Breakdown of genes in archaeal and bacterial genomes by the taxonomic affinity of the most similar homolog. The data for the hyperthermophilic bacterium Thermotoga maritima and for the mesophilic archaeon Methanosarcina mazei are shown. The results were obtained by searching all protein sequences from each genome against the NCBI database of protein sequences using the BLASTP program (Altschul et al., 1997).

A key problem in "horizontal genomics" is the relationship between lineage-specific gene loss and HGT. A fundamental observation that reveals the complex and "nontrivial" character of evolution in prokaryotes is the patchy phyletic pattern seen in numerous COGs (see Figure 5-5). The emergence of these patterns can be explained by either HGT or gene loss, or any combination of the two. The simplest (or the most parsimonious, as it is traditionally called) evolutionary scenario can be identified when the relative rates of HGT and gene loss are known, but this ratio (which undoubtedly differs among prokaryotic groups—as discussed later in this chapter and in Chapter 6) is one of the big unknowns of prokaryote genomics. Several global reconstructions of prokaryotic evolution have been reported, all of them based on some version of the parsimony principle and either exploring scenarios with varying gain/loss rate ratios or attempting to estimate the optimal value of this ratio (Kunin and Ouzounis, 2003; Mirkin, et al., 2003). The conclusions of these analyses are that HGT might be almost as common or moderately less common (perhaps approximately twice less common) than gene loss during prokaryotic evolution. Accordingly, at least one HGT event is likely to have occurred during the evolution of most COGs, even within the limited sets of organisms that were analyzed. Of course, these studies were based on grossly oversimplifying assumptions, such as uniform rates of HGT and gene loss across the prokaryote groups, the notion that highly complex ancestral forms are unlikely (a strongly intuitive but apparently wrong idea—see Chapter 8 on the evolution of complexity), and the very concept of an underlying species tree. Although the results did not strongly depend on the species tree topology, the basic notion of a tree with distinct clades representing evolution of the compared organisms is indispensable for any reconstruction. Herein is the fundamental problem that literally reaches philosophical heights: to meaningfully speak of HGT, one *must* define the "main," vertical direction of evolution. However, if organisms exchange genes at high rates—in the extreme, freely and uniformly—the concept of vertical evolution makes no sense, nor does the orthogonal concept of HGT. Hence, a web (network) representation of the evolution of prokaryotes seems to be a logical necessity (see Figure 5-6). Having said this, I must immediately correct myself: Although it is not necessary for a preferred, tree-like component of the evolutionary process to exist, such a central trend in the evolution of prokaryotes is actually detectable; Figure 5-6 shows this pattern, which is one of the main messages of Chapter 6.

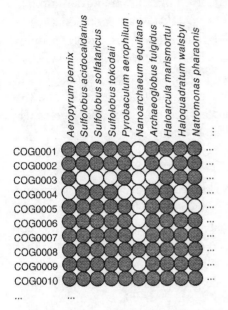

Figure 5-5 Phyletic patterns of COGs. Filled circles indicate the presence of a COG member in a genome; white circles indicate absence.

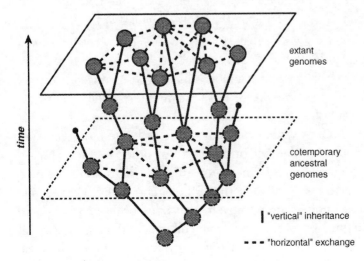

Figure 5-6 The web representation of prokaryote evolution.

Here we continue to speak of HGT under the understanding that a tree-like pattern exists as an important central trend in the evolution of prokaryotes (see Chapter 6). It is widely believed that "informational" genes coding for proteins involved in translation, transcription, and replication are much less prone to HGT than operational genes that encode metabolic enzymes and transport systems and other "operational" proteins. The rationale behind this view is the so-called complexity hypothesis (Jain, et al., 1999). Under this hypothesis, the cause of the low rate of HGT among informational genes is that the products of these genes typically are parts of complex molecular machines (unlike the products of most of the operational genes) that are strongly coadapted and thus cannot be easily displaced with orthologs from distant organisms (known as xenologs). However, the validity and general applicability of the complexity hypothesis remain open questions, as many clear-cut cases of HGT have been discovered among informational genes. Perhaps surprisingly, these include not only most, if not all, aminoacyl-tRNA synthetases (aaRS), enzymes that function in relative isolation, but also many ribosomal proteins, components of the paradigmatic molecular machine, the ribosome (Makarova, et al., 2001b). Strong evidence of HGT has been presented also for such traditional markers of vertical phylogeny as DNA-dependent RNA polymerase subunits (Iyer, et al., 2004a). The difference in the modes of evolution of informational and operational genes has to do both *with the much lower incidence of NOGD and with the generally reduced rate of HGT among the informational genes.*

It has been suggested that HGT between closely related organisms (as judged by the sequence similarity of rRNAs and other conserved genes) is more common than HGT between distant organisms, and this gradient of HGT might substantially contribute to the apparent phylogenetic coherence of prokaryotic groups (Gogarten, et al., 2002). A systematic study of the ability of bacterial genes to functionally complement orthologs from other bacteria showed that complementation became less efficient with the increasing sequence divergence between the orthologous genes (Diaz, et al., 2011). These findings are intuitively plausible because a transferred gene has to function within a different cell milieu; thus, statistically, it can be expected that the less divergence there is between the gene repertoires and orthologous gene structures, the more likely it is that a transferred gene succeeds and is fixed. It stands to reason that this nonuniformity of the fixation of transferred genes could be a cohesive factor in the prokaryotic world in the face of extensive HGT. We return to this issue in Chapter 6 in the context of the Tree of Life concept.

Finally, in this brief discussion of the different faces of HGT in the prokaryotic world, we must return to the selfish operon hypothesis, which posits that "the organization of bacterial genes into operons is beneficial to the constituent genes because proximity facilitates horizontal co-transfer of all genes required for a selectable phenotype" (Lawrence, 1999). There is no contradiction between the functional and selfish aspects of operon evolution: An operon is a "prepackaged" functional unit, often coming together with its own regulator. In that capacity, operons are more likely than single genes to be fixed after HGT. Whereas the initial fixation of an operon is affected by the benefits of coregulation of functionally linked genes, their maintenance and spread through the prokaryotic world are mediated by HGT, an evolutionary modality that does confer on operons some (but certainly not all) of the properties of selfish, mobile elements. Moreover, the selfish character of operons can be seen as a way of overcoming the constraints imposed by the complexity hypothesis, considering that the most common operons encode subunits of protein complexes (see the discussion earlier in this chapter). Packaging all subunits of a complex in one operon provides for the transferability of the requisite complexity. An excellent case in point is the evolutionary history of membrane proton and sodium-translocating ATP synthases, during which operons encoding multiple (up to eight) subunits of these elaborate molecular machines were repeatedly transferred between archaea and bacteria (Mulkidjanian, et al., 2008).

So what is the take-home message on the prevalence and role of HGT in the prokaryotic world? In my view, it is no longer a matter of sensible dispute *that HGT is a defining process in the evolution of prokaryotes that affects all aspects of bacterial and archaeal biology.* Attempts to dismiss HGT as a marginal phenomenon seem outdated and hopeless; the web metaphor of evolution (see Figure 5-6) is here to stay. At the quantitative level, however, the HGT issue is far from being settled, and it is also far from being clear which are the main factors constraining the HGT of individual genes and operons. These problems are central to our understanding of evolution among prokaryotes; Chapter 6 confronts those head on.

The prokaryotic mobilome

As noted in the preceding section, hardly any COG is refractory to HGT in principle, but some genes are much more equal than others in this respect. A substantial part of the prokaryotic genetic material consists of selfish elements for which horizontal mobility is the dominant mode of dissemination and that have been aptly termed the mobilome (Frost, et al., 2005). We also discuss the mobilome in the context of the virus world (see Chapter 10), but to sketch the emerging coherent view of the evolution of prokaryotes, we must briefly summarize here the salient features of this class of genetic elements. The mobilome consists of bacteriophages, plasmids, transposable elements, and genes that are often associated with them and regularly become passengers such as restriction-modification (RM) and toxin-antitoxin (TA) systems. It seems natural that, inasmuch as viruses and plasmids are mobile by definition, so are the defense systems. The mobilome is inextricably connected with the "main" prokaryotic chromosomes. Viruses (bacteriophages) and many plasmids systematically integrate into chromosomes, either reversibly, in which case they often mobilize chromosomal genes, or irreversibly, when a mobile element becomes "domesticated," giving rise to resident genes—initially, ORFans (see the earlier section "The fractal genome space-time, pangenomes, and clustering of prokaryotes"). Since the classic experiments of Jacob and Wollman in the 1950s, it has been well known that conjugative plasmids can mediate the transfer of large segments of bacterial chromosomes, whereas viruses (bacteriophages) have been long known to mediate transduction (Bushman, 2001). The discovery of the GTAs, which seem to be specialized HGT vectors, further emphasizes the existence of channels of regular exchange of genetic material between the mobilome and the chromosomes (Paul, 2008).

Transfer of antibiotic resistance and secondary metabolic capabilities on plasmids are textbook examples of bacterial mobilome dynamics, but the role of plasmids extends far beyond such relatively narrow areas of biology. Actually, the boundary between chromosomes and plasmids is quite fuzzy. Plasmids are replicons (typically circular, but in some cases, linear) that, similarly to prokaryotic chromosomes, carry a replication origin site and encode at least some of the proteins involved in the plasmid replication and partitioning (distribution of replicated plasmids between daughter cells during division). The key proteins involved in plasmid and chromosome partitioning, particularly ATPases of the FtsK/HerA family, are homologous throughout the prokaryotic world (and also found in numerous viruses—see Chapter 10), a fact that emphasizes common evolutionary origins and strategies of diverse prokaryotic replicons (Iyer, et al., 2004b; McGeoch and Bell, 2008).

The "canonical" genomes of numerous bacteria and archaea include, in addition to the "main" chromosome(s), one or more relatively stable, essential, large extrachromosomal elements, often described as megaplasmids. Megaplasmids can be remarkably persistent during evolution. For instance, the single megaplasmid of *Thermus thermophilis* is homologous to one of the two megaplasmids of *Deinococcus radiodurans* and, by implication, derives from the common ancestor of these related but highly diverged bacteria (Omelchenko, et al., 2005). However, over the course of evolution of this ancient bacterial group, the megaplasmids have accumulated (relative to their size) many more differences in their gene repertoires than the chromosomes. Moreover, the megaplasmids carry numerous horizontally transferred genes, including genes from thermophilic organisms that apparently were acquired by the *Thermus* lineage and appear to be important for the thermophilic lifestyle. Thus, although megaplasmids can persist in prokaryotic lineages over long evolutionary spans, they display greater genomic plasticity than chromosomes and appear to serve as reservoirs of HGT.

Nearly all sequenced prokaryotic genomes contain traces of integration of multiple plasmids and phages. It is particularly notable that most of the archaeal genomes possess multiple versions of the *herA-nurA* operon that encodes key components (ATPase and nuclease) of the plasmid partitioning machinery. Each of these operons is a remnant of a distinct replicon, so *replicon fusion* is likely to be a common event in prokaryotes. Over the course of evolution, such fusions might have been a major factor that shaped the observed architectures of prokaryotic chromosomes (Iyer, et al., 2004b; McGeoch and Bell, 2008).

Defense and stress response systems, particularly restriction-modification and toxin-antitoxin systems, can be considered special parts of the mobilome. Comparative analysis of these systems reveals rapid evolution and frequent HGT, and they are often found in plasmid and bacteriophage genomes. Despite their enormous molecular diversity, these systems function on the same principle: Each contains a toxin, a protein that destroys the chromosomal DNA (restriction enzymes), blocks translation (RNA endonuclease toxins), or kills the cell by making holes in the membrane (Kobayashi, 2001; Van Melderen and Saavedra De Bast, 2009). The toxin-induced cell death is prevented by specific methylation of the DNA, in the case of restriction-modification systems, or by neutralization of the toxin by the antitoxin, in the case of toxin-antitoxin systems, either through protein-protein interaction between the toxin and the antitoxin or through abrogation of the translation of the toxin mRNA by the antitoxin antisense RNA. These systems possess properties of selfish elements that have evolved to make the host cells dependent on them ("addicted" to the selfish element). When the respective genes are lost from a cell, the cell typically dies either because the toxin is more stable than the antitoxin, so that its activity is unleashed once the antitoxin degrades but cannot be replenished, or because of the differential effects of dilution on the restriction and modification enzymes. Because of this property of toxin-antitoxin systems, plasmids that carry toxin-antitoxin genes and so ensure plasmid "addiction" by killing cells that have lost the plasmid enjoy a strong selective advantage over plasmids that lack toxin-antitoxin systems. The currently known toxin-antitoxin systems probably represent the proverbial tip of the iceberg, as bacterial and archaeal genomes carry a great variety of operons that mimic the properties of toxin-antitoxin operons (a pair of genes encoding small proteins and occurring as a stable combination in diverse genomes and genomic neighborhoods) but have not been experimentally characterized (Makarova, et al., 2009a).

Recently, a highly unusual class of mobile defense systems has been shown to exist in the majority of the archaea and about one-third of bacteria with sequenced genomes (Deveau, et al., 2010; Koonin and Makarova, 2009). This system is centered on arrays of *Clustered Regularly Interspaced Short Palindromic Repeats* (CRISPR) repeats and includes about 50 distinct families of CRISPR-associated *(cas)* genes; remarkably, it comes across as the second-largest array of connected gene neighborhoods in prokaryotic genomes (after the ribosomal superoperon; Rogozin, et al., 2002). The CRISPR-Cas system protects prokaryotic cells against phages and plasmids via a "Lamarckian route" (we return to this issue at greater length in Chapter 9), whereby a fragment of a phage or plasmid gene is integrated into the CRISPR locus on the bacterial chromosome and is subsequently transcribed and utilized, via still poorly characterized mechanisms, to abrogate the selfish agent's replication. The CRISPR-Cas system shows extreme plasticity, even among closely related isolates of bacteria and archaea, and shows strong evidence of extensive HGT.

The selected examples discussed here point to enormous, still incompletely understood diversity of the prokaryotic mobilome and emphasize the major contribution that the mobilome makes to the evolution of the prokaryotic genome space-time.

The indispensability of HGT for the evolution of prokaryotes

It seems not widely realized that HGT is essential for the evolution of prokaryotes and can be legitimately viewed as a necessary condition of the long-term survival of archaea and bacteria. Any asexual population is headed for eventual extinction because it does not possess effective means to eliminate the inevitably accumulating deleterious mutations. Usually, the advantage of sexual populations over asexual ones is attributed to the mechanism known as Müller's ratchet (Müller, 1964). Under Müller's ratchet, the accumulation of deleterious mutations in the absence of recombination (sex) leads to the gradual loss of fitness and decline of an asexual population. The effect of Müller's ratchet is most severe in small populations, due to the power of genetic drift. Michael Lynch and coworkers developed a more elaborate model of the decline of asexual population, known as mutational meltdown (Lynch, et al., 1993). Given that the majority of mutations are (at least) slightly deleterious, an asexual population goes into a "downward spiral" of mutational meltdown when Müller's ratchet acts in conjunction with genetic drift. In this case, the population size drops as the result of the elimination of deleterious mutations by purifying selection, resulting in an increased drift and a

greater chance of random fixation of additional deleterious mutations. Thus, mutational meltdown seems to set limits on genome size
and the survival span of populations of asexual organisms.

Most prokaryotes do not engage in regular sexual relationships,
although a mechanism known as bacterial sex, conjugation, is a thoroughly characterized process. However, this bacterial sex requires the
presence of a specialized conjugative plasmid (described as the F factor in the classic early experiments of Wollman-Jacob, Lederberg,
and Cavalli-Sforza) or the so-called integrative and conjugative elements resident in the chromosome, and is far from ubiquitous in
prokaryotes (Bushman, 2001). Actually, among the currently well-
characterized bacteria, conjugation is known to exist only in a minority, whereas in archaea, conjugation apparently has not been reported
(Frost and Koraimann, 2010; Wozniak and Waldor, 2010). Bacteria
that engage in conjugation frequently form large panmictic populations and perhaps species resembling the classical species of eukaryotes. In these cases, sex absolves the bacteria from mutational
meltdown. However, if conjugation is absent or very infrequent, as
seems to be the case in archaea and many bacteria, there is no way to
avoid the meltdown catastrophe other than HGT, which may be considered a form of illegitimate recombination. In the long term, an
asexual population (of prokaryotes) can survive *only if it receives, via
HGT, at a sufficient frequency, functional versions of genes that accumulate deleterious mutations* (see Figure 5-7). This population-
genetic perspective on the role of HGT in the evolution of
prokaryotes strongly suggests that *selection operates to maintain an
optimal level of HGT,* a rate that is sufficiently high to prevent mutational meltdown and to provide opportunities for potentially adaptive
innovation, but low enough to avoid frequent disruption of functionally important gene associations (operons). A clear prediction of the
HGT optimization hypothesis is that functionally important genes
that evolve fast and are often lost in the course of evolution should
also experience a high rate of HGT. In Chapter 6, we shall see that
this prediction is indeed validated by comparative analysis of phylogenetic trees for prokaryotic genes. This perspective allows us to rationalize the evolution of GTAs as specialized HGT vehicles that keep the
rates of gene transfer above the meltdown threshold. Furthermore,
DNA pumps involved in transformation (Chen, et al., 2005) also may

be considered devices for HGT rather than simple scavengers of nucleotides, as sometimes suggested.

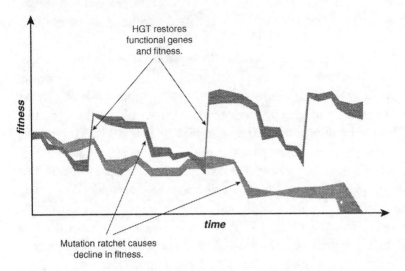

Figure 5-7   The inevitability of HGT: the fates of asexual populations in isolation and in the presence of HGT.

Conversely, any asexual population that is (virtually) isolated from HGT is headed for decline and eventual extinction. This indeed appears to be the case for obligate bacterial parasites, especially intracellular ones. The intracellular symbionts with the smallest genomes, such as the aforementioned *Hodgkinia cicadicola* or *Carsonella rudii*, with its slightly larger genome, have approached or even reached the status of organelles in the host eukaryotic cells (McCutcheon, et al., 2009) and, in all likelihood, have travelled far along the path to extinction. As often happens (more on this in Chapter 8), tension exists between the global pressure exerted by population dynamics and local adaptations. Some of the insect endosymbionts with small but not tiny genomes (typically, 500 genes or so), such as *Wolbachia* or *Wigglesworthia*, retain certain metabolic pathways that supply essential metabolites, particularly amino acids, to the host (Wu, et al., 2006). This adaptation could allow these organisms to maintain a relatively large effective population size and, hence, at least temporarily, avoid the meltdown. In the long run, however, it still appears likely that such bacteria have a relatively short lifespan (on the evolutionary scale).

Horizontal gene transfer, the universal "laws" of genomics, and the well-mixed prokaryotic gene pool

In the preceding chapter, we discussed several universal dependencies between genomic variables (the "laws" of genomics), particularly the inversely linked scaling of functional classes of genes and gene family size. We now cannot avoid the question: What is the relationship between these "genomic laws" and HGT that is so prominent in the prokaryote world? Indeed, comparative genomic analysis shows that the gene families in prokaryotes are shaped to a greater extent by HGT than by duplication (Treangen and Rocha, 2011). Most of the genes that come across as paralogs in the analysis of a single genome are actually pseudoparalogs (Makarova, et al., 2005). Despite this route of origin, the family size distributions are reproduced by Birth, Death, and Innovation Models with a high accuracy (see Figure 4-7). The only explanation for this fit seems to be that the rates of gene birth and death are actually proportional not to the size of a family in the given genome, but to the size of the family in the donor gene pool. Given that the power law distributions of family size are closely similar in all genomes, the donor pool is actually equivalent to the entire genome universe of prokaryotes. In other words, this aspect of the structure of the genome universe can be described with a single power law distribution of gene family sizes (this distribution is obviously quite different from the CSC structure because it relates to broadly defined gene families rather than orthologous sets).

Similar logic pertains to the scaling of functional classes. Given that HGT is a key contribution to the gene composition of prokaryotic genomes, (nearly) universal scaling laws require that the prokaryote genome universe be treated as a single gene pool. The explicit theory of HGT-dominated evolution remains to be developed. However, the universality of the scaling laws and the previous qualitative considerations suggest that, on average, the *prokaryotic genome universe is a well-mixed gene pool*. Certainly, there are significant local nonhomogeneities and "highways" of HGT (see Chapter 6), but, on average, the rate of gene mixing is sufficiently high to provide for the universal scaling laws.

Genomic signatures of distinct lifestyles of bacteria and archaea and the nonisomorphous mapping of the gene and functional spaces

One of the greatest hopes associated with comparative genomics is the possibility, at least, in principle, to delineate "genomic signatures" of distinct organismal lifestyles (phenotypes)—that is, sets of genes that are necessary and sufficient to support these lifestyles. In the current, rapidly growing collection of prokaryotic genomes, a particular lifestyle is often represented by multiple, diverse genomes, so the time seems ripe for the study of the genome-phenotype links to start in earnest. So far, only very modest success can be claimed. When a life style is linked to a well-defined biochemical pathway, such as in methanogens or photosynthetic organisms, identification of a genomic signature can be a relatively straightforward task. Even so, the analysis of the genes for proteins involved in photosynthesis, for example, illustrates the complex intertwining of lifestyle- and lineage-specific features. The most complete set of "photosynthetic" genes was detected in cyanobacteria, whereas the other groups of photosynthetic bacteria possessed various subsets of these genes (Mulkidjanian, et al., 2006).

Genomic signatures of more complex phenotypes, such as thermophily or radiation resistance, turned out to be much more elusive. The most sustained effort, perhaps, has been dedicated to the search for signs of thermophilic adaptation. Remarkably, a single gene is found in all sequenced hyperthermophilic genomes, but not in any of the mesophiles, and this gene encodes a protein that is strictly required for DNA replication at extreme high temperatures, the reverse gyrase (Forterre, 2002). Moreover, the genome of a moderate thermophile *Thermus thermophilus* (strain HB27) contains a reverse gyrase pseudogene, whereas the related strain HB8 contains an intact reverse gyrase gene, demonstrating an ongoing process of reverse gyrase elimination after the probable switch from hyperthermophilic to moderate thermophilic lifestyle (Omelchenko, et al., 2005). However, a search for other thermophile-specific genes yielded limited information, with no genes other than reverse gyrase showing a clean pattern of presence-absence correlated with (hyper)thermophily and only a few showing significant enrichment in hyperthermophilic compared to mesophilic archaea and bacteria (Makarova, et al., 2003). In addition, there have been many attempts to identify determinants of the thermophilic phenotype at the level of nucleotide and protein sequences and structures. Although these studies have revealed several potentially distinctive features of thermophilic proteins, such as a high charge density and over-representation of disulphide bridges, the ultimate significance of each of these features remains uncertain (Beeby, et al., 2005). In phylogenetic trees of highly conserved genes (see Chapter 6), thermophiles often cluster with mesophiles, such as proteins from *Thermus* with homologs from the mesophilic bacterium *Deinococcus* (consider the famous Taq polymerase, an essential tool of genetic engineering). These findings indicate that common evolutionary history leaves a much stronger imprint in the protein sequences than thermophilic (and other) adaptations. The overall conclusion from these studies is that, so far, comparative genomics has failed to reveal "secrets" of the (hyper)thermophilic lifestyle. (Intuitively, one would suspect that there must be major genome-encoded differences between organisms whose optimal growth temperature exceeds 95°C and those that optimally grow at 37°C.)

The story of the search for genomic correlates of extreme radiation resistance and desiccation resistance might be even more illuminating. Some bacteria and archaea, of which the best characterized is the bacterium *Deinococcus radiodurans,* show extreme radiation resistance that is thought to be a side benefit of their adaptive desiccation resistance. Extensive genome analysis of *D. radiodurans* did not immediately reveal any unique features of the genome or of DNA repair systems that could explain the exceptional ability of this organism to survive radiation damage, although homologs of plant proteins implicated in desiccation resistance and, at the time, not found in any other bacteria, have been identified (Cox and Battista, 2005; Makarova, et al., 2001a). *Deinococcus radiodurans* is a model experimental system, so subsequently, transcriptomic and proteomics studies have been undertaken to characterize the response of this bacterium to high-dose irradiation. These studies have generated some excitement because substantial up-regulation was observed for several uncharacterized genes whose products were implicated in potentially relevant processes such as double-strand break repair (Liu, et al., 2003). However, knockout of these genes failed to affect radiation resistance, whereas knockouts of a few genes that did not encode any recognizable domains and were not upregulated upon irradiation did render the organism radiation sensitive (Blasius, et al., 2008; Cox and Battista, 2005; Makarova, et al., 2007a). The comparative analysis of two related, radiation-resistant bacteria, *D. radiodurans* and *D. geothermalis*, conspicuously failed to resolve and even further confounded the problem of genomic determinants of radiation resistance (Makarova, et al., 2007a). No genes with a clear relevance to radiation resistance were discovered that would be unique to these radioresistant bacteria. Moreover, orthologs of many of the genes that are strongly upregulated in *D. radiodurans* upon irradiation are missing in *D. geothermalis*. The careful comparison of operon structure and predicted regulatory sites in the two *Deinococcus* genomes led to the prediction of a putative radiation-resistance regulon (Makarova, et al., 2007a). However, for most of the genes that comprise this putative regulon, the relevance for radiation and desiccation resistance is uncertain. The principal determinants of radioresistance remain elusive, and growing evidence shows that important roles could belong to genes that mediate resistance in unexpected, indirect ways, such as through regulation of the intracellular concentrations of divalent cations that affect the level of protein damage resulting from irradiation or desiccation (Daly, 2009).

The only possible conclusion on the current state of understanding of the genome-phenotype connections in prokaryotes is that these links are multifaceted and that distinct sets of genes responsible for complex phenotypes are not readily identifiable, despite the existence of clear signatures of certain lifestyles (such as the reverse gyrase, in the case of hyperthermophily). The complexity of the genome-phenotype relationship can be represented as a *nonisomorphous, many-to-many mapping between the genome and functional spaces of prokaryotes* (Koonin and Wolf, 2008b). Each gene is pleiotropic (linked to multiple functions), and each function is multigenic (linked to multiple genes; see Figure 5-8). We arrived at this crucial conclusion through the analysis of prokaryotic genomes, but there is no doubt that it reflects the general lack of determinism in the genotype-phenotype mapping (see Chapter 13).

genes   functions

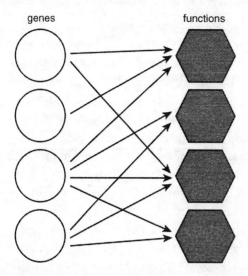

Figure 5-8   The nonisomorpohous, many-to-many mapping of the genomic and functional spaces.

Archaea and bacteria in light of comparative genomics: Whither prokaryotes?

The very validity of the term and concept of a prokaryote has been challenged as outdated and based on a negative definition: the absence of the eponymous organelle of the "higher" organisms (eukaryotes), the nucleus (Pace, 2009b, 2006). Instead of the purportedly inadequate notion of a prokaryote, the proposal has been vigorously propounded to classify life forms solely on the basis of phylogenetic divisions that have been derived primarily from rRNA trees and supported by trees for a few other (nearly) universal informational genes (Pace, 2009a). The argument on the negative definition of prokaryotes has been countered by defining positive characters such as transcription-translation coupling (Martin and Koonin, 2006b). Regardless of the relative merits of these arguments, comparative genomics throws its own light on the prokaryote conundrum. As discussed in this chapter, there is very little universal conservation in the repertoires of genes (COGs) between archaea and bacteria, and even less in the organization of specific operons. Indeed, in trees built by comparing the gene repertoires or conserved pairs of adjacent genes, the split between bacteria and archaea is unequivocal (Wolf, et al., 2002).

In stark contrast, the overall genome organization of bacteria and archaea is remarkably uniform. Some exceptions notwithstanding, this general principle of genome organization can be easily captured in a succinct description: *Bacteria and archaea have compact genomes with short intergenic regions so that many genes form directons that tend to function as operons.* The formation of directons, many of which become operons, can be considered a direct consequence of genome contraction (see Chapter 8 for further discussion). The persistence of operons is subsequently ensured by a combination of purifying selection and frequent HGT, as captured in the selfish operon concept. The uniform principle of organization of the genomes of bacteria and archaea emerges as a direct consequence of the selective pressures that operate in the evolution of these life forms. These selective factors themselves are gauged through population dynamics (see Chapter 8). Considering this unity, I am compelled to conclude that the concept of *prokaryotes as life forms characterized by a distinct mode of evolution that involves extensive and essential HGT, which creates a well-mixed gene pool and leads to a common type of genome organization,* is valid. Whether or not *prokaryotes* is a good term to describe this part of the biosphere remains a debatable issue (this is brought to focus in the discussion of eukaryogenesis in Chapter 7) but, arguably, one of secondary importance.

Synopsis and perspective

By any account, the progress in our knowledge of the prokaryotic world brought about by comparative genomics has been enormous. Many of the major trends and patterns discussed here, such as the clear distinction between archaea and bacteria, along with fundamental similarities in the mode of evolution and ensuing genome organization, the operonic organization of bacterial and archaeal genes, and the existence of HGT, have been noticed in the pregenomic era, but more as anecdotes than as general patterns. Comparative genomics allows one to actually determine how common a particular pattern is, and the strength of such inference increases with the growth of the genome collection. In the early days of genomics, there was hope for a new suite of "laws of genomics." Certain striking, nearly universal quantitative regularities have been revealed by comparing prokaryotic genomes. The best candidates for "laws of genomics" seem to be the scaling of different functional classes of genes with the genome size, the power law distribution of gene family size, and the universal distribution of the evolutionary rates in orthologous gene sets (see Chapter 4).

On the whole, however, 15 years into the comparative-genomic enterprise, it seems more appropriate to speak of regularities, constraints, and perhaps principles, not laws carved in stone. Indeed, in terms of general organization, the great majority of the archaeal and bacterial genomes are notably similar and are built according to the same simple "master plan," with wall-to-wall protein-coding and RNA-coding genes, preferentially organized in directons, typically with a single origin of replication. Most of the archaeal and bacterial genes are simple units, with uninterrupted coding sequence and short regulatory regions. There seems to be a nontrivial connection between gene functions and genome complexity: Scaling of the number of genes of different functional classes appears to be (nearly) the same across the wide range of the available genomes, with the nearly constant, "frozen" set of genes involved in translation and a steep increase in the number of regulators and signaling proteins with genome size. This increased "burden of bureaucracy," along with energetic constraints, is likely to be one of the important factors that set the upper limit for prokaryotic genome size and, accordingly, complexity. These regularities come as close to "laws of genomics" as one can imagine, although, as always in biology, there are multiple exceptions to any rule.

More important, within these simple constraints, we observe the enormous diversity and intricacy of the content, operation, and history of prokaryotic genomes. Cases in point abound. The demonstration that a substantial majority of genes in each genome are not ORFans, but rather have orthologs is the very cornerstone of comparative genomics, which underlies all functional annotation of the sequenced genomes, as well as evolutionary reconstructions. However, the flip side of the coin, the patchy distribution of clusters of orthologous genes in the gene space, is no less fundamental. This distribution is the product of the major factors that shape prokaryotic evolution: HGT; gene loss that often reflects genome streamlining; and NOGD, which reflects the nonisomorphous mapping between the genome space and the functional space. The effectively unlimited flexibility of the architecture of prokaryotic genomes owing to extensive rearrangements, which create diverse variations on the themes of conserved operons, and the discovery of previously unsuspected signaling, regulatory, and defense systems (only a few of which are briefly discussed in this chapter) add to the complexity of the prokaryotic genome space that is revealed by comparative genomics.

Arguably, the primary conceptual novelty brought about by genomics is the demonstration of the ubiquity of HGT in the prokaryote world, even as the extent of gene movement between distantly related organisms remains an issue of study and debate. Regardless of the course these debates take in the years to come, the wide spread of HGT and the apparent absence of impenetrable barriers means that the prokaryote world is a single, connected gene pool. This pool has a complex, compartmentalized structure, with its distinct parts exchanging genes at widely varying rates. Horizontal gene transfer affects different classes of genes to different extents, at least partly according to the complexity hypothesis, but no gene seems to be completely refractory to HGT. It is critical to realize that a sufficient level of HGT is essential for the long-term survival of any asexual prokaryotic population; otherwise, such a population is extinguished by mutational meltdown. Thus, a sufficient rate of HGT is a condition *sine qua non* for the continuous survival and evolution of the prokaryotic world. Moreover, considering the demonstrated formative role of HGT in the evolution of prokaryotic genomes jointly with the universal scaling laws leads to the conclusion, even if only a qualitative one so far, that the prokaryotic gene pool is well mixed overall, all local nonhomogeneities notwithstanding.

Importantly, a substantial fraction of most prokaryotic genomes belongs to the mobilome, the vast set of genes that come and go at striking rates and generally are selfish genetic elements devoid of any adaptive value for the host organisms, even if occasionally recruited by the hosts for specific biological functions.

Taken together, these findings amount to a new picture of the dynamic prokaryote world that is best represented as a complex network of genetic elements that exchange genes at widely varying rates. In this network, the distinction between the relatively stable chromosomes and the mobilome is a difference in degree (of stability and mobility) rather than in kind. The remarkably uniform general organization of prokaryotic genomes appears to be determined by the dynamic character of the prokaryotic genome space-time, along with the intensive purifying selection underpinned by the large effective population size of most prokaryotes that, considering the looming and otherwise inevitable mutational meltdown catastrophe, is itself contingent on the extensive gene exchange (more on this in Chapter 8).

The paradox of today's state of the art is that, despite the tremendous progress (but also owing to these advances), the emerging complexity of the prokaryotic world is currently beyond our grasp. We still have no really fitting language, in terms of theory or tools, to describe the workings and histories of the genomic web. Developing an adequate conceptual framework for understanding the evolution of prokaryotes is the major challenge for the next stage in the evolution of prokaryotic genomics. Chapter 6 describes some modest steps in this direction.

Recommended further reading

Davids W., and Z. Zhang. (2008) "The Impact of Horizontal Gene Transfer in Shaping Operons and Protein Interaction Networks—Direct Evidence of Preferential Attachment." *BMC Evolutionary Biology* 8: 23.

A quantitative study of the rates of HGT between core and shell genes in bacteria. The shell genes, particularly those involved in defense, have higher HGT rates.

Doolittle, W. F. (1999) "Lateral Genomics." *Trends in Cell Biology* 9: M5–8.

An early discussion of the impact of HGT on the Tree of Life concept.

Doolittle, W. F., and O. Zhaxybayeva. (2009) "On the Origin of Prokaryotic Species." *Genome Research* 19: 744–756.

A conceptual discussion of prokaryotic species, with the focus on the plurality of evolutionary processes in archaea and bacteria that lead to different forms of clustering of organisms, some of which resemble speciation in eukaryotes and others that do not.

Frost, L. S., R. Leplae, A. O. Summers, and A. Toussaint. (2005) "Mobile Genetic Elements: The Agents of Open Source Evolution." *Nature Reviews Microbiology* 3: 722–732.

A review of the evolutionary impact of mobile elements.

Gogarten, J. P., and J. P. Townsend. (2005) "Horizontal Gene Transfer, Genome Innovation and Evolution." *Nature Reviews Microbiology* 3: 679–687.

Review asserting the key role of HGT in evolution, at least in the prokaryotic world.

Jain, R., M. C. Rivera, and J. A. Lake. (1999) "Horizontal Gene Transfer Among Genomes: The Complexity Hypothesis." *Proceedings of the National Academy of Sciences USA* 96: 3,801–3,806.

The key study presenting the hypothesis that genes involved in multiple interactions (complexes) are less prone to HGT than genes with fewer partners.

Koonin, E. V., and Y. I. Wolf. (2008) "Genomics of Bacteria and Archaea: The Emerging Dynamic View of the Prokaryotic World." *Nucleic Acids Research* 36: 6,688–6,719.

A broad conceptual overview of the evolution of prokaryotes, with an emphasis on the dynamic character of evolution caused by HGT, NOGD, lineage-specific gene loss, and the activity of mobile elements. The tripartite organization of the prokaryote gene universe is reported.

Lapierre, P., and J. P. Gogarten. (2009) "Estimating the Size of the Bacterial Pan-Genome." *Trends in Genetics* 25: 107–110.

An attempt to estimate the size of the entire pangenome based on the tripartite model of the gene spread among prokaryotes, similar to the model in the preceding article.

Lawrence, J. (1999) "Selfish Operons: The Evolutionary Impact of Gene Clustering in Prokaryotes and Eukaryotes." *Current Opinion in Genetics & Development* 9(6): 642–648.

Overview of the data in support of the Selfish Operon Hypothesis that posits that operons are maintained primarily by HGT rather than by the benefits of gene coregulation and coexpression.

Sapp, J. (2005) "The Prokaryote–Eukaryote Dichotomy: Meanings and Mythology." *Microbiology and Molecular Biology Reviews* 69: 292–305.

A historical analysis of the Woeseian revolution and the conceptual inadequacy of prokaryotes.

Woese, C. R. (1994) "There Must Be a Prokaryote Somewhere: Microbiology's Search for Itself." *Microbiology Reviews* 58:1–9.

A fascinating discussion of the "Woeseian revolution" in (micro)biology, by the architect of the revolution himself. The emphasis is on the "decoherence" of prokaryotes and the relationship between archaea and eukaryotes.

6

The phylogenetic forest and the quest for the elusive Tree of Life in the age of genomics

A very brief history of the TOL

The concept of the Tree of Life (TOL) in its modern meaning was introduced by Darwin in his notebooks as early as 1838. Twenty years later, Darwin captured it in the single illustration of *The Origin of Species*. Certainly, he did not invent the idea of depicting genetic relationships in the form of a tree. For centuries, trees have been used to represent genealogies, such as actual histories of families (for example, royal ones). However, Darwin was the first to come up with the seminal idea that *different species were related by a tree*, with the leaves corresponding to extant species and the internal nodes corresponding to extinct ancestral forms. Moreover, Darwin formulated the sweeping hypothesis that ultimately the entire history of life could be presented in the form of a single huge tree:

> The affinities of all the beings of the same class have some-
> times been represented by a great tree. I believe this simile
> largely speaks the truth. The green and budding twigs may
> represent existing species; and those produced during each
> former year may represent the long succession of extinct
> species. ...The limbs divided into great branches, and these
> into lesser and lesser branches, were themselves once, when
> the tree was small, budding twigs; and this connexion of the
> former and present buds by ramifying branches may well

represent the classification of all extinct and living species in groups subordinate to groups. (Darwin, 1859)

In the sixth edition of *Origin* (Darwin, 1872), Darwin went further and explicitly introduced the TOL:

As buds give rise by growth to fresh buds, and these, if vigorous, branch out and overtop on all sides many a feebler branch, so by generation I believe it has been with the great Tree of Life, which fills with its dead and broken branches the crust of the earth, and covers the surface with its everbranching and beautiful ramifications.

For Darwin's day, this was an incredibly bold proposition as no hard evidence supported the common origin of all life forms, not to mention that Darwin and other biologists of the 19th century had no idea of the true diversity of life on earth. The Universal Common Ancestry hypothesis nevertheless caught on. Several years after the publication of *Origin*, Ernst Haeckel populated Darwin's conceptual TOL with real life forms that included almost exclusively animals, with "MAN" on top, and some amoebae and "Monera" (the nineteenth-century name for bacteria) at the roots (Haeckel, 1997). Since then, the TOL has become the centerpiece of evolutionary biology and, in a sense, of biology in general.

For nearly 140 years after Darwin and Haeckel, phylogenetic trees (initially constructed using phenotypic characters, but, following the seminal work of Emile Zuckerkandl and Linus Pauling in the early 1960s, increasingly reliant upon molecular sequence comparison) have been construed as a generally accurate depiction of the evolution of the respective organisms. In other words, a tree built for a specific character or a gene was, by default, equated with a *"species tree."* The adoption of the 16S rRNA, a molecule that is universal in cellular life forms, as the gold standard for phylogenetic reconstruction yielded the three-domain TOL of Woese and coworkers. This was a fitting culmination to the heroic period of phylogenetics (Pace, 1997, 2006; Woese, 1987; Woese, et al., 1990). The 16S tree included parts with excellent resolution of the branches, and although many other parts remained poorly resolved, especially deep in the tree, further improvement of phylogenetic methods, along with the analysis of several additional universal genes, was expected to reveal the detailed, definitive topology of the TOL in a not-so-remote future (Pace, 1997).

The trouble for the TOL concept started even before the advent of genomics, as it became clear that certain common and essential genes of prokaryotes experienced multiple horizontal gene transfers (HGT). J. Peter Gogarten and colleagues then proposed the metaphor of a "net of life" as a potential replacement for the TOL (Hilario and Gogarten, 1993). However, these ideas did not get much traction in the pregenomic era, and HGT was generally viewed as a minor evolutionary process, crucial in some areas (such as the spread of antibiotic resistance), but secondary in the overall course of evolution—and a minor complication to the construction of the overarching TOL. In the late 1990s, comparative genomics of prokaryotes dramatically changed this picture by showing that the patterns of gene distribution across genomes are typically patchy (members of most COGs are scattered among diverse organisms), and topologies of phylogenetic trees for individual genes are often incongruent. These findings suggested that HGT was extremely common among bacteria and archaea, and was important also in the evolution of eukaryotes, especially in the context of endosymbiotic events (see Chapter 7). Thus, a perfect TOL turned out to be a chimaera because extensive HGT prevents any single gene tree from being an accurate representation of the evolution of entire genomes. The realization that HGT among prokaryotes is the dominant mode of evolution rather than an exceptional process led to the idea of "uprooting" the TOL—above all, in several influential review articles by W. Ford Doolittle (Doolittle, 1999a, b, 2000). The purported demise of the TOL received a huge amount of attention, not just in professional publications, but also in (semi) popular literature (Pennisi, 1999). This is often viewed as a paradigm shift in evolutionary biology, if not biology as a whole (O'Malley and Boucher, 2005; see Appendix A).

The views of evolutionary biologists on the status of the TOL in the face of the pervasive HGT span the entire range, from (i) continued denial of a significant role of HGT in the evolution of life, to (ii) a "moderate" revision of the TOL concept, to (iii) genuine uprooting, whereby the TOL is declared meaningless as a representation of the evolution of organisms or genomes (O'Malley and Boucher, 2005). With the accumulation of comparative genomic data, the anti-HGT stance is quickly becoming more of a psychological oddity than a defendable scientific position. The real debate seems to be between the "revisionist" and "radical uprooting" views (ii and iii). The moderate approach maintains that, all the differences between individual gene trees notwithstanding, the TOL remains relevant as a central trend that, at least in principle, can be revealed through a comprehensive comparison of gene tree topologies (Wolf, et al., 2002). The radical view counters that massive HGT obliterates the very distinction between the vertical and horizontal routes of genetic information transmission, so the TOL concept should be abandoned in favor of a (broadly defined) network representation of evolution (Doolittle and Bapteste, 2007; Gogarten et al., 2002).

The TOL controversy finds a striking illustration in the debate surrounding the automatically produced "highly resolved tree of life" that Peer Bork and coworkers generated from a concatenation of sequence alignments of 31 highly conserved proteins, primarily those involved in translation (Ciccarelli, et al., 2006). It did not take long for this purported TOL to be summarily dismissed as a "tree of 1%" (of the genes in any given genome) that does not actually reflect the history of genomes. To me, at least, the eloquent argument of Tal Dagan and Bill Martin (Dagan and Martin, 2006) sounds compelling and worth a quote:

> When chemists or physicists find that a given null hypothesis can account for only 1% of their data, they immediately start searching for a better hypothesis. Not so with microbial evolution, it seems, which is rather worrying. Could it be that many biologists have their heart set on finding a tree of life, regardless of what the data actually say?

In this chapter, I present a comprehensive dissection of the evolution of prokaryotes into the tree-like and web-like components that I believe has the potential to objectively determine the role and place of trees in our understanding of evolution, and go some way to actually settle the TOL controversy. However, before turning to this quantitative analysis, we examine the roots of tree thinking at a conceptual level.[1]

The fundamental units of evolution and its intrinsic tree-like nature

As discussed in Chapter 2, replication of the genetic material, a process that is intrinsically error prone, is both the condition and the direct cause of evolution. A critical point for defining the status of trees in biology is that replication and the evolution that necessarily follows are inherently tree-like processes (Koonin and Wolf, 2009a). Indeed, a replicating molecule gives rise to two copies (in the case of semiconservative replication of dsDNA that occurs in all cellular organisms and many viruses) or multiple copies (in the case of the conservative replication of viruses with ssDNA or ssRNA genomes) with errors, resulting in a tree-like process of divergence (see Figure 6-1). In graph-theoretical terms, such a process can be isomorphously represented by a specific form of a directed acyclic graph known as arborescence, a generalized tree in which multifurcations are allowed and all edges are directed away from the root (see Figure 6-1). Although the occasional extinction of one or both progeny molecules yields vertices that emit no edges, the graph remains an arborescence; the definition of this class of graphs does not require the leaves to be at the same level (see Figure 6-1; hereafter, instead of referring to arborescences, I use the common term "tree").

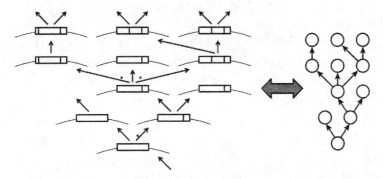

Figure 6-1   A tree (arborescence) as an isomorphous representation of the error-prone gene replication process. A schematic of the replication history of a genetic element that includes both bifurcations and a multifurcation (shown by asterisks). Fixed mutations are shown by strikes. Adapted from Koonin and Wolf, 2009a.

A potential major complication to the tree-like character of evolution is recombination. If common, recombination would turn the tree-like representation of the history of a replicating lineage (see Figure 6-1) into a network (or, worse, a morass). Is it possible to determine a fundamental, "atomic" level of genetic organization at which recombination is negligible? This does not seem to be feasible in the case of homologous recombination that is extensive during coreplication of closely related sequences, particularly in eukaryotes that engage in regular sex and in "quasi-sexual" prokaryotes. Essentially, the unit of homologous recombination is a single base pair. However, homologous recombination cannot occur between distantly related sequences, so HGT between diverse prokaryotes involves only nonhomologous (illegitimate) recombination complemented by more specific routes such as dissemination via bacteriophages and plasmids (see Chapter 5). In contrast to homologous recombination, a strong preference for evolutionary fixation of nonhomologous recombination events outside of genes or between parts of genes, encoding distinct domains of multidomain proteins, should be expected; preservation of gene integrity after nonhomologous recombination within genes is extremely unlikely. The prevalence of intragenic recombination in the course of HGT between distantly related prokaryotes has not been studied in sufficient detail. Nevertheless, at least one study shows that regions encoding relatively small protein domains are significantly avoided by recombination (Chan, et al., 2009). Hence, the conclusion appears important and plausible, even if not yet sufficiently supported by data: *The evolutionary history of a gene or domain is reticulate on the micro scale due to homologous recombination but is largely tree-like on the macro scale* (see Figure 6-2).

W. Ford Doolittle and Eric Bapteste have argued and demonstrated in compelling examples that a tree can well describe relationships that have nothing to do with common descent, so "tree thinking" was deemed not to be *a priori* relevant, or at least not central in biology (Doolittle and Bapteste, 2007). Although valid in itself, this argument seems to miss the crucial point discussed earlier, that *a tree is a necessary formal consequence of the descent history of replicating nucleic acids and the ensuing evolution.* Thus, trees cannot be banished from evolutionary biology for a fundamental reason: They *are intrinsic to the evolutionary process.* The main pertinent question then becomes this: What are the fundamental genetic units whose evolution is best represented by trees? In the practice of evolutionary biology, trees are most often built for individual genes or for sets of genes that are believed to evolve coherently. However, it is usually implied (or even stated explicitly) that the ultimate goal is a species (organismal) tree. The lack of clarity about the basic unit to which tree analysis applies seems to be an important (if not the main) source of the entire TOL controversy.

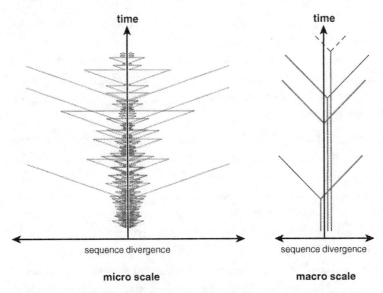

micro scale　　　　　**macro scale**

Figure 6-2　Evolution of a gene is reticulate on the micro scale but tree-like on the macro scale. The figure schematically depicts the evolution of four genes. The divergence history of each gene was simulated under the model of random homologous recombination, with the probability of recombination exponentially decreasing with sequence divergence. At each simulation step, the two daughter genes diverge by a constant amount (clock-like divergence) and either undergo homologous recombination (which brings the difference between the two back to zero) or do not, preserving the existing state of divergence. After a number of short periods of divergence and recombination, the genes stochastically diverge far enough for (homologous) recombination to become extremely unlikely. After that point, they continue diverging without recombination. At a macro scale, this pattern appears as a simple bifurcation in the tree-graph. Adapted from Koonin and Wolf, 2009a.

Conceptually, the answer to the previous question seems clear: *The fundamental unit of evolution can be most adequately defined as the smallest portion of genetic material with a distinct evolutionary trajectory*—that is, one that evolves independently of other such units through a substantial duration of evolution. In practice, given the dynamical character of the evolution of prokaryotes described in Chapter 5, *a genomic locus that encodes an RNA or protein molecule (or a distinct evolutionary domain) meets the criteria of a fundamental unit of tree-like evolution.* (Obviously, this unit corresponds to a gene, except for the case of multidomain proteins.) Indeed, as first explicitly stated by Richard Dawkins in his eponymous book (Dawkins, 2006), genes are substantially selfish; that is, subject to selection that is partially independent of other genes. Under conditions of extensive HGT, a gene or an operon has the potential to sweep a broad range of organisms. Of course, this typically happens when the gene in question confers a selective advantage to the organisms that harbor it, so the evolution of genes and the evolution of organisms are tightly linked.

The realization that individual genes, as opposed to genomes, are the "atomic" units of evolution undermines the very idea of a TOL. However, as shown earlier, trees are inalienable from any description of evolution, for the simple reason that replication of the genetic material is an intrinsically tree-like process. Together these two fundamental observations lead to a clear conclusion on what should replace the TOL: the *Forest of Life (FOL), or the collection of phylogenetic trees for all genes* (with the obvious exception of ORFans). The reconstruction of the history of life (obviously, not the entire history, but its "skeleton"), then, is not as simple as an analysis of the topology of the TOL. Instead, such a reconstruction requires charting the FOL in search of "groves" of similar trees that might reflect long-term trends of coherent (vertical) evolution of gene ensembles and "vines" of HGT. Arguably, comprehensive exploration of the FOL is the primary goal of phylogenomics. In the following sections, I mostly discuss the results of recent analysis of the FOL performed by my colleagues Pere Puigbo and Yuri Wolf and myself (Puigbo, et al., 2009, 2010). These are by no means the only studies that compare phylogenetic trees and try to distinguish vertical from horizontal trends in evolution. However, this work is up-to-date, and I believe that we have found useful ways to present the relationships between trees of numerous genes, so a summary of these results provides a good idea of the structure of the FOL. (The presentation in the next two sections is somewhat more technical than most of this book; some readers might decide to skip to the concluding paragraphs of each section and then to the synopsis of the chapter.)

The Forest of Life and the nearly universal trees

In principle, the FOL includes trees for "all" genes. In practice, however, working with all 1,000 or so sequenced prokaryote genomes (this number will increase by a few hundred by the time this book comes out) is technically hard because the maximum likelihood methods for tree construction that provide the best resolution are computationally expensive. Fortunately, using all genomes does not seem to be particularly important. The dynamic picture of the evolution of prokaryotes notwithstanding, core and shell genes in closely related organisms (defined, for example, by the high sequence similarity of their rRNAs or other core genes) evolve congruently most of the time (and only core and shell genes are widespread enough to yield meaningful trees). Thus, a carefully selected representative set of organisms should be sufficient to reveal major trends in the FOL. For the studies discussed here, we constructed such a representative set of 100 prokaryotic genomes, 41 archaeal and 59 bacterial (in the rest of this chapter, we refer to these prokaryotes as species—with the full understanding of the limitations of this concept pointed out in Chapter 5). Trees were built for all sets of orthologs with more than four members (the minimum number of sequences required to make an unrooted tree), so we obtained almost 7,000 trees altogether. Predictably, given the core-shell-cloud structure of the prokaryotic gene space described in the Chapter 5, most of these trees are small: Only 2,040 trees included more than 20 species, and only a small set of 102 Nearly Universal Trees (NUTs) included more than 90% of the analyzed prokaryotes.

Usually, phylogeneticists attempt to identify HGT by comparing trees of individual genes to a predefined "species tree." However, as we have seen in the preceding section, the very concept of a "species tree" is invalidated by the pervasive HGT and the selfishness of individual genes that are the fundamental units of tree-like evolution. We sought to dissect the FOL without any preconceived idea of a standard tree against which to compare the rest of the trees. To this end, we analyzed the complete matrix of the topological distances between the trees; this is quite a large matrix that includes almost 24 million pair-wise tree comparisons, although many cells in the matrix are empty because the respective trees consist of nonoverlapping sets of species.

In Figure 6-3, the FOL is represented as a network in which each node is a tree. We see that the group of NUTs occupies a rather special position in this network: About 40% of the trees are highly similar to at least one NUT. (Two trees are considered "topologically similar" when there is only a small difference in the connections between their branches; topological differences are used to calculate distances between trees. We skip the details of these calculations.) In sharp contrast, using the same similarity cutoff, 102 randomized NUTs were connected only to about 0.5% of the trees in the FOL. Thus, there is a high and nonrandom topological similarity between the NUTs and a large part of the FOL.

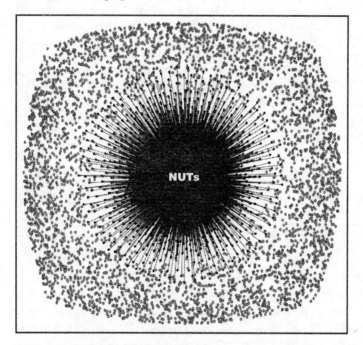

Figure 6-3   The Forest of Life as a network of trees. Each node in the network denotes a tree. The 102 Nearly Universal Trees (NUTs) are shown as solid circles in the middle, and the rest of the trees are shown as empty circles. The NUTs are connected to trees with similar topologies—trees with at least 50% of similarity to at least one NUT. Adapted from Puigbo, et al., 2009.

Knowing all the distances between the trees in the FOL, we can apply statistical methods for data clustering—that is, determining whether the FOL is simply a cloud of randomly scattered points (trees in the topology space) or contains distinct clusters of trees with similar topologies. The applied statistical procedure partitioned the FOL into seven clusters of trees. Notably, all the NUTs formed a compact group within one of the clusters (see Figure 6-4). The seven clusters showed considerable differences in the distribution of the trees by the number of species, the distribution of archaea and bacteria, and the functional classification of the respective genes. So the clustering results indicate that the FOL can be partitioned into large, distinct groups of topologically similar trees; however, at this stage, it remains unclear how much of this clustering is due to "vertical" evolutionary processes and how much to "horizontal" ones. The key observation is that all the NUTs occupy a compact and contiguous region of the tree space, are not partitioned into distinct clusters (in contrast to the rest of the FOL), and are separated by approximately the same distance from all clusters of trees (see Figure 6-4).

The results of this first part of our expedition into the thicket of the FOL suggest an important conclusion: *The topologies of the NUTs are highly similar to each other and seem to represent a central evolutionary trend in the FOL.* The claim about the central trend might seem vague but actually reflects very simple, straightforward observations:

1. The topologies of the NUTs are similar to the topologies of many other trees in the FOL.

2. The NUTs are approximately equidistant from clusters of other trees. In a sense, they occupy a central position in the FOL.

So far, we have spoken of the NUTs in the abstract, without considering the actual genes that contribute to this set of big trees. In fact, the identities of the NUTs are all too predictable: These are the genes that encode ribosomal proteins and other highly conserved proteins involved in translation, along with a few core subunits of the DNA-dependent RNA polymerase. These are the genes that are expected to be least prone to HGT, according to the complexity hypothesis (Jain, et al., 1999). Somewhat paradoxically, this set of nearly universal genes also encompasses some of the most spectacular examples of HGT, particularly among the aaRS, some of which are responsible for antibiotic resistance, but also among quite a few ribosomal proteins. Nevertheless, the observations reported here unequivocally show that the group of NUTs is internally topologically consistent and, moreover, is linked by topological similarity to numerous other trees in the FOL.

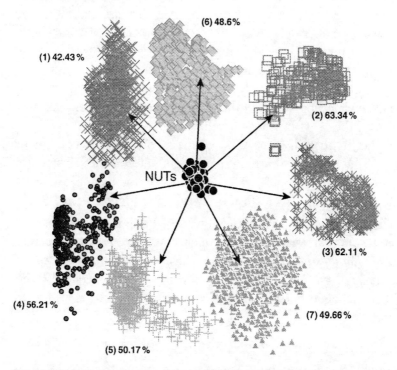

Figure 6-4   Clustering of the Forest of Life in the tree topology space. The clusters were obtained using classical multidimensional scaling (a clustering method that essentially is a more sophisticated version of the popular principal component analysis approach). The NUTs are arbitrarily placed in the center, and the mean similarity between the NUTs and each of the clusters is shown. Adapted from Puigbo, et al., 2009.

In light of the near-ubiquitous HGT, nothing can revive the TOL in its old glory. However, if we were to seek the best meaningful approximation of a TOL, the consensus topology of the NUTs would look like the best candidate. But before we inaugurate the NUTs in this capacity, deeper aspects of evolution need to be discussed.

Deep into the Forest of Life: Big Bang or compressed cladogenesis?

We saw in the preceding section that the NUTs occupy a special position in the FOL. They are topologically similar to each other and to numerous other trees in the FOL, and so might represent a central trend of vertical, tree-like evolution. However, before we conclude that these trees for nearly universal, essential genes indeed reflect a central trend in the FOL, a crucial question needs to be addressed: Does the tree signal permeate the entire history of life, or is it limited to relatively recent evolution?

We have good reason to ask this question. Many phylogenetic studies, including an examination of a supernetwork of the NUTs (glossing over technical details, the supernetwork is a consensus tree produced by "averaging" the topologies of the NUTs), clearly show that deep internal nodes in phylogenetic trees tend to be poorly resolved, compared to the nodes located closer to the leaves (see Figure 6-5A). This pattern recurs at many different levels in the history of life: For instance, poor resolution among the deepest branches is seen both in the phylogenetic tree of mammals that spans about 100 million years and in the purported TOL that covers more than 3.5 billion years (Rokas and Carroll, 2006). Thus, the intervals of evolution that involve the primary radiation of major groups of organisms appear to be special, different from "normal" epochs of evolution (an analogy with punctuated equilibrium, even if superficial, is tempting—see Chapter 2). Two models have been proposed to account for this pattern:

1. Compressed Cladogenesis (see Figure 6-5A; Rokas and Carroll, 2006)

2. The more radical "Biological Big Bang" model[2] (see Figure 6-5B; Koonin, 2007a)

Under the Compressed Cladogenesis model, evolution—or, more precisely, the emergence of new groups of organisms (clades, or distinct monophyletic branches in trees) in the transformational epochs—occurs rapidly, resulting in very short internal branches. Accordingly, these branches can be extremely hard to reliably resolve using any phylogenetic method. Nevertheless, in principle, under the

Compressed Cladogenesis mode, there is a unique branching order throughout the TOL (regardless of the exact interpretation of the TOL idea). The Biological Big Bang model posits that the transitional epochs are qualitatively different from "normal" tree-like phases of evolution: The model postulates that, as a result of rampant HGT, including massive gene fluxes caused by endosymbiosis and other processes, these stages of evolution are completely devoid of the tree signal. Somewhat simplifying the model, it can be said that, in the transitional phases, the memory of the preceding tree-like evolution is obliterated, and the evolving genomes form a single gene pool from which new clades emerge. Whatever lengths are assigned to the respective internal branches by tree construction procedures, under the Big Bang model, these are artifacts; the true length of all these branches is exactly zero (see Figure 6-5B). The question of whether or not a discernible phylogenetic signal exists in the deepest nodes of the trees is obviously relevant for the validity of a central tree-like trend in the FOL that potentially could be approximated by the topologies of the NUTs. Fortunately, the two models can be tested by a deeper analysis of the trends in the FOL.

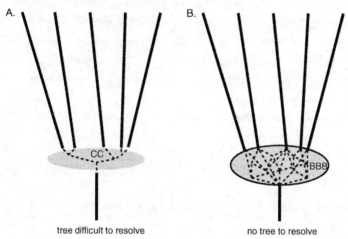

Figure 6-5   The two models for the transitional epochs in evolution: (A) Compressed cladogenesis (CC); (B) Biological Big Bang (BBB).

We introduced a new measure, the Inconsistency Score, which determines how representative the topology of the given tree is of the entire FOL (the score is simply the inverse of the fraction of the times the splits from a given tree are found in all trees of the FOL

(Puigbo, et al., 2009). Using the Inconsistency Score, we objectively examine trends in the FOL, without relying on the topology of a pre-selected "species tree." The plots in Figure 6-6 show the dependence of the inconsistency score on the phylogenetic depth for the trees in the entire FOL and in the NUTs alone. Again leaving technicalities aside, to generate these plots, one has to cut the trees within a certain depth interval (a special procedure, the details of which are irrelevant, was developed to estimate the depth on the scale from 0 to 1) and take the mean of the inconsistency scores for that interval only. The two plots and the difference between them are quite remarkable. The plot for the entire FOL resembles plots that describe phase transitions in various physical processes: At a certain depth, the value of a particular variable (in our case, the inconsistency score) changes abruptly (see Figure 6-6). The plot for the NUTs is quite different: It shows significantly lower values of the inconsistency score (that is, the topologies of the NUTs are, on average, much more congruent with each other than those of other trees in the FOL) and a less abrupt change at the critical depth that would not qualify as a phase transition (see Figure 6-6).

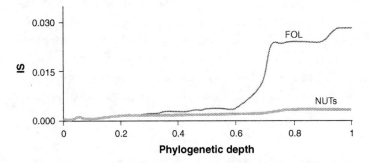

Figure 6-6   The dependence of the tree inconsistency score on the phyloge-netic depth in the Forest of Life. Note the sharp phase transition in the entire FOL and the much smoother transition among the NUTs. Adapted from Puigbo, et al., 2009.

The appearance of a phase transition suggests a distinct possibility that the deep parts of the FOL are best described by the Biological Big Bang model (indeed, in modern cosmology, the Big Bang is literally considered a phase transition, as explained in Appendix B). To address this possibility, we designed a computer model of evolution that simulated a Big Bang (that is, complete randomization of the

branching order in trees) at different phylogenetic depths and reproduced the plots shown in Figure 6-6 with different rates of additional
HGT. Much to our surprise, we failed to find a combination of
parameters (the Big Bang depth and the HGT rate) that would produce a plot closely resembling that in Figure 6-6. A curve with a good
fit to the empirically observed one was obtained only in a simulation
without a Big Bang at the point of or after the radiation of the bacterial phyla—and a Big Bang (or any other event) antedating that split is
beyond our "event horizon" in this analysis. Thus, the comparison of
the trees in the FOL seems to favor the Compressed Cladogenesis
model, although, given the difficulty of the problem, additional analysis is certainly required.

If the Compressed Cladogenesis model holds, we have to conclude that the NUTs indeed represent *a central tree-like trend that
persisted throughout the evolution of cellular life.* Couched in more
biological terms, about 100 protein-coding genes that comprise the
translation and the core of the transcription systems (along with the
universal rRNAs and tRNAs) evolved mostly in concert since the Last
Universal Common Ancestor (LUCA) of all cellular life (see Chapter
11). *Thus, the evolution of this set of genes seems to be the best possible reflection of the history of organisms that can be obtained from
molecular phylogenies.* As for the transitional epochs in the evolution
of life, these are probably best depicted as phases of extremely rapid,
explosive evolution triggered by extinction of the preceding diversity
of life forms and severe population bottlenecks among the few sur-
vivors (see Chapter 9).

Dissection of the evolution of prokaryotes into the tree-like and web-like components

As shown in the preceding section, the signal of tree-like evolution,
defined as the consensus topology of the NUTs, seems to reflect a
central trend of evolution in the FOL and is traceable throughout the
entire range of phylogenetic depths despite the substantial rate of
HGT. By contrast, the sum total of all evolutionary patterns that
appear incompatible with the consensus NUTs topology, whether
caused by HGT or by other processes (such as parallel gene losses
that are also common among prokaryotes), can be denoted the

web-like signal. We developed a quantitative measure to directly estimate (on a 0–1 scale) the tree-like and web-like contributions to the
evolutionary distances between species (Puigbo, et al., 2010). The
lower the score (that is, the closer to the distance expected by chance,
under the assumption that genes are freely mixed), the more the relationship between the given pair of species is determined by web-like
evolutionary processes. The tree-net map of the NUTs was dominated by the tree-like signal (dark in Figure 6-7A): The mean score
for the NUTs was 0.63, so the evolution of the nearly universal genes
of prokaryotes appears to be nearly two-thirds tree-like. The exceptions are the radioresistant bacterium *Deinococcus radiodurans* that
showed primarily web-like relationships with most of the archaea and
several bacterial taxa (*Thermotogae, Aquificales, Cyanobacteria,
Actinobacteria, Chloroflexi, Firmicutes*, and *Fusobacteriae*), each of
which formed a strongly connected network with other bacteria (see
Figure 6-7A).

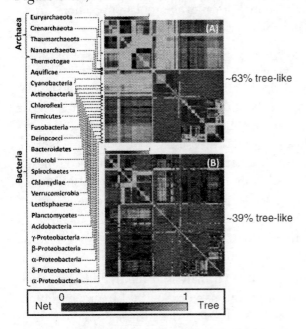

Figure 6-7   The tree-like and web-like signals in the evolution of prokaryotes:
(A) The 102 NUTs; (B) The FOL without the NUTs (6,799 trees). The tree-like
signal increases from dark (web-like evolution) to light (tree-like evolution). The
species are ordered in accord with the topology of the supertree of the 102
nearly NUTs, which is taken to represent the vertical (tree-like) signal. In (A),
the major groups of Archaea and Bacteria are denoted. Adapted with permission from Puigbo, et al., 2010.

In a stark contrast to the NUTs, the rest of the FOL is dominated by web-like evolution, with the mean score of 0.39 (about 60% web-like). Remarkably, areas of tree-like evolution are interspersed with areas of web-like evolution across different parts of the FOL (see Figure 6-7B). The major web-like areas observed among the NUTs recurred in the FOL, but additional ones became apparent, including Crenarchaeota, which showed a pronounced signal of a non-tree-like relationship with diverse bacteria, as well as some Euryarchaeota (see Figure 6-7B). A more detailed dissection of the FOL shows that the web-like signal dominates the evolution of genes that are present in a small number of prokaryotes, whereas the evolution of more widespread genes is more tree-like and more closely resembles the pattern seen among the NUTs (Puigbo, et al., 2010). This trend is clearly compatible with the HGT optimization hypothesis (see Chapter 5), according to which genes that are frequently lost during evolution should be also frequently transferred if the extinction of these genes and the overall mutational meltdown of microbial populations are to be avoided (see Chapter 5).

Different functional classes of genes showed major differences with respect to the tree-like or web-like trends in their evolution, from the dominance of the tree-like signal among genes for translation machinery components and proteins involved in intracellular trafficking, to almost fully web-like evolution of genes for ion transport, signal transduction, and defense system components (see Figure 6-8). This pattern is generally compatible with the complexity hypothesis but also reveals a more nuanced picture, with substantial differences, for instance, between enzymes of nucleotide metabolism that evolve mostly in a tree-like fashion and proteins involved in amino acid or carbohydrate metabolism and transport, for which the web-like signal was much more prominent (see Figure 6-8).

To summarize, the quantitative analysis of the tree-like and web-like signals reveals an apparent paradox of prokaryote evolution: Although the tree-like evolution is by far the strongest single trend in the FOL, quantitatively, evolution of prokaryotes is dominated by the combination of the web-like processes, such as HGT and lineage-specific gene loss. The tree-like pattern accounts for most of the evolution among the NUTs; however, because the FOL consists mostly of small trees among which the tree signal is barely detectable, the web-like processes that govern the evolution of relatively small gene families are quantitatively dominant.

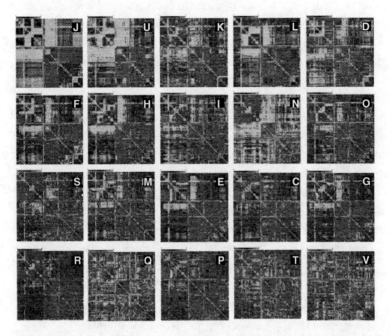

Figure 6-8   The signals of tree-like and web-like evolution for different functional classes of prokaryotic genes. The order and numbering of the species are as in Figure 6-7. The functional classification of genes was from the COG system (Tatusov, et al., 2003). The designations are: J: Translation, ribosomal structure and biogenesis; U: Intracellular trafficking, secretion, and vesicular transport; K: Transcription; L: Replication, recombination, and repair; D: Cell cycle control, cell division, and chromosome partitioning; F: Nucleotide transport and metabolism; H: Coenzyme transport and metabolism; I: Lipid transport and metabolism; N: Cell motility; O: Posttranslational modification, protein turnover, and chaperones; S: Function unknown; M: Cell wall/membrane/ envelope biogenesis; E: Amino acid transport and metabolism; C: Energy production and conversion; G: Carbohydrate transport and metabolism; R: General function prediction only; Q: Secondary metabolites biosynthesis, transport, and catabolism; P: Inorganic ion transport and metabolism; T: Signal transduction mechanisms; V: Defense mechanisms. Adapted with permission from Puigbo, et al., 2010.

Tree-like evolution or/and biased horizontal gene transfer?

Gogarten, Lawrence, and Doolittle proposed a heretical (and ingenious) hypothesis to explain away the tree signals that are observed in phylogenetic analyses of individual genes or ensembles of genes (Gogarten, et al., 2002). According to this proposal, the tree-like pattern of evolution actually might be a consequence (one might provocatively say, an artifact) of nonuniform, biased HGT, whereby organisms that appear "close" in phylogenetic trees actually exchange genes frequently, and organisms that seem "distant" in trees are those between which HGT is rare. As already pointed out in Chapter 5, this possibility certainly makes biological sense: Given that HGT moves a gene into a non-native intracellular environment, one would expect that the less divergent that environment is from the native one (the source of the transferred gene), the greater the chance for the transferred gene to be fixed. Although anecdotal, there is already some experimental support for this conjecture (Diaz, et al., 2011).

We used the FOL framework to simulate evolution with varying declining gradients of HGT rate from close to distant organisms, and to assess the possibility that the tree-like pattern we observed was a simple consequence of biased HGT. In each series of the simulations, we tested whether the observed characteristics of the NUTs, such as the mean distance between trees and the degree of separation of archaea from bacteria, could be reproduced under different models of evolution. The first series of simulations started with the topology of the supertree of the NUTs, which was taken to represent the signal of tree-like evolution, and measured the characteristics of the resulting trees, depending on the slope of the HGT gradient. We did indeed find that a moderate HGT gradient from the tips to the center of the tree reproduced the empirically observed features of the NUTs. The second series of simulations started from "star" trees, under the assumption that tree-like evolution is an artifact, and then gradually evolved the HGT gradient by assigning increased HGT rates to randomly joined branches. This simulation failed to reproduce the observed characteristics of the NUTs, even at extremely high rates of HGT. Although these simulations undoubtedly are oversimplified models of evolution, they seem to suggest that the tree-like trend and biased HGT coexist and interact in the course of evolution of prokaryotes. Indeed, the high rate of HGT between organisms whose core genes are closely related via tree-like evolution translates into a self-reinforcing process that maintains coherent clusters of prokaryotes at different phylogenetic depths.

Synopsis and perspective

When Darwin introduced the TOL metaphor, his argument came from observations on the evolution of animals. However, he generalized the tree pattern of evolution to life in general with considerable confidence. In the narrow sense, Darwin was correct: No one denies that evolution of animals is tree-like. However, this is not a TOL, but only a description of the evolution of a single, relatively small, tight group of eukaryotes. The generalization to the entirety of cellular life on Earth fails because of the complex net of extensive HGT that is most common among prokaryotes but that also prominently contributed to the evolution of eukaryotes, particularly via endosymbiosis (see Chapter 7).

However, notwithstanding the newly discovered web-like character of evolution, Darwin's metaphor reflects a deeper truth: Trees remain the natural representation of the histories of individual genes, given the fundamentally bifurcating character of gene replication and the substantially low frequency of intragenic recombination compared to intergenic recombination at long evolutionary distances. Thus, although no single tree can fully represent the evolution of complete genomes and the respective life forms, the realistic picture of evolution necessarily combines trees and networks. These components can be revealed through the analysis of the Forest of Life (FOL), the complete collection of phylogenetic trees for individual genes.

The quantitative dissection of the FOL reveals a complex landscape of tree-like and web-like evolution. The signals from these two types of evolution are distributed in a highly nonrandom fashion among different groups of prokaryotes and among functional classes of genes. Overall, the web-like signal is quantitatively dominant, a finding that (almost literally) vindicates the concepts of "lateral genomics" or "net of life." These results are decidedly incompatible with the representation of prokaryote evolution as a TOL adorned with thin, random "cobwebs" of HGT (Ge, et al., 2005; Kunin, et al., 2005). However, the tree-like signal compatible with the consensus topology of the NUTs is also unmistakably detectable and strong; by our measurement, up to 40% of the evolution in the prokaryote world follows this pattern. *The crucial, even if somewhat paradoxical, feature of prokaryote evolution appears to be that, although web-like processes are quantitatively dominant, the single strongest trend is the tree-like evolution reflected in the consensus tree topology of the NUTs that also largely recapitulates the rRNA tree.* In principle, one could speak of this trend as a "statistical" or "weak" TOL, although I tend to think that this terminology is counterproductive: The proper object of phylogenomics is the FOL and evolutionary patterns that can be discerned in it rather than an illusory Tree of Life.

The tree-like trend of evolution seems to be related to the gradient of HGT from closely related (via tree-like evolution) to distant life forms. The interaction between tree-like evolution and biased HGT may create a self-reinforcing evolutionary process that accounts for the coherence of groups of prokaryotes at different phylogenetic depths.

To conclude this chapter, I have to echo the conclusion of the preceding one: Although the approaches to the quantitative analysis of the FOL outlined here are informative and illustrative, they certainly are not the final word in phylogenomic methodology. A truly adequate conceptual apparatus and technical tools for the simultaneous, comprehensive analysis of tree-like and web-like evolutionary processes remain to be developed. Once such methods are at hand, we will start discerning the real seascape of evolution.

Recommended further reading

Bapteste, E., M. A. O'Malley, R. G. Beiko, M. Ereshefsky, J. P. Gogarten, L. Franklin-Hall, F. J. Lapointe, J. Dupré, T. Dagan, Y. Boucher, and W. Martin. (2009) "Prokaryotic Evolution and the Tree of Life Are Two Different Things." *Biology Direct* 4: 34.

Another paper on the interface of philosophy and biology, with the main emphasis that "the belief that prokaryotes are related by ... a tree has now become stronger than the data to support it."

Ciccarelli, F. D., T. Doerks, C. von Mering, C. J. Creevey, B. Snel, and P. Bork. (2006) "Toward Automatic Reconstruction of a Highly Resolved Tree of Life." *Science* 311: 1,283–1,287.

The culmination of the classic tradition of TOL studies. A computational pipeline is described that automatically generates a Tree of Life from alignments of 31 universally conserved proteins (naturally, all of these are proteins involved in translation).

Dagan, T., and W. Martin. (2006) "The Tree of One Percent." *Genome Biology* 7: 118.

A spirited refutation of the "brave new TOL" of Cicarelli, et al: According to Dagan and Martin, a tree of about 1% of the genes in compared genomes, even if free of internal contradictions, in principle cannot represent genome evolution and, thus, cannot claim to be a TOL.

Doolittle, W. F. (2009) "The Practice of Classification and the Theory of Evolution, and What the Demise of Charles Darwin's Tree of Life Hypothesis Means for Both of Them." *Philosophical Transactions of the Royal Society of London Biological Sciences* 364: 2,221–2,228.

The abstract of this brief review of the TOL controversy is so aphoristic that it is worth quoting almost in full by way of annotation: "Debates over the status of the tree of life (TOL) often proceed without agreement as to what it is supposed to be: a hierarchical classification scheme, a tracing of genomic and organismal history or a hypothesis about evolutionary processes and the patterns they can generate. I will argue that for Darwin it was a hypothesis, which lateral gene transfer in prokaryotes now shows to be false."

Doolittle, W. F. (1999) "Phylogenetic Classification and the Universal Tree." *Science* 284: 2,124–2,129.

The first principled revision of the TOL concept, on the basis of the discovery of extensive HGT among prokaryotes. The classic image of the three-domain TOL is replaced with a web-like image that still contains the three major trunks but also shows numerous horizontal connections—so many of these that the trunks are barely discernible.

Doolittle, W. F., and E. Bapteste. (2007) "Pattern Pluralism and the Tree of Life Hypothesis." *Proceedings of the National Academy of Sciences USA* 104: 2,043–2,049.

A remarkable blend of philosophy and biology. The paper illustrates how the same (tree-like) pattern emerges through different processes, some of which are not tree-like at all. Doolittle and Bapteste concluded: "Pattern pluralism (the recognition that different evolutionary models and representations of relationships will be appropriate, and true, for different taxa or at different scales or for different purposes) is an attractive alternative to the quixotic pursuit of a single true TOL."

Koonin, E. V., and Y. I. Wolf. (2009) "The Fundamental Units, Processes, and Patterns of Evolution, and the Tree of Life Conundrum." *Biology Direct* 4: 33.

An extended argument that tree-like patterns are central to evolution derived from the fundamental bifurcating replication process. Inasmuch as homologous recombination within genes is much less frequent than between genes, evolution of individual genes appears fundamentally tree-like, and the description of genome evolution should be sought by searching for trends in the "forest" of gene trees.

O'Malley, M. A., and Y. Boucher. (2005) "Paradigm Change in Evolutionary Microbiology." *Studies in History and Philosophy of Biological and Biomedical Sciences* 36: 183–208.

A card-carrying philosopher and an evolutionary biologist present a classification of the "postmodern" views of the TOL, from the denial of any substantial contribution of HGT to the summary dismissal of tree thinking.

Puigbo, P., Y. I. Wolf, and E. V. Koonin. (2010) "The Tree and Net Components of Prokaryote Evolution." *Genome Biology and Evolution* 2: 745–756.

A quantitative dissection of the evolution of prokaryotes into the tree-like and web-like components vindicating the lateral genomics concept, but also revealing the tree-like central trend in prokaryote evolution.

Puigbo, P., Y. I. Wolf, and E. V. Koonin. (2009) "Search for a 'Tree of Life' in the Thicket of the Phylogenetic Forest." *Journal of Biology* 8: 59.

A comprehensive comparison of the topologies of about 7,000 trees that comprise the prokaryotic FOL. The analysis reveals an objective central trend in the FOL and shows that the Compressed Cladogenesis model for major evolutionary transitions is better compatible with the data than the Biological Big Bang model.

7

The origins of eukaryotes: Endosymbiosis, the strange story of introns, and the ultimate importance of unique events in evolution

Organisms with large, complex cells are known as eukaryotes—that is, possessing bona fide nuclei. These organisms include the three kingdoms of multicellular life forms, plants, brown algae, and animals, as well as a huge variety of unicellular forms (also known as protists). Eukaryotic cells typically are orders of magnitude bigger than prokaryotic cells and possess complex intracellular organization with diverse membrane-bounded organelles, including the eponymous nucleus and the mitochondria that evolved from endosymbiotic bacteria. Thus, by any reasonable criteria, eukaryotic cells are dramatically more complex than bacteria and archaea. How this complexity evolved is a major enigma of evolutionary biology, which is, of course, all the more tantalizing, thanks to the inescapable parochial interest. After all, when we explore the origin of eukaryotes, we are looking into *our own* origins.

The fundamental differences in cellular organization between eukaryotes and prokaryotes make a rather paradoxical counterpoint to the differences in genome architecture that we already touched upon in Chapter 3. While eukaryotic cells possess a far more elaborate, ordered, and complex organization than prokaryotic cells, the genomes of eukaryotes are by far less optimized and more haphazard than the genomes of prokaryotes. Understanding the evolutionary underpinnings of this apparent paradox is a major challenge, and the solution is likely to hold the key to the origin of eukaryotes—or, more precisely, the origin of the eukaryotic cell organization. The problem is far from being solved and remains a matter of heated (sometimes, perhaps, overheated) disputes. In this chapter, we address this conundrum and, more generally, the evolutionary relationships between archaea, bacteria, and eukaryotes as objectively and logically as possible, to gain insight into the emergence of the "eukaryotic state." By the end of this discussion, I hope to show that, although numerous key details remain to be elucidated, the contours of a specific, plausible scenario for eukaryote origin are becoming apparent, and this scenario accounts for at least some of the unusual features of the eukaryotic genomes and the remarkable complexity of the eukaryotic cellular organization. Furthermore, this chapter comes to the conclusion that Woese's three-domain scheme is not a proper depiction of the history of life and ponders general implications of the eukaryote story that are germane to the central theme of this book, the interplay of chance and necessity in evolution.

The eukaryotic cell, its internal architecture, and the chasm between prokaryotic and eukaryotic cellular organizations

It is neither practical nor necessary in this book to delve into the innumerable fine details of biological structures. However, to discuss the origin of eukaryotes (hereafter *eukaryogenesis*) in earnest, we need to fully appreciate the nature and depth of the gulf that separates the cells of eukaryotes from prokaryotic cells. Indeed, there is a sharp divide in the organizational complexity of the cell between eukaryotes and prokaryotes: A typical eukaryotic cell is about a thousand-fold larger by volume than a typical bacterium or archaeon and possesses elaborate intracellular compartmentalization that is not seen even in the most sophisticated prokaryotes. Later in this chapter, we briefly discuss some interesting exceptions, such as giant prokaryotic cells and prokaryotic cells containing intracellular compartments; nevertheless, a careful examination of these cases supports the fundamental eukaryote-prokaryote dichotomy in cellular organization.

The compartmentalization of eukaryotic cells relies on an elaborate, diversified endomembrane system and the actin-tubulin-based cytoskeleton. A striking consequence of the intracellular compartmentalization is that eukaryotic cells are physically distinct from the cells of prokaryotes. In prokaryotes, the content of the cell is a solution, even if a viscous one, so macromolecules (proteins and nucleic acids) diffuse more or less freely and reach their cellular destinations through the combination of stochastic movements with trapping in specific complexes. By contrast, in eukaryotes, macromolecules are largely prevented from free diffusion and instead reach their target sites through complex trafficking systems. This difference is clearly demonstrated by a simple experiment in which the membranes of prokaryotic and eukaryotic cells are opened (permeabilized): Proteins and nucleic acids ooze out of permeabilized bacterial cells, but not (typically) from eukaryotic cells (Hudder, et al., 2003). So the cytosol of eukaryotic cells has much lower entropy than prokaryotic cells—it is hard to think of a more fundamental difference.

The nucleus, the eponymous eukaryotic organelle, encloses the genomic DNA that is organized into chromatin and partitioned among multiple chromosomes; it is the site of transcription, splicing, and ribosome assembly. The nucleus itself is part of the endomembrane system: The nuclear envelope is continuous with the membranes of the endoplasmic reticulum. Obviously, for a eukaryotic cell to function, the nucleus must constantly communicate with the cytosol. Indeed, the nuclear envelope is perforated by pores, extremely complex structures that are responsible for both passive and active trafficking of all kinds of molecules (and even macromolecular complexes such as the ribosomal subunits) to and from the nucleus. Note that the confinement of the chromatin and transcription to the nuclear compartment eliminates transcription-translation coupling, a hallmark feature of gene expression in prokaryotes. Later in this chapter, we examine the fundamental implications of this decoupling.

Inside, the nucleus is filled with a highly structured matrix and, in this respect, resembles the eukaryotic cytosol. The eukaryotic chromatin contained within the nucleus is by no means just a DNA molecule covered with proteins and regularly packed in three dimensions. Instead, the chromatin is an extremely complex, dynamic system of molecular machines that consist of numerous specialized proteins that regulate and coordinate the processes of replication and expression largely through the so-called chromatin remodeling—that is, modification of the chromatin structure that changes the pattern of accessible regions (Clapier and Cairns, 2009). Although the picture of prokaryotic expression regulation is becoming increasingly complex and is now a far cry from the simple Jacob-Monod scheme (see Chapter 5), nothing in prokaryotes can compare with the complexity of the eukaryotic chromatin.

Qualitative distinctions between eukaryotes and prokaryotes are numerous and span various aspects of cell biology, particularly those that are involved with information processing, signal transduction, and intracellular trafficking (see Box 7-1). The organizational complexity of the eukaryotic cells is complemented by extremely sophisticated, cross-talking signaling networks. The main signaling systems in eukaryotes are the kinase-phosphatase machinery that regulates protein function through phosphorylation and dephosphorylation; the ubiquitin network that governs protein turnover and localization through reversible protein ubiquitylation; regulation of translation by microRNAs; and regulation of transcription at the levels of individual genes and chromatin remodeling.

In Chapter 3, we discussed some of the major differences in the genome architectures of prokaryotes and eukaryotes. Later in this chapter, we consider in greater detail the evolution and possible origins of one of the most fascinating signatures of eukaryotes, the exon-intron structure of genes. Note that the differences are manifest at all levels of genome organization, from gross features such as the partitioning of the genome into multiple linear chromosomes, to the fine details such as the size and structure of untranslated regions in protein-coding genes (see Box 7-1).

No direct counterparts to the signature eukaryotic organelles, genomic features, and functional systems exist in archaea or bacteria. Hence, the very nature of the evolutionary relationships between prokaryotes and eukaryotes becomes a cause of bewilderment. Indeed, sequence comparisons between genome-wide gene sets show beyond doubt that a few thousand eukaryotic genes responsible for key cellular functions (translation, transcription, and replication) share common origins with homologs from archaea and/or bacteria (Koonin, et al., 2004). This evolutionary unity of cellular life forms makes it an extremely hard and fascinating challenge to explain how largely common components give rise to cells that are so dissimilar in so many respects.

Box 7-1: A brief comparison of major structural and functional features of eukaryotic and prokaryotic cells

Feature/system	Eukaryotes	Prokaryotes
Endosymbiosis	Mitochondria-related endosymbionts in all eukaryotes Plastids in all Plantae and many Chromalveolata	Apparently extremely rare, but bacterial endosymbionts of other bacteria reported
Intracellular membranes/ compartmentalization	Advanced endomembrane system: endoplasmic reticulum, Golgi complex Membrane-bounded organelles: nucleus, vacuoles, peroxisomes Fully compartmentalized cytosol	Typically, no membrane-bounded organelles Limited endomembrane systems and intracellular compartmentalization Intracellular compartments in some groups (Verrucomicrobia-Planctomycetes) and in specialized cell forms (spores, cyanobacterial heterocysts)
Chromatin organization	Highly complex chromatin with nucleosome organization Hundreds of associated protein complexes and diverse modifications Multiple linear chromosomes	Relatively simple chromatin organization Typically, a single or a few circular chromosomes
Cytoskeleton	Complex cytoskeleton consisting of tubulin-based microtubules and actin filaments that interact with numerous protein complexes and ancillary cytoskeleton proteins	Transient cytoskeleton structures, like the FtsZ ring that is formed during cell division Microtubules in Prosthecobacteria that possess a tubulin horizontally transferred from eukaryotes Probable actin filaments in Thermoproteales

Feature/system	Eukaryotes	Prokaryotes
Transcription-translation coupling	No	Yes
Intracellular trafficking of proteins and nucleic acids	Primarily highly organized trafficking mediated by cytoskeleton and intracellular membrane system	Primarily free diffusion
Cell wall	None in most eukaryotes Cellulose cell walls in Plantae	Peptidoglycan cell walls in most bacteria Proteinacious cell walls (S layers) in archaea Many derived wall-less forms.

The preceding section outlined several fundamental distinctions between prokaryotic and eukaryotic cells (see Box 7-1). However, the most striking of these differences merits a separate section because it might well hold the key to the entire problem of the origin of eukaryotes. This major hallmark of the eukaryotic cell is the presence of mitochondria, which play the central role in energy transformation and perform many additional roles in eukaryotic cells, such as involvement in diverse forms of signaling and programmed cell death. Mitochondria are compartments of a characteristic ("lady's shoe–like") shape bounded by a double membrane; the inner membrane contains the electron transport chain, an array of strictly ordered protein complexes. Strikingly, mitochondria harbor their own genomes, typically circular DNA molecules that vary in size between eukaryotic kingdoms (very small, only about 10 Kb, in animals, and larger, on the order of 100 Kb up to about 1 Mb, in other eukaryotes) and encode a small number of proteins (only 13 in most animals, largely subunits of the electron transfer complexes, along with 34 rRNA and tRNA molecules). The larger mitochondrial genomes of plants, fungi, and protists may contain more functional genes up to 100 in the excavate *Reclinomonas americana*), but mostly the large mitochondrial genomes consist of inserted mobile elements

(Barbrook, et al., 2010). Furthermore, mitochondria possess their own transcription and translation systems that mediate the expression of the mitochondrial genome. In every respect, these systems resemble the prokaryotic counterparts more than those of eukaryotes. Many eukaryotic cells contain numerous mitochondria, and under the electron microscope, a eukaryotic cell looks almost as if it is packed with multiple parasitic or symbiotic bacteria. And, in fact, this is exactly the case.

These days, biologists have no reasonable doubts that mitochondria originated from bacteria that were endosymbionts of an ancestral eukaryote and have undergone reductive evolution that molded them into organelles that are fully dependent on the host cell, yet have retained some signature prokaryotic features. It was relatively easy to identify the bacterial ancestors of the mitochondria (Yang, et al., 1985): Phylogenetic analysis of the mitochondrial rRNA and some protein-coding genes confidently placed them within α-proteobacteria, a distinct branch of the double-membrane Proteobacteria that, interestingly, includes, along with a huge variety of free-living bacteria, a number of intracellular parasites (like *Rickettsia*) and endosymbionts (like *Wolbachia*). So at least conceptually, the path from α-proteobacteria to mitochondria seems straightforward. However, at the molecular level, the transition is anything but trivial. Indeed, most of the mitochondrial genomes have shrunk to a bare minimum, and the shrinkage has been accompanied by the transfer of hundreds of former bacterial genes into the host genome (for now, let us stick to this neutral description of the endosymbiont's host—later in this chapter, we discuss the nature of the host in detail). The protein products of most of these genes—including, among others, all proteins that constitute the mitochondrial translation system—are targeted back to the mitochondria, where they perform their functions (see Figure 7-1). For this circuit to work, the genes transferred to the host chromosomes need to be transcribed by the host, which requires the proper regulatory signals, and translated in the cytosol, which requires the full complement of the eukaryotic translation signals; finally, these proteins have to be imported into the mitochondrion, which requires specific import signals and specialized protein machinery in the outer membrane of the mitochondria. Adaptation of the transferred endosymbiont genes for the proper passage through this circuitous path seems almost like a problem unsolvable in its complexity. However, there seems to be an obvious enough solution; I keep the suspense for now and address it in the section on the origin of the eukaryotic cell, later in this chapter.

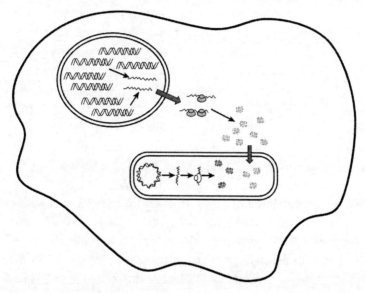

Figure 7-1   A schematic of the mitochondrion, with its genome and translation system and the routing of mitochondrial proteins in the eukaryotic cell.

Endosymbiosis, mitochondria, hydrogenosomes, and plastids

A crucially important, perhaps somewhat underappreciated discovery of the first years of the twenty-first century is the finding that *all* eukaryotes that have been studied in sufficient detail possess mitochondria or related organelles (Shiflett and Johnson, 2010; van der Giezen, 2009). A variety of unicellular eukaryotes (protists), such as many amoebas, microsporidia, some anaerobic fungi, and diverse excavates, lack typical mitochondria and, for a long time, have been considered primitive, primary amitochondrial eukaryotic forms (often collectively denoted *archezoa*). However, more recent ultrastructural studies revealed previously unknown tiny organelles resembling mitochondria in all of these organisms. These protists are anaerobes, so the organelles that became known as hydrogenosomes, or mitosomes, or simply mitochondria-like organelles (MLOs) are not involved in aerobic respiration like mitochondria. However, they still possess the iron-sulfur clusters that are the main catalytic centers of the mitochondrial electron transfer chains. In MLOs, these clusters, along with a variety of other enzymes, catalyze diverse anaerobic redox reaction; in particular, one major pathway yields molecular hydrogen that is consumed by the cytosolic metabolic systems. Despite their diversity, MLOs contain a number of proteins in common with the typical mitochondria. They also use protein import machineries that closely resemble the mitochondrial one. What these tiny MLOs conspicuously lack is the genome and the translation system that are always present in bona fide mitochondria. However, the genes for several hallmark proteins that the mitochondria and the MLOs share have been detected in the nuclear genomes of the respective organisms. All things considered, there is no reasonable doubt that the MLOs are degraded mitochondrial derivatives that probably lost their genomes upon the transition of the respective organisms to their anaerobic lifestyle. In all likelihood, this reduction of the mitochondria occurred on several independent occasions during the evolution of eukaryotes. Hence, the crucial conclusion: *We are currently unaware of any true amitochondrial eukaryotes*. It is not impossible that, as I am typing these words, some archezoa are quietly reproducing, for example, in a little pond nearby. But with each new eukaryote that is found to possess mitochondria or MLOs, this proposition is becoming increasingly unlikely.

The story of endosymbiosis in eukaryotes certainly is not limited to mitochondria. The second key endosymbiosis was the acquisition of a cyanobacterium by the common unicellular ancestor of green algae and land plants. That cyanobacterial endosymbiont evolved into a plastid that subsequently diversified into chloroplasts and chromoplasts. After the cyanobacterial endosymbiosis, a variety of protists went on a veritable rampage of engulfing green algae and other evolving plastid-carrying cells and, as a result, acquiring complex endosymbionts that consist of a plastid (probably the selective factor behind the evolutionary fixation of the endosymbiosis) and the remains of a eukaryotic cell (often called nucleomorph), the original host of the plastid. Endosymbiosis seems to have been the main factor in the diversification of Chromalveolata, one of the eukaryotic supergroups (see later) (Bhattacharya et al., 2007; Lane and Archibald, 2008).

In several notable articles and a provocative book co-authored with Dorion Sagan, Lynn Margulis, the founder of the endosymbiosis concept in its modern incarnation, propounded endosymbiosis as the single dominant evolutionary process in eukaryotes (Margulis, 2009; Margulis et al., 2006; Margulis and Sagan, 2003). Margulis not only proposed that some organelles, in addition to mitochondria and plastids, such as centrioles and flagella, evolved via endosymbiosis, but even argued that endosymbiosis underlies all speciation events in eukaryotes. However, in a sharp contrast to the cases of mitochondria and plastids, these ideas have little, if any, empirical support. Other known cases of bacterial endosymbiosis in eukaryotes are quite numerous but mostly transient, although there are notable exceptions, such as the long-term, mutualistic endosymbiosis in various insects (Gibson and Hunter, 2010).

The supergroups of eukaryotes and the root of the eukaryotic evolutionary tree

As noted in Chapter 6, the tree-like evolutionary process is a much better depiction of the evolution of eukaryotes than it is in the case of prokaryotes, the main reason being the partial suppression of indiscriminate HGT that dominates the prokaryotic world but that eukaryotes have replaced with regular sex (see the discussion later in this chapter). However, the suitability of the tree metaphor in principle does not mean that the correct tree can be easily reconstructed. Several eukaryotic kingdoms, such as animals, fungi, plants, and ciliates, are well defined and monophyletic beyond a reasonable doubt; in addition, the evolutionary relationships within them mostly fit well-resolved trees. However, deciphering the evolutionary relationships between these kingdoms and numerous other groups of unicellular eukaryotes (protists) is a daunting task, and the primary radiation of eukaryotes from the Last Eukaryote Common Ancestor (LECA) stage arguably is the hardest phylogenetic problem, as far as the evolution of eukaryotes is concerned (Koonin, 2010a).

The problem of the primary eukaryotic radiation is linked to the ubiquity of mitochondria and MLOs in eukaryotes, which we discussed in the preceding section. For many years, most evolutionary biologists favored the so-called crown group phylogeny (aka "archezoan" tree), whereby the tree of eukaryotes consisted of a "crown" that included animals (Metazoa), plants (Viridiplantae), fungi, and various assortments of protists, depending on the methods used for the tree construction (Cavalier-Smith, 1998; Patterson, 1999; Roger, 1999). The rest of the protists that lacked typical mitochondria, such as microsporidia, diplomonads, and parabasalia, collectively comprised "archezoa" and were considered "early branching eukaryotes" that branched off the tree prior to the mitochondrial symbiosis (see Figure 7-2A). This topology of the eukaryotic tree was compatible with most of the phylogenies of rRNA and various conserved proteins. In trees rooted with prokaryotic outgroups, the archezoan lineages typically fell outside the "crown," as would be expected if the root indeed was between the archezoa and the crown. However, during the first decade of the twenty-first century, the archezoan scenario crumbled and fell apart (Embley and Martin, 2006). The main cause of its demise was the discovery of mitochondria or MLOs in all modern eukaryotes, a discovery that destroyed the biological underpinning of the near-root positions of the (former) early branching groups of protists. Simultaneously, the greatly improved taxon sampling resulting from extensive genome sequencing, together with the new, more robust methods for phylogenetic analysis, has shown that the deep placing of the "early branching" groups of protists seen in many trees was a long-branch attraction artifact caused by the fast evolution of the respective groups (Brinkmann and Philippe, 2007).

So there is no reason to consider any group of eukaryotes a primitive, presymbiotic archezoan. Instead, taking into account the small genomes and the high rate of evolution characteristic of the protist groups previously thought to be early branching (the former archezoa) and their parasitic lifestyle, it is becoming increasingly clear that most, or perhaps all, of them evolved from more complex ancestral forms by reductive evolution (Brinkmann and Philippe, 2007; Koonin, 2010a). Parasites tend to lose genes, organelles, and functions that are provided by the host (reductive evolution), and they tend to evolve rapidly, owing to the perennial arms race with the host defenses (more about that with respect to viruses in Chapter 10). Thus, the archezoan (crown group) phylogeny has been effectively refuted, and the study of the deep phylogeny and the origin of eukaryotes had to start from scratch (Embley and Martin, 2006).

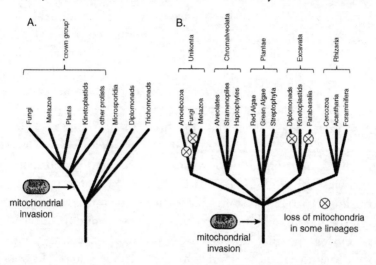

Figure 7-2   Phylogeny of the eukaryotes: (A) The hypothetical archezoan tree; (B) The consensus "star" tree of five supergroups.

This time, phylogenomic approaches were mainly used—that is, phylogenetic analysis of genome-wide sets of conserved genes. The key accomplishment at this new stage was the delineation of "supergroups" each of which combines highly diverse groups of eukaryotic organisms in a monophyletic clade (Adl, et al., 2005; Keeling, 2007; Keeling, et al., 2005). Most of the phylogenomic analyses published so far converge on five supergroups (or six, if the Amoebozoa and Opisthokonts do not form a single supergroup, the Unikonts; Figure 7-2B). Although proving monophyly is nontrivial for each of the supergroups, with the possible exception of Plantae, the general structure of the tree, with a few supergroups forming a star-like phylogeny, is reproduced consistently, and the latest studies seem to support the monophyly of each of the five supergroups. Examination of the composition of the supergroups is most instructive and can have a major effect on our perception of eukaryotes. Of the five supergroups in the star tree (see Figure 7-2B), only three—Unikonts, Plantae, and Chromalveolata (the brown algae, in the latter case)—include complex multicellular organisms, and even in these three supergroups, multicellular organisms form only the "crowns" (or just one branch in

Chromalveolata), whereas the numerous deep branches in these three supergroups and the entire remaining two supergroups consist of protists. All appearances notwithstanding, *eukaryotic life is mainly defined by the enormous diversity of unicellular forms*, whereas the conspicuous, large multicellular creatures are only offshoots in three eukaryote branches founded by protists.

The relationship between the supergroups is a formidable problem. The internal branches are extremely short, suggesting that the radiation of the supergroups occurred rapidly (on the evolutionary scale), perhaps resembling an evolutionary Big Bang (see Chapter 6). Two detailed phylogenetic studies each analyzed more than 130 conserved proteins from several dozen eukaryotic species and, after exploring the effects of removing fast-evolving taxa, arrived at a three-megagroup structure of the eukaryotic tree (Burki, et al., 2008; Hampl, et al., 2009). The megagroups consist of Unikonts, Excavates, and the assemblage of Plantae, Chromalveolata, and Rhizaria (see Figure 7-2B). Still, the support for the megagroups is hardly decisive, and for now, the supergroup level is the deepest robust phylogeny of eukaryotes.

Several attempts have been made to infer the position of the root of the eukaryotic tree (see Figure 7-2B). Phylogenetic approaches in themselves yield no information on the root, and using prokaryotic outgroups leads to a loss of resolution, so independent information is required. A popular idea is to identify so-called *derived shared characters* (synapomorphies) that would split the tree into two subtrees and thus point to the position of the root. The problem is to find such characters that are highly unlikely to emerge in two or more lineages independently. The first rooted alternative to the crown group tree was proposed by Tom Cavalier-Smith and coworkers (Richards and Cavalier-Smith, 2005; Stechmann and Cavalier-Smith, 2003), who used distinct rare genomic changes (RGCs), such as the fusion of the genes for two enzymes (dihydrofolate reductase and thymidylate synthase) and later the domain architecture of myosins, to place the root between the Unikonts and the rest of eukaryotes, the Heterokonts (see Figure 7-2B). This separation seems biologically plausible because Unikont cells have a single cilium, whereas all other eukaryotic cells have two. Nevertheless, this conclusion could be suspect because the use of only a few RGCs makes it difficult to rule out parallel emergence of the same RGC, such as gene fusion or fission, in different lineages (the phenomenon known as homoplasy).

. Igor Rogozin and colleagues used a different RGC approach based on rare replacements of highly conserved amino acid residues that require two nucleotide substitutions and inferred the most likely position of the root between Plantae and the rest of eukaryotes (see Figure 7-2B; Rogozin, et al., 2009). Again, this seems to make sense biologically because the cyanobacterial endosymbiosis that gave rise to plastids occurred on the Plantae lineage and, under this scenario, might have been the event that triggered the primary radiation of eukaryotes. More generally, endosymbiosis is a crucial factor of evolutionary innovation and diversification in eukaryotes. Several major lineages of the Chromalveolata were "seeded" by engulfment of unicellular algae by ancestral, plastid-less unicellular eukaryotes (see Figure 7-2B).

Another potential position of the eukaryote root comes directly from the analysis of mitochondrial genomes. As mentioned earlier in this chapter, the excavate *Reclinomonas americana* has by far the most complex of the known mitochondrial genomes, with about 100 functional genes, as opposed to less than 20 in other eukaryotes. One could surmise that *Reclinomonas* represents the earliest branching eukaryotic lineage that split from the trunk of the eukaryote tree before the ultimate degradation of the endosymbiont genome. This scenario would place the root within the excavate supergroup. However, a viable, perhaps more likely alternative is that the final steps of genome degradation occurred only after the divergence of the major lineages of eukaryotes and independently proceded along convergent routes in different lineages. The latter scenario implies a powerful evolutionary process that leads to the loss (or transfer to the host genome) of all endosymbiont genes with the exception of those few that are strictly required to remain in the mitochondrial genome for the mitochondrion to be functional; later in this chapter, we address the possible nature of this process and the requirements to the mitochondrial genome.

The lack of consensus about the root position and the monophyly of at least some of the supergroups, let alone the megagroups, indicate that, despite the emerging clues, the deep phylogeny of eukaryotes currently should be considered unresolved. In a sense, given the likely Big Bang of early eukaryote radiation, the branching order of the supergroups, in itself, might be viewed as relatively unimportant. However, the biological events that triggered these early radiations are of major interest, so earnest attempts to resolve the deepest branches of the eukaryotic tree will undoubtedly continue with larger and further improved data sets and methods.

The complex, expanding LECA—and the dark ages of eukaryote evolution

Reconstructing LECA

The conservation of the major features of cellular organization and, more importantly, the existence of a large set of genes that are conserved across all or most of the diverse eukaryotic forms leave no doubt that all extant eukaryotes evolved from a Last Eukaryote Common Ancestor (LECA). As discussed in the beginning of this chapter, all eukaryotes that have been studied in sufficient detail possess either mitochondria or MLOs. The simplest (most parsimonious) inference from this pattern is that LECA already possessed mitochondria—and, again, the likelihood of this inference increases with each newly characterized group of eukaryotes in which mitochondria-related organelles are found.

Reconstruction of the evolution of the eukaryote gene repertoire is based on the same principles and methods as the reconstruction of prokaryote evolution outlined in Chapter 5—essentially, maximum parsimony and maximum likelihood. To describe the gist of these approaches very simply, when it comes to the evolution of eukaryotes, genes that are represented in diverse extant representatives of the major eukaryotic lineages, even though apparently lost in some lineages, can be mapped back to LECA. The results of all these reconstructions consistently point to a complex LECA, in terms of both the number of ancestral genes and, perhaps even more importantly, the ancestral presence of the signature functional systems of the eukaryotic cell. Maximum parsimony reconstructions based on phyletic patterns in clusters of orthologous genes of eukaryotes map approximately 4,100 genes to the LECA (Koonin, et al., 2004). Such estimates are highly conservative, as they do not account for lineage-specific loss of ancestral genes, a major aspect in the evolution of eukaryotes. Indeed, even animals and plants, the eukaryotic king-doms that seem to be the least prone to gene loss, appear to have lost about 20% of the putative ancestral genes identified in the free-living excavate *Naegleria gruberi* (Fritz-Laylin, et al., 2010; Koonin, 2010b). Thus, the reconstructions suggest that *the genome of the LECA was at least as complex as those of typical extant free-living unicellular eukaryotes* (Koonin, 2010a).

This conclusion is supported by comparative-genomic recon-structions of the ancestral composition of the key functional systems of LECA, such as the nuclear pore (Mans, et al., 2004), the spliceo-some (Collins and Penny, 2005), the RNA interference machinery (Shabalina and Koonin, 2008), the proteasome and the ubiquitin sig-naling system (Hochstrasser, 2009), the endomembrane apparatus (Field and Dacks, 2009), and the cell division machinery (Makarova, et al., 2010). The outcomes of all these analyses are straightforward and consistent, even when different topologies of the phylogenetic tree of eukaryotes are used as the scaffold for the reconstruction: The LECA already possessed all these structures in their fully functional state, possibly as complex as the counterparts in modern eukaryotes.

Reconstruction of other aspects of the genomic composition and architecture of the LECA similarly points to a highly complex ances-tral genome. Comparative-genomic analysis of intron positions in orthologous genes within and between supergroups suggests high intron densities in the ancestors of the supergroups and in LECA, at least as dense as in modern free-living unicellular eukaryotes, but most likely closer to the intron-rich genes of animals and plants (later in this chapter we return to the fascinating history of eukaryotic introns in some detail). A systematic analysis of widespread paralo-gous genes in eukaryotes indicates that hundreds of duplications antedate LECA, especially duplications of genes involved in protein turnover such as molecular chaperones (Makarova, et al., 2005). Taken together, these results clearly show that the *LECA was a typi-cal, fully developed eukaryotic cell*. The subsequent evolution of eukaryotes has shown no consistent trend toward increased cellular complexity, except for lineage-specific embellishments found in the multicellular groups (animals, plants, and brown algae), as well as some protists, like green algae or ciliates.

The stem phase: The dark age of eukaryotic evolution

The demonstration that LECA possessed highly complex cells implies a key "stem" phase of the evolution of eukaryotes (see Figure 7-3), after the emergence of eukaryotic cells but before LECA. Among other developments, the stem stage included extensive duplication of numerous genes so that the set of ancestral genes approximately doubled (Makarova, et al., 2005). How long was the stem phase in the evolution of eukaryotes? Given that we are unaware of any eukaryotic diversity prior to LECA, intuition suggests a very short stem phase, with the implication that the events between the emergence of the first eukaryotic cell and LECA unfolded in rapid succession, perhaps in an explosive manner (see Figure 7-3A). There is, however, a perfectly legitimate and logically valid alternative: The stem phase was long and involved substantial diversification, but LECA (once again, the last common ancestor of all *extant* eukaryotes) is a survivor of a single lineage, whereas all others have gone extinct (see Figure 7-3B). Some of the attempts to date the primary radiation of eukaryotes—or, in other words, estimate the age of LECA—yield results compatible with a long stem phase.

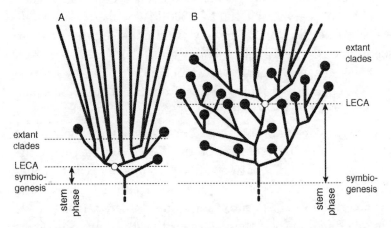

Figure 7-3 Evolution of eukaryotes after and before LECA: (A) An "explosion" scenario with a short stem phase of evolution; (B) A scenario with an extended stem phase and significant extinct diversity antedating LECA.

Molecular dating of evolutionary divergence events is a highly specialized research field, with many difficult technical problems (Bromham and Penny, 2003; Graur and Martin, 2004). We must skip most of the technicalities and get right to the results. Note that the principle is to map molecular divergence data (that is, the results of sequence comparison linked to a phylogenetic tree) onto the fossil record by using several accurately dated fossils as calibration points (for instance, the earliest indisputable mammalian fossils date to about 120 million years ago, so this is the minimum age of the mammalian radiation). Assuming either a strict or a relaxed molecular clock, it is possible to obtain the time estimate for any divergence event relative to the calibration points and given the particular tree topology. The estimates may be quite reliable when they involve only interpolation (that is, inference of divergence times within the interval spun by the calibration points) but much less robust when it comes to extrapolation (dates outside the calibration interval). Unfortunately, for ancient dates such as the age of LECA, extrapolation is inevitable. The time estimates obtained by different researchers span an extremely broad range of dates, between 1,000 million years ago (MYA) and 2,300 MYA. Several more recent, independent, and advanced dating approaches that employed relaxed molecular clock models or adapted RGCs with similar behavior have independently converged on "young LECA" estimates that date the primary radiation of eukaryotes at about 1,100 to 1,300 MYA (Chernikova, et al., 2011; Douzery, et al., 2004). Certainly, the problem is not solved but this seems to be the best available time estimate for LECA. This estimate implies a long stem phase of several hundred million years (see Figure 7-3B) because unequivocal fossil remains of eukaryotes date to more than 1,500 MYA (Knoll, et al., 2006).

This conclusion seriously affects our assessment of the current knowledge of the early evolution of eukaryotes. On one hand, the reconstruction results that portray LECA as a modern-type unicellular eukaryote with fully developed signature functional systems of the eukaryotic cell become less surprising: Indeed, there seems to have been ample time to evolve these advanced features after the emergence of the (primitive) eukaryotic cell. Ditto for the numerous gene duplications that map to LECA: Under the long-stem scenario, they did not have to occur in a single burst; there was plenty of time to duplicate genes gradually. On the other hand, the stem phase is a veritable dark age of eukaryote evolution about which we know next to nothing and can hope to learn precious little. Indeed, LECA is effectively the "event horizon" for comparative genomics: Using genome comparisons only, we cannot peer into the stem phase. We can get a glimpse of what was going on through a detailed study of ancient eukaryotic gene duplications, but that is about the only source of information on the stem phase. We have no idea about the actual diversity of the dark age eukaryotes and little hope to be able to evaluate it in the future. The fossil data reveals limited variety, but the record is never complete, and it is hard to tell just how incomplete it is. With few exceptions, the early and midproterozoic eukaryotic fossils do not seem to represent any extant taxa, an observation that has to be taken cautiously, but at least at face value is compatible with a young LECA and the existence of some extinct diversity that is currently inaccessible to us (see Figure 7-3B). An obvious possibility is that the LECA was the breakthrough eukaryote that captured the mitochondrial endosymbiont and that endosymbiosis triggered the radiation of the extant eukaryotes. This scenario implies that the pre-LECA, Proteozoic diversity of eukaryotes represents an extinct archezoan (primary amitochondrial) biota. However, a distinct and probably more plausible alternative is that the mitochondrial endosymbiosis actually triggered eukaryogenesis itself, so the dark age eukaryotes already harbored mitochondria or MLOs. We discuss this dilemma and the arguments in support of endosymbiosis-driven eukaryogenesis later in this chapter, after examining the comparative genomic evidence of evolutionary connections between eukaryotes and prokaryotes.

The archaeal and bacterial roots of eukaryotes

Search for the archaeal and bacterial "parents" of eukaryotes

All eukaryotes are hybrid (chimeric) organisms, in terms of both their cellular organization and their gene complement. Indeed, as pointed out earlier in this chapter, all extant eukaryotes seem to possess mitochondria or MLOs derived from α-proteobacteria, whereas Plantae and many groups of Chromalveolata additionally possess cyanobacteria-derived plastids. The gene complement of eukaryotes is an uneven mix of genes of apparent archaeal origin, genes of probable bacterial origin, and genes that so far seem eukaryote-specific, without convincing evidence of ancestry in either of the two prokaryote domains. Paradoxical as this might appear, although trees based on rRNA genes and concatenated alignments of information-processing proteins, such as polymerases, ribosomal proteins, and splicesosome subunits, put archaea and eukaryotes together, genome-wide analyses consistently and independently show that there are three or more times more genes with closest bacterial homologs than with closest archaeal homologs (see Figure 7-4; Esser, et al., 2004; Koonin, et al., 2004; Makarova, et al., 2005). The archaeal subset is strongly enriched in information processing functions (translation, transcription, replication, splicing), whereas the bacterial subset consists largely of metabolic enzymes, membrane proteins and components of membrane biogenesis systems, various signaling molecules, and other "operational" proteins (see later in the chapter for more details).

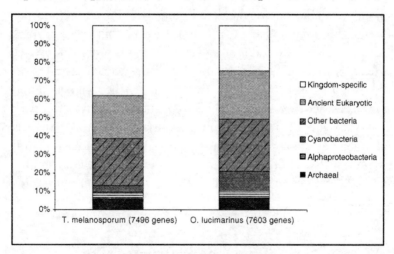

Figure 7-4 Breakdown of the genes in two distantly related eukaryotes, according to their probable origins: archaeal, bacterial, or eukaryote specific. The sequences of all encoded proteins from the fungus Tuber melanosporum (black truffle) and the green alga Ostreococcus lucimarinus were compared to the NCBI RefSeq database using the BLASTP program (Altschul, et al., 1997), and the probable phylogenetic affinity of each protein-coding gene was determined using a custom script. Note the similar, relatively small fractions of genes of apparent alpha-proteobacterial origin and the higher fraction of cyanobacterial genes in the alga.

At a coarse-grain level, these observations are best compatible with genome fusion scenarios in which the eukaryotic genome emerged through a fusion between two ancestral genomes, an

archaeal or archaea-related one, and a bacterial (most likely α-proteobacterial) one, given the well-established ancestry of the mitochondrial endosymbiont (Embley and Martin, 2006; Rivera and Lake,
2004). Genome fusion is most easily interpreted as a reflection of
symbiogenesis. However, attempts to pinpoint the specific archaeal
and bacterial "parents" do not lead to clear-cut results and seem to
suggest complicated evolutionary relationships. Although many of the
bacterial-like genes in eukaryotes have α-proteobacterial homologs,
these are far from dominant among the bacterial-like genes that show
apparent evolutionary affinities with a variety of bacterial groups (see
Figure 7-4). An important cause of this complicated breakdown of
the bacterial-like component of the eukaryotic gene complement is
the large size of the alpha-proteobacterial pangenome (see Chapter 5).
Thus, without knowing the actual identity of the alpha-proteobac-
terium that gave rise to the eukaryotic mitochondria, it is hard to
delineate its genetic contribution (Martin, 1999; Esser, et al., 2007).
Apart from this uncertainty about the gene complement of the
endosymbiont, it is impossible to rule out multiple sources of the
bacterial-like genes in eukaryotes, which might have come from
sources other than the genome of the α-proteobacterial endosymbiont that gave rise to mitochondria. In particular, whatever the
actual nature of the archaeal-like ancestor, it probably lived at moderate temperatures and non-extreme conditions, and was consequently in contact with a diverse bacterial community. Modern
archaea with such lifestyles (for example, *Methanosarcina*) have
numerous genes of diverse bacterial origins, indicating extensive
horizontal acquisition of genes from bacteria (see Chapter 5). Thus,
the archaeal-like host of the endosymbiont could have already had
many bacterial genes, partly explaining the observed prevalence and
diversity of "bacterial" genes in eukaryotes.

Identifying the archaeal(-like) parent of eukaryotes is even more
difficult than identifying the bacterial ancestor(s) because there is no
unequivocal data on the ancestral archaeal lineage that would parallel
the unambiguous origin of mitochondria from α-proteobacteria. Phylogenomic studies that use different methods point to different major
archaeal lineages (Crenarchaeota, Euryarchaeota, Thaumarchaeota,
or an unidentified deep branch) as the best candidates for the eukaryote ancestor (Cox et al., 2008; Kelly et al., 2010; Pisani et al., 2007;

Yutin et al., 2008). Unequivocal resolution of such deep evolutionary relationships is extremely difficult. Moreover, some of these analyses explicitly suggest the possibility that the archaeal heritage of eukaryotes is genuinely mixed, with the largest contribution coming from a deep lineage, followed by the contributions from Crenarchaeota (Thaumoarchaeota) and Euryarchaeota (see Figure 7-5; Yutin, et al., 2008). A tempting speculation suggested by these findings is that the archaeal parent of eukaryotes belonged to a (probably extinct) deep lineage of archaea with a highly complex genome (Makarova et al., 2010). This conjecture seems to be congruent with the results of comparative genomic reconstructions that point to complex archaeal ancestors (Csuros and Miklos, 2009; Makarova et al., 2007b; see discussion later in this chapter).

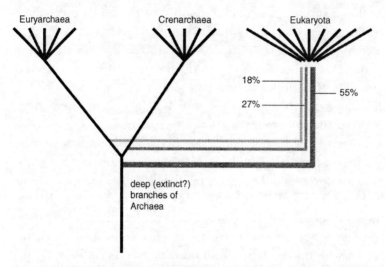

Figure 7-5   The contributions of different groups of archaea to the origin of eukaryotes. The percentage of genes of apparent archaeal descent apparently derived from Euryarchaeota, Crenarchaeota, and deep archaeal branches is indicated. The data comes from Yutin, et al., 2008.

Origin of the key functional systems of the eukaryotic cell

Another major theme emerging from comparative-genomic studies is the interplay between the archaeal and bacterial contributions to the origin of eukaryote-specific functional systems, particularly the mixed archaeao-bacterial origin of some of these systems. Conceptually, there seem to be two types of relationships between functional systems of eukaryotes and prokaryotes (see Box 7-2):

1. Eukaryotic systems that evolved from homologous and functionally analogous systems of prokaryotes

2. Eukaryotic systems that evolved by assembly from components that, in prokaryotes, are involved in functionally distinct, often multiple processes, sometimes along with additional proteins that appear to be eukaryote-specific

Box 7-2: Inferred origins of some key functional systems and molecular machines of eukaryotes

System/Complex/ Function	Inferred Origins	Duplication and Other Complex Features in Eukaryotes
	Type 1: Origin from functionally analogous prokaryotic ancestors	
DNA replication and repair machinery	Archaeal, with either crenarchaeotal or euryarchaeotal affinities for DNA polymerases and other central replication proteins A mix of archaeal and bacterial for repair enzymes	Several early duplications of DNA polymerase and other replication proteins
Transcription machinery	Archaeal; at least two RNA polymerase subunits of crenarchaeotal/korarchaeotal origin	Pre-LECA duplications yielding 3 RNA polymerases
Translation apparatus, including ribosomes	Mostly archaeal; some aaRS displaced with bacterial homologs	Minimal duplication
Proteasome: regulated proteolysis	Archaeal	Extensive duplication of subunits and substantial structural embellishment
Ubiquitin signaling: regulated proteolysis and protein topogenesis	Archaeal	Massive duplication throughout the evolution of eukaryotes

System/Complex/ Function	Inferred Origins	Duplication and Other Complex Features in Eukaryotes
Exosome: regulated RNA degradation	Archaeal	Extensive pre-LECA and some lineage-specific duplication
Type 2: assembly from diverse prokaryotic ancestors		
Nuclear pore complex: nucleocytosolic transport	Bacterial; some key proteins of the nuclear pore complex repetitive and of uncertain origin	Extensive pre-LECA duplication in eukaryotes and lineage-specific duplication
Chromatin/nucleosomes	Complex mix of archaeal and bacterial	Extensive duplications throughout the evolution of eukaryotes (including pre-LECA duplication of histones) and addition of ESPs
RNA interference	Hybrid of archaeal and bacterial components	Extensive lineage-specific duplications
Endomembrane system/endoplasmic reticulum	Complex mix of archaeal and bacterial components	Extensive duplication throughout the evolution of eukaryotes
Programmed cell death machinery	Bacterial	Extensive lineage-specific duplication in eukaryotes

Many of the Type 1 systems have been long viewed as eukaryotic innovations. However, a remarkable trend is that the more carefully we look into the rapidly expanding comparative genomic data for archaea and bacteria, the more evolutionary antecedents for the signature eukaryotic systems are being discovered. For instance, it has long been known that the protein degradation machinery of eukaryotes, the proteasome, has a simpler counterpart in Archaea, and an even more primitive version exists in bacteria (Groll, et al., 2005).

The indisputable evolutionary relationship and functional analogy between archaeal and eukaryotic proteasomes has been well established even before genome sequences became available. Comparative genomic analysis led to a parallel prediction: Highly conserved arrays of predicted archaeal operons were identified that encoded proteins homologous to subunits of eukaryotic exosomes, the molecular machines that degrade RNA in eukaryotes (Koonin, et al., 2001b). Sure enough, this predicted archaeal exosome has been experimentally discovered within a few years (Hartung and Hopfner, 2009).

For a long time, the ubiquitin signaling network that governs degradation and topogenesis of proteins within eukaryotic cells through the conjugation of a small, extremely conserved protein named ubiquitin (Ub) and much less common Ub paralogs to the target proteins has been considered a quintessential eukaryote-specific functional system, a unique eukaryotic signature (Hochstrasser, 2009). Later, thanks to the increasing diversity of the sequenced archaeal and bacterial genomes, and the improved methods for detecting sequence and structural similarity between proteins, prokaryotic Ub homologs have been detected. These small proteins are particularly abundant in archaea but have been thought to function in reactions of sulfur insertion that are required for the biosynthesis of certain coenzymes. However, a detailed comparative-genomic analysis has led to the discovery, in a variety of bacteria, of operons that combine genes for Ub homologs with genes for homologs of two ubiquitin ligase subunits and a deubiquitinating enzyme. Although these proteins are only distantly related to their eukaryotic homologs, the colocalization of all these genes was highly suggestive of the possibility that the bacterial ancestry of the Ub system had been discovered (Iyer, et al., 2006). Then in 2010, experiments were reported demonstrating that, in at least some Archaea, a particular group of Ub homologs function similarly to the classic eukaryotic Ub—that is, these small proteins are conjugated with various other proteins and target them for degradation (Humbard, et al., 2010).

This is not where the Ub story ends. In December 2010, when this book was already essentially complete and at the editing stage, a startling finding was published. It comes from the newly sequenced genome of Candidatus, *Caldiarchaeum subterraneum*, an archaeon that was isolated from a gold mine and might represent a novel archaeal phylum or a new group within the Crenarchaeota (Nunoura, et al., 2010). The genome of this organism contains an operon that encodes four proteins that, in a sequence database search, come across as typically eukaryotic—with numerous, highly conserved eukaryotic homologs but with no comparable similarity to any proteins from other Archaea or bacteria. These proteins are Ub and three subunits of the Ub ligase (known as E1, E2, and E3). Moreover, a eukaryotic-type deubiqutinating enzyme is encoded next to the Ub operon in the opposite DNA strand. Thus, this new archaeal genome encodes the complete suite of proteins required for the reversible protein ubiquitylation in eukaryotes. Interestingly, when I performed an additional database search with protein sequences from the same genomic neighborhood, I managed to identify yet another E3 subunit, so even the proliferation of E3 that reaches a dramatic scale in eukaryotes seems to have already started in Archaea. The degree of similarity between these proteins and their eukaryotic homologs is unexpectedly high (much greater than for bacterial proteins encoded in similar operons), suggesting the unusual possibility of HGT from eukaryotes to Archaea. However, this does not seem to be the most parsimonious scenario, given the distinct operonic organization of these genes in *Caldiarchaeum subterraneum*. What remains is to conclude that this archaeon encodes the ancestral ubiquitin system. If that is the case, we will be compelled to conclude that this system evolved in Archaea to a fully formed state so that eukaryotes received it "ready-made" and what happened to the Ub network during the evolution of eukaryotes amounts to diversification and embellishment. It is quite striking that it took more than 100 archaeal genomes to be sequenced for this putative ancestral Ub system to be discovered; this shows that the ancestral versions of some key eukaryotic functionalities are quite "exotic" among the Archaea. I described this discovery in such detail not only because of its obvious importance for understanding the origin of the Ub network, but even more for its general implications for the evolution of eukaryotes, which I emphasize later in this chapter.

The Type 1 molecular machines and systems generally followed the major trend of eukaryotic evolution—namely, serial gene duplication with subsequent diversification: Where an archaeal complex consists of multiple copies of one or two proteins, the evolved eukaryotic version includes instead diversified paralogous subunits (Makarova, et al., 2005; see Box 7-2).

Type 2 systems can be exemplified by the quintessential eukaryotic molecular machine, the nuclear pore complex, for which there is no functional analogs in prokaryotes. Notably, the nuclear pore complex does not show any indications of archaeal ancestry, but rather is built of several proteins of apparent bacterial origin, combined with proteins consisting of simple repeats whose provenance is difficult to ascertain (Mans, et al., 2004). By contrast, the RNA interference (RNAi) machinery, a system of antivirus defense (innate immunity) and expression regulation in eukaryotes that attracted a lot of attention in the last decade, partly thanks to its outstanding utility as an experimental tool, has a readily demonstrable chimeric, archaeo-bacterial origin (Shabalina and Koonin, 2008). Specifically, one of the key RNAi proteins, the endonuclease Dicer, consists of two bacterial RNAse III domains and a helicase domain of apparent euryarchaeal origin; the other essential RNAi protein, Argonaute, also shows a euryarchaeal affinity (Shabalina and Koonin, 2008). Another signature molecular machine of eukaryotes, the spliceosome, is, to some extent, intermediate between the first and second types of eukaryotic systems (Collins and Penny, 2005). The Sm proteins that constitute the core of the spliceosome have readily identifiable archaeal orthologs but these are involved in a different kind of RNA processing reactions; indeed, there are no bona fide spliceosomes outside eukaryotes.

Taken together, phylogenomic findings suggest that the archaeal ancestor of eukaryotes combined a variety of features that are found separately in diverse extant archaea. Evolutionary reconstructions using Maximum Parsimony and especially advanced Maximum Likelihood methods point to a genetically complex common ancestor of all extant archaea—in the least, comparable to the typical extant forms, but quite possibly containing an even greater diversity of genes (Csuros and Miklos, 2009; Makarova, et al., 2007b). The currently existing archaeal lineages probably evolved by differential streamlining or reductive evolution of the complex ancestral forms (more about this route of evolution in Chapter 8), whereas eukaryotes largely retained the ancestral complexity (Makarova, et al., 2010). The diversity of the origins of different functional systems of the eukaryotic cells has major implications for the models of eukaryogenesis which we discuss later in this chapter.

Eukaryogenesis: The origins of the distinctive eukaryotic cellular organization

Symbiogenesis versus archezoan scenarios

The numerous phylogenomic observations that are briefly summarized in the preceding section do not explain where the eukaryotic cell came from, but they do provide the essential groundwork for constructing scenarios of eukaryogenesis. Box 7-3 lists the key observations that must be included in any evolutionary account of the origin of eukaryotes (eukaryogenesis) and the early stages of their evolution. Given these observations, the main issue now revolves around the role of endosymbiosis: Was it the cause of the entire chain of events that led to the emergence of LECA (the stem phase of evolution), as in the symbiogenesis scenario, or was it a step in the evolution of the already-formed eukaryotic cell, as in the archezoan scenario? In other words, was the host of the α-proteobacterial symbiont (the future mitochondrion) a prokaryote or an amitochondrial eukaryote, an archezoan?

Box 7-3: Key points to consider for modeling eukaryogenesis

- All extant eukaryotes have mitochondria or related organelles, so endosymbiosis must predate LECA.

- LECA was a highly complex organism that already had all signature functional systems of eukaryotes and was probably a typical eukaryotic cell, so all key innovations of eukaryogenesis must have occurred at the stem phase of evolution before LECA. Among others, these innovations include introns and the spliceosome.

- The duration of the stem phase is unknown, but there is a distinct possibility that it was long and that a considerable diversity of pre-LECA eukaryotes existed.

- Highly conserved genes of eukaryotes are a chimeric set: A minority of genes encoding information transmission systems and some other key molecular machines, such as the cell division apparatus, are of archaeal origin, whereas the majority of metabolic enzyme genes originate from bacteria.

- Some of the key functional systems of the eukaryotic cell, such as RNA interference or repair pathways, are archaeo-bacterial chimeras. Other essential molecular machines of the eukaryotic cell, such as the nuclear pore complex, seem to be primarily of bacterial provenance.

- Ancestors of eukaryotic genes are scattered among archaeal and bacterial lineages.

Given that eukaryogenesis most likely was a unique event and that intermediate evolutionary stages between the emergence of the first eukaryotic cells and the advent of LECA are barely accessible to comparative genomic or any other conceivable methods, one would doubt that these questions will ever be answered with full certainty. However, this is not a reason to stay agnostic. The ubiquity of mito-chondria/MLO in extant eukaryotes is often invoked as an argument in support of the symbiogenesis scenario, but this argument loses much, if not all, of its strength if there was a long stem phase of eukaryotic evolution. Indeed, that dark age could belong to extinct archezoa, of which only one lineage that domesticated an α-pro-teobacterium survived and gave rise to all extant diversity of eukary-otes. Nevertheless, the symbiogenesis scenario does seem to be more plausible than the archezoan scenario, for three principal reasons.

1. Under the archezoan scenario, there are no plausible selective factors behind the evolution of the nucleus and, in particular, the elaborate nuclear pore complex. The nucleus disrupts the transcription-translation coupling that is typical of bacteria and archaea and necessitates the evolution of the time- and energy-consuming mechanism of nucleocytosolic transport of mRNA.

The symbiogenesis hypothesis offers a plausible selective factor: defense against the invasion of the host genome by Group II self-splicing introns (these are actually selfish genetic elements—more about them in Chapter 8), the evolutionary precursors of the spliceosomal introns, that are abundant in α-proteobacteria but not in Archaea. The repeated exposure of the archaeal host genome to the bacterial endosymbiont DNA from disrupted endosymbiont cells could lead to activation of Group II introns and their massive insertion into the host genome. These inserted introns would fatally disrupt gene expression unless transcription and translation were decoupled and compartmentalized, hence the driving force for the evolution of the nucleus (see details in the next section).

2. As outlined in the preceding section, a combination of comparative genomic, ultrastructural, and functional studies in prokaryotes, particularly archaea, show that not only the molecular components of the numerous signature eukaryotic systems, but also their actual structures and functions have evolved in archaea and, thus, antedate eukaryogenesis (Type 1 systems discussed previously; see Box 7-2). However, the endomembrane system and the nucleus, as well as the mitochondria themselves, the introns interrupting protein-coding genes, and the spliceosome that mediates exon splicing (intron excision) are Type 2 systems (features). These systems do not have functional analogs in prokaryotes, although they seem to have been assembled from prokaryotic components. Thus, a single causal chain of events seems plausible (see Figure 7-6): *Eukaryogenesis was triggered by endosymbiosis, and the endomembrane systems, including the nucleus, evolved as a defense against the invasion of Group II introns and perhaps bacterial DNA in general* (Martin and Koonin, 2006a; Lopez-Garcia and Moreira, 2006). It does not seem accidental that many key components of these endomembrane systems appear to be of bacterial origin, whereas others are repetitive proteins that might have evolved *de novo*.

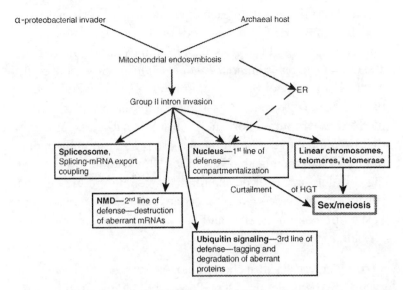

Figure 7-6   The emergence of eukaryotic cellular organization as a multilevel system of defense against intron invasion: a hypothetical single chain of causation. Adapted from Koonin, 2006.

3. Simple estimates made by Nick Lane and Bill Martin suggest that the emergence of large, complex cells like those of eukaryotes does not appear energetically feasible without the acquisition of multiple energy-producing organelles capable of autonomous reproduction and regulation (Lane and Martin, 2010). In prokaryotes, protein complexes that comprise the electron transfer chain and membrane ATP synthases that convert proton or sodium gradient into ATP are located on the plasma membrane. The biogenesis of these complexes is intrinsically coupled to the synthesis of their subunits, highly hydrophobic proteins that insert into the membrane cotranslationally. Given that the surface of a cell is proportional to the square of the diameter, whereas the volume is proportional to the cube of the diameter, the increase in the cell size at some point makes this model of bioenergetics inefficient and hence unsustainable. We are currently aware of only two routes to efficient bioenergetics that had the potential to trigger the evolution of large cells. The first route was taken by the eukaryotes, with multiple energy-producing, endosymbiont-derived

organelles residing within each cell. The second version is realized in some recently discovered giant bacteria that contain numerous copies of the genomic DNA in each cell (more than 100,000 in the fish symbiont *Epulopiscium* sp.; Mendell, et al., 2008). Each copy of the chromosome seems to be attached to the membrane, so the synthesis of membrane proteins, particularly those required for energy transformation, is probably tightly coupled to the insertion of these proteins into the membrane. Unlike the eukaryotic solution, this second invention of big cells has not involved the dramatic genome reduction, the key feature of mitochondria that contributes to the efficiency of eukaryotic cell energetics, and has not spawned diverse complex life forms.

Several objections against the symbiogenesis scenario have been put forward (Kurland et al., 2006; Poole and Penny, 2007). First, prokaryotic endosymbionts in prokaryotic hosts are not widespread, prompting the view that phagocytosis, which is apparently unique to eukaryotic cells, should be essential for the acquisition of the mitochondria. This argument does not appear compelling, for four reasons:

1. Eukaryogenesis is extremely rare, probably unique, in the history of life. As a (nearly) unique event, it would not necessarily require a mechanism that routinely operated in the host of the primary endosymbiont.

2. Endosymbiotic bacteria within other bacteria are not common but do exist (von Dohlen, et al., 2001). Intracellular bacterial predation also might be a route to endosymbiosis (Davidov and Jurkevitch, 2009).

3. Observations on membrane remodeling systems and actin-like proteins in archaea (Makarova, et al., 2010; Yutin, et al., 2009) suggest the possibility of still unexplored mechanisms for engulfment of other prokaryotes, perhaps resembling primitive phagocytosis.

4. A computer modeling study suggests that differentiation of cellular life forms into predators and prey is an intrinsic feature of cell evolution and so would emerge soon after the appearance of the first cells (de Nooijer, et al., 2009).

A second, potentially strong argument against the symbiogenesis scenario could be the existence of a substantial number of eukaryote signature proteins (ESPs), proteins found only in eukaryotes. The provenance of the ESPs is an intriguing question. However, as already discussed in part in the preceding section with regard to eukaryotic signature functional systems, careful sequence and structure searches lead to the identification of an increasing number of archaeal and/or bacterial homologs of proteins originally considered ESPs, or else the existence of such homologs becomes obvious with the appearance of new genomes. The discovery of prokaryotic homologs of tubulin, actin, and ubiquitin are well-known examples; more recent cases include the so-called GINS subunits of eukaryotic DNA replication complexes (Marinsek, et al., 2006), the ESCRT-III systems (Makarova, et al., 2010) and the subunits of the TRAPP complex (Barrowman, et al., 2010) that play key roles in eukaryotic vesicle trafficking. Under the symbiogenesis scenario, the former and remaining ESPs result primarily from acceleration of the evolution of genes whose functions have substantially changed during eukaryogenesis; differentiation of simple repetitive protein structures into distinct folds could be another important route of ESP evolution (Aravind et al., 2006).

A third serious objection against the symbiogenesis scenario could be that neither archaeal-like nor bacterial-like genes of eukaryotes can be traced to a single prokaryotic lineage (although the origin of the mitochondria from alpha-proteobacteria is well established). However, the pangenomes of prokaryotes are large, whereas the gene composition of individual organisms is extremely malleable, so reconstruction of the actual partners of the endosymbiosis that led to eukaryogenesis might not be feasible from the limited available set of extant genomes (Martin, 1999; Esser, et al., 2007). Moreover, many, if not most, modern archaea and bacteria might have evolved by streamlining (see Chapters 5 and 8), so eukaryogenesis could have been triggered by symbiosis between two prokaryotes with complex genomes.

It is currently impossible to strictly rule out the possibility that the key eukaryotic innovations evolved independently from and prior to the mitochondrial endosymbiosis. Thus, in principle, the host of the endosymbiont might have been an archezoan. However, the archezoan scenario does not provide plausible staging of events during the evolution of the complex internal organization of the eukaryotic cell, does not offer a *raison d'être* for the nucleus, and does not account for the presence of signature functional systems of eukaryotes in different archaeal lineages. In contrast, the symbiogenesis scenario can tie all these diverse lines of evidence into a coherent, even if still woefully incomplete and admittedly speculative, narrative.

The symbiogenesis scenario of eukaryogenesis: The origin of the key eukaryotic innovations triggered by endosymbiosis

We have already touched upon many aspects of the hypothesis of symbiogenesis-triggered eukaryogenesis. This section integrates various lines of evidence and sketches a coherent scheme of eukaryogenesis. In doing so, we should not forget that the specifics of what actually happened might not be decipherable, so if we engage in elaborate speculation about the details, we are doomed to end up with a "just so story." Nevertheless, if we keep the discussion at a relatively coarse-grained level, it might be possible to discover some logic even behind unique events in evolution, such as eukaryogenesis.

We try to link the scenario of eukaryogenesis (see Figure 7-7) to specific stages and times in the history of life and Earth itself (Kasting and Ono, 2006). The time and place is approximately two billion years ago (Paleoproterozoic), moderate temperatures and salinity, probably ocean floor at a shallow depth. The Earth atmosphere (and, accordingly, the ocean) in the first 1.5 billion year of the history of life were strictly reducing. However, around the time we are focusing on, microoxygenation of the Earth began, thanks to the emergence of oxygenic photosynthesis in cyanobacteria. The concentration of oxygen was probably two to three orders of magnitude lower than it is today, but aerobic respiration already might have been possible. The diversity of the microbial biota in the biosphere was comparable to that in the extant biosphere, with the exception of the paucity (near lack) of aerobic organisms. All major groups of archaea and bacteria we are aware of already existed, and quite likely there were others, now extinct. The ecological setting: An important point that is sometimes overlooked in discussions of endosymbiosis is that the action most likely took place in microbial mats, widespread and literally tightly-knit communities of diverse bacteria and archaea (Allen and

Banfield, 2005). In microbial mats, the level of HGT presumably is as high as it gets, and so should be the frequency of engulfment of one prokaryotic cell by another that potentially leads to endosymbiosis, rare as this process appears to be among prokaryotes.

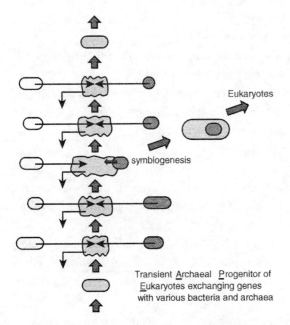

Figure 7-7   The Archaeal Progenitor of Eukaryotes (APE) as a transient, complex archaeal form prone to HGT and the symbiogenesis scenario of eukaryogenesis.

The principal player, the host of the future endosymbiont (let us denote it APE, for Archaeal Progenitor/Parent of Eukaryotes): As already mentioned, this would be a mesophilic archaeon, conceivably with a large genome that could consist of as many as 5,000 to 6,000 genes. Our knowledge of the extant mesophilic archaea is woefully incomplete, compared to other groups of prokaryotes. Nevertheless, what we do know is compatible with the possibility that genetically complex organisms with many horizontally transferred genes are common in this ecological group. Indeed, the largest known archaeal genomes, the only archaea with more than 5,000 genes, are found among mesophilic archaea (namely, certain *Methanosarcina*), and genes of relatively recent bacterial origin might account for up to 20% of these genomes. Other known mesophilic archaea, such as the well-characterized Thaumarchaeota, an archaeal phylum that until recently remained hidden under the nondescript name of mesophilic Crenarchaeota, have smaller genomes but are similarly enriched with "bacterial" genes (Brochier-Armanet et al., 2008). None of the currently known mesophilic archaea looks like a viable candidate for the glorious role of the APE. As discussed earlier in this chapter, the archaeal heritage of today's eukaryotes is mixed, with subsets of genes shared with different groups of Crenarchaeota, Thaumarchaeota, or Euryarchaeota. An interesting possibility is that the elusive APE combined many (perhaps most) of these genes within a single genome/cell prior to the endosymbiosis, although subsequent acquisition via HGT could have been important as well. Driving from the dynamic picture of the prokaryotic world outlined in Chapter 5, it might be productive to consider *transient* rather than *extinct* groups (lineages) of archaea and bacteria as potentially important evolutionary intermediates. Such relatively short-lived life forms with highly complex, mosaic genomes might emerge rather often (on the evolutionary scale), but most of the time, would gradually lose large fractions of their genes and degrade into more stable, familiar forms. However, some of these putative complex transient states could spawn bursts of new diversity (see Figure 7-7).

A startling discovery reported while this chapter was being written illustrates the unexplored archaeal diversity and reinforces the possibility that close relatives of the elusive APE might have survived to this day. Olivier Gros and colleagues have reported two species of Thaumarchaeota that inhabit shallow-water marine habitats (Muller, et al., 2010). These archaea possess giant cells that form distinct macroscopic filaments. Moreover, the cells of one of these species are covered with symbiotic γ-proteobacteria. The bacteria are ectosymbionts rather than endosymbionts; nevertheless, this type of archaeobacterial association could well create the conditions facilitating endosymbiosis. These newly discovered archaea might not be close relatives of the APE (then again, the chance that they are might not be negligible), but at any rate, this discovery speaks volumes to the plausibility of the endosymbiotic scenario of eukaryogenesis.

The history of eukaryotes (see Figure 7-7) begins with the APE engulfing an α-proteobacterium whose precise identity is hard to pinpoint. The APE might have been specifically prone to internalizing other prokaryotic cells, although it certainly was not a bona fide phagocyte like the modern amoebae. However, it seems likely that the APE was a wall-less archaeon, similar in that respect to the extant thermophilic archaea of the genus *Thermoplasma*. Moreover, it probably possessed some form of cytoskeleton formed by actinlike proteins related to those discovered in another group of thermophilic archaea (a group of Crenarchaeota known as *Thermoproteales*); comparative analysis of the sequences of these archaeal homologs of actins (which, unfortunately, still have not been studied experimentally) even suggested the possibility that they form branched filaments, a key structure involved in the eukaryotic phagocytosis (Yutin, et al., 2009). So it is not unreasonable to propose that the APE "grazed" on a bacterial mat, from time to time internalizing bacterial cells. Most of the consumed bacteria would end up as food; other bacteria would kill the predator, and some might become transient symbionts. The fixation of an evolutionarily stable endosymbiont is an extremely tall order because many hurdles have to be cleared to establish such a stable symbiosis. It appears inevitable that, although the initial engulfment of the future endosymbiont would occur by sheer chance, the fixation of the endosymbiont would become possible only inasmuch as it was associated with a distinct selective advantage of the emergent chimeric organism.

What could be the selective factor(s) behind the emergence of the archaeo-bacterial chimeric system? Given the probable microaerophilic conditions at the time of eukaryogenesis (or possibly even anaerobic conditions in the specific environ where eukaryogenesis took place), the selective advantage most likely was not the aerobic respiration–based bioenergetics. Instead, the initial "rationale" behind the endosymbiont stabilization could have been metabolic integration of the host and the symbiont that gradually became mutualistic. A specific model of such a metabolic association, the so-called Hydrogen Hypothesis, has been proposed by Bill Martin and Miklos Müller (Martin and Müller, 1998). Under the Hydrogen Hypothesis, the metabolism of the archaeal host was based on the utilization of molecular hydrogen that was a waste product of the anaerobic, heterotrophic metabolism of the symbiont. Anaerobic production of ATP, as well as facultative aerobic respiration, might have been additional benefits of the endosymbiosis.

Endosymbiosis would create a peculiar intracellular environment in the hybrid organism. Obviously, to be inherited, the bacteria-turned-endosymbionts had to divide. All modern aerobic eukaryote cells contain numerous mitochondria, and it stands to reason that this relationship—multiple endosymbionts within a single chimeric cell—stems from a very early phase of evolution, effectively, from the emergence of the chimera. The endosymbiont copy number could grow gradually, with the increasing dependence of the cell on the symbiont's metabolism. In this situation, the endosymbionts inevitably would be subject to lysis, leading to the release of the symbiont DNA into the surrounding (host) cytosol of the chimeric cell. Remarkably, even in modern plants and animals, where the chromosomes are partly protected from alien DNA by the nuclear envelope and fixation of any inserted DNA is complicated by the necessity of integrating into the germ line and surviving recombination during meiosis, inserts of large pieces of mitochondrial DNA into the nuclear genome are rather common (Hazkani-Covo, et al., 2010). In the chimeric proto-eukaryote cell shortly after the endosymbiosis, the unprotected host DNA would be subject to a veritable barrage of endosymbiont DNA. Note that the situation is inherently asymmetrical because, first, genomes of viable endosymbionts are protected from invasion of the host DNA by the bacterial membrane, whereas the host DNA is exposed; and, second, because the sheer amount of free endosymbiont DNA is much greater. Hence, *a ratchet of gene transfer from the endosymbiont to the host results* (Martin and Koonin, 2006a).

The exposure of the host genome to the endosymbiont DNA has several major consequences. Inserting a piece of endosymbiont DNA within host genes with functions important for cell survival typically would be deleterious and, most often, would not be fixed in the proto-eukaryote population. Prokaryotes possess "wall-to-wall" genomes, mostly consisting of protein-coding genes (see Chapter 5), and there is no reason to believe APE was exceptional in that regard. Thus, the propagation of the endosymbiont accompanied by occasional lysis would put enormous pressure on the chimeric cell population, probably leading to an extended population bottleneck. Such a bottleneck has the potential to dramatically increase the rate of genetic drift and thus the role of chance in evolution, while lowering the intensity of selection (see Chapter 8 for a more detailed discussion of this important phenomenon). One could perceive a paradox in the present scenario of eukaryogenesis: Endosymbiosis is considered beneficial by virtue of the metabolic cooperation, but at the same time deleterious because of the release of the endosymbiont DNA and other effects of the intracellular propagation of the endosymbiont. I submit that this situation creates strong tension but not a paradox: The chimeric cell could survive the onslaught of alien DNA without shedding the endosymbiont if the *mutualistic relationship between the host and the symbiont was established very shortly after the symbiont invasion.* This tension between the necessity to maintain the endosymbiont and the burden that it exerted on the chimeric cell could be a necessary condition for the emergence of the eukaryotic innovations.

A distinct class of sequences potentially causes limited damage when inserted into genes, even functionally important ones. These are the so-called Group II self-splicing introns, a class of reverse-transcribing selfish genetic elements that "jump around" genomes of many bacteria and some mesophilic archaea, as well as fungal and plant mitochondria (Lambowitz and Zimmerly, 2004). These elements have a very interesting, unusual life cycle: Using RNA (ribozyme) catalysis, they excise themselves from the transcripts of the respective host genes and then insert into new sites on the host chromosome after making their own DNA copies using the reverse transcriptase they encode. It is now considered well established that Group II introns, which in the eukaryotic world are present only in some endosymbiont-derived organelles, are ancestors of the spliceosomal introns that interrupt eukaryotic protein-coding genes (Keating, et al., 2010; Toor, et al., 2008). Indeed, the terminal structures of Group II introns that are responsible for the intron excision closely resemble the canonical terminal structures of spliceosomal introns. What is more, the small RNA molecules in the spliceosome that catalyze splicing in all eukaryotes also are derived from Group II introns. Most bacteria keep Group II introns in check with only a few copies (if any) per bacterial chromosome because of the intense purifying selection in bacterial populations (see Chapter 8). Interestingly, α-proteobacteria are relatively enriched for these elements, with up to 30 copies per bacterial genome. The highest content of Group II

introns is seen in fungal and plant mitochondria, where they constitute a significant fraction of the genome. This propagation of Group
II introns in the endosymbiont genome might have started shortly
after the symbiosis was established, conceivably triggered by the
inevitable drop in the effective population size of the symbiont and
the consequent inability to effectively purge selfish elements.

Thus, Group II introns might have been a significant presence in
the endosymbiont DNA that bombarded the host genome. Moreover,
these elements possess the ability to actively integrate into other
DNA molecules, so they would aggressively attack the host chromosome by inserting into genes and then moving to additional locations
in the genome (Martin and Koonin, 2006a). Although after transcription Group II introns autocatalytically excise from the transcript so
that the surrounding exons are spliced together, massive infestation of
host genes would be a severe hazard. Indeed, splicing is a relatively
slow process, much slower than translation. Given that, in prokaryotes, transcription and translation are coupled, transcripts with
inserted Group II introns on many occasions would be translated
before there is time for splicing to occur. The consequences could be
dramatic, probably fatal, if intron insertions were numerous: Aberrant proteins would accumulate, with a severe detrimental effect on
the affected cell. Even more serious would be the consequences of
inactivation of the open reading frame encoding the Group II reverse
transcriptase (RT) that acts *in cis* as a splicing cofactor (not an
enzyme at this stage). Splicing of genes containing introns with inactivated RT genes would have to occur *in trans*. This is known to be an
inefficient reaction, so such introns would effectively abolish the production of the respective functional proteins. Thus, the invasion of
the host genes by Group II introns would create a powerful driving
force for a cascade of evolutionary innovations (Koonin, 2006):

1. A splicing machinery capable of efficient *in trans* action

2. A "defense" device that would decouple translation from
   transcription, allowing the relatively slow splicing process to
   occur before translation begins

3. Additional "lines of defense" against the accumulation of
   aberrant polypeptides

Indeed, all three types of adaptations to the intron invasion have evolved during the pre-LECA stage of evolution of prokaryotes: the spliceosome, the nucleus, and the additional quality-control systems such as Nonsense-Mediated Decay (NMD), the machinery that eliminates immature transcript and the Ub-dependent system of protein degradation that destroys aberrant proteins directly (see Figure 7-6).

So in principle, the onslaught of retroelements from the endosymbiont on the host genome provides the selective pressure for the emergence of several defining innovations of the eukaryotic cell, above all, the endomembrane system of which the nucleus is a major component. When one looks closer, however, the problem of evolving these systems still uncomfortably reminds one of "irreducible complexity." Specific explanations are needed, and these are not easy to come up with. For instance, the elaborate nuclear pore complex cannot function and, accordingly, cannot be selected for in the absence of the nuclear envelope, but the latter cannot communicate with the cytosol without nuclear pore complexes. It appears most likely that the evolution of the endomembrane systems and the nucleus, even if rapid on the geological/evolutionary scale, did go through intermediate stages. The proliferation of the endosymbionts within the evolving chimeric cells could have been sufficiently gradual to allow the proto-eukaryotes to survive long enough for the fixation of innovations with a limited beneficial effect. Conceivably, the series of innovations would start with the formation of vesicles from the endosymbiont membrane. These vesicles would form a primitive endomembrane system including a protonucleus—that is, a compartment enclosing the chromosome(s) that did not contain modern-type pores, but only gaps between flattened vesicles; each of those then remained connected to the endomembrane system. The holes in the protonuclear membrane would allow passive transport of proteins and nucleic acids into and out of the evolving protonuclear compartment but would prevent the access of ribosomes to the sites of transcription, thus disrupting the transcription-translation coupling typical of prokaryotes and minimizing the damage from the inserting retroelements (Group II introns). This would allow further proliferation of the (proto)mitochondria and release of more DNA and retroelements from them, and these would push for further elaboration of the nucleus, culminating in the modern-type pore complex that actively controls the nucleocytosolic traffic and couples splicing of pre-mRNAs with the extrusion of mature mRNAs from the nucleus. The proliferation of the endomembranes eventually led to the complete overhaul of the membrane system of the protoeukaryotic cell, with the ancestral archaeal plasma membrane replaced with the bacterial counterpart, probably from within, by the expanding symbiont-derived endomembranes.

A similar scenario can be imagined for the evolution of the spliceosome, starting with an RNA-only system in which both the introns and the catalytic small RNA involved in splicing are derived from the retroelements. The next stage of evolution would involve recruiting the Sm protein that stabilizes RNA duplexes involved in splicing (Veretnik, et al., 2009), followed by elaborating the ribonu-cleoprotein spliceosome. Remarkably, recent observations indicate that one of the key protein components of the spliceosome, Prp8, is an inactivated derivative of the Group II intron RT (Dlakic and Mushegian, 2011). This unexpected discovery further attests to the multiple contributions of Group II introns to the origin of both spliceosomal introns and the spliceosome itself. More generally, such stepwise *evolutionary bootstrapping* might partly account for the evolution of the signature complex systems of the eukaryotic cell.

Of course, many important aspects of the eukaryotic cellular organization cannot be easily linked to immediate outcomes of endosymbiosis. Consider the eukaryotic chromatin, with the multiple linear chromosomes replacing circular chromosomes that are most common in bacteria and archaea. The extremely intricate organization of the eukaryotic chromatin, with its regular nucleosome structure, at least outwardly is dramatically different from the much simpler structure of prokaryotic chromosomes (Branco and Pombo, 2007) although archaea (euryarchaeota) possess simple nucleosomes comprised of histones (Bailey, et al., 2002). Added to this is the fundamental change in genome architecture, whereby the operonic organization, which is the guiding architectural principle of prokaryotic genomes, is abandoned. This series of dramatic changes associated with the eukaryogenesis is hard to attribute to specific effects of endosymbiosis. Nevertheless, some interesting connections can be traced. Linear chromosomes face the difficult problem of replicating the ends, given that all known DNA polymerases require a primer and cannot start from the first nucleotide of the template. Unless a special mechanism to restore the ends is in operation, the ends shorten in each replication cycle, making replication unsustainable. All eukaryotes employ the enzyme named telomerase, which restores an array of repeats at chromosome ends by reverse transcription of a small RNA molecule associated with the enzyme (Autexier and Lue, 2006). Strikingly, the telomerase is another (after Prp8) evolutionary derivative of the Group II intron RT, in this case retaining the enzymatic activity, a link that might bring the transition to linear chromosomes within the framework of endosymbiosis-related and, more specifically, retroelement-stimulated cascade of innovations (Koonin, 2006; see Figures 7-6 and 7-7).

For all subsequent evolution of eukaryotes, an apparently inevitable and critically important consequence of the emergence of the nucleus is the drastic, even if not complete, curtailment of HGT. Although multiple acquisitions of bacterial genes by unicellular eukaryotes have been reported, the level of HGT is hardly comparable to that seen in no-parasitic bacteria and archaea (Keeling and Palmer, 2008). Most of the DNA that enters a eukaryotic cell is destroyed without ever entering the nucleus and reaching the chromatin. This precipitous drop in HGT suggests a natural answer to the otherwise puzzling question: Why have eukaryotes lost all the operons of their prokaryotic ancestors? (The host archaeon undoubtedly had a regular operonic organization of genes, and so did the endosymbiont.) Recalling the selfish operon concept, once the HGT effectively ceases, a ratchet is put into action, whereby once an operon is disrupted, it is extremely unlikely to be re-created through recombination and then retained by selection. Effectively, then, the operon is irreversibly lost in the given lineage. Apparently, this ratchet mechanism wiped out all prokaryotic operons in early stages of the evolution of eukaryotes. The operons that do exist in some eukaryotes, such as nematodes, have nothing to do with prokaryotic operons; apparently, they evolved *de novo* and are not conserved in divergent eukaryotic lineages.

This scenario yields a straightforward yet unexpected prediction: Those genes that can function only within operons but exert deleterious effects when taken out of the operon context will be completely

the logic of chance

lost in eukaryotes. Strikingly, this is precisely the case with the toxin-antitoxin and restriction-modification systems that are extremely common in bacteria and archaea (see Chapter 5) but seem to be completely absent in eukaryotes.

The near-elimination of HGT also provides an evolutionary incentive for the extensive gene duplication that is the primary route of innovation in eukaryotes (Lespinet, et al., 2002). The population bottleneck caused by the propagation of endosymbionts allowed an explosion of duplications during the stem phase (Makarova, et al., 2005; see also Chapter 8), but more generally, duplication substitutes for HGT as the main source of novelty throughout the evolution of eukaryotes.

Last, but certainly not least, the low rate of HGT in eukaryotes can be considered the principal factor behind the evolution of meiotic sex, one of the defining biological processes in eukaryotes. Indeed, *in eukaryotes deleterious mutations typically cannot be complemented by horizontally acquired genes, hence the pressure to evolve a system of regular recombination that would preclude accumulations of such mutations and the eventual mutational meltdown.* Such a system countering Müller's ratchet evolved in the form of meiosis and sex. This is not necessarily the only factor that drove the evolution of sex, but it certainly appears to be an important one (we do not have the opportunity to discuss in detail this problem that is enormously popular among evolutionary biologists [de Visser and Elena, 2007]). Given that the curtailment of HGT largely results from the evolution of the nucleus, the "invention" of meiosis and sex—on the basis of archaeal repair and cell division systems—seems to be part of the chain of defense and damage-control adaptations triggered by intron invasion (see Figure 7-6).

The other major consequence of the proto-mitochondrial endosymbiont proliferation within the chimeric proto-eukaryotic cell that is, in a way, complementary to the curtailment of HGT from outside sources is the release of random pieces of bacterial DNA (as opposed to selfish elements) through the lysis of endosymbionts. Such DNA fragments also have the potential to insert into the host chromosome, albeit at a lower rate than selfish elements. On many occasions, such insertions will be fatal. However, when an entire gene from the endosymbiont is inserted in an intergenic region of the host chromosome, there might be no significant deleterious effect. More-over, the inserted gene can be expressed if necessary regulatory elements are available next to the insertion site. Fragments of mito-chondrial DNA occasionally insert into the nuclear genomes of plants and animals even now (Hazkani-Covo, et al., 2010), notwithstanding the protection provided by the nucleus and by the systems of defense against cytosolic DNA. Undoubtedly, the rate of insertion was much greater during eukaryogenesis, before the eukaryotic cellular organi-zation was fully established. The ratchet of gene transfer results in gene doubling when functional copies of the same gene are present both in the endosymbiont and in the nuclear genome. Some of the nuclear genes that evolved via this route diverged and were recruited for cellular functions outside the endosymbiont. However, on other occasions, the inserted gene would be preceded by a sequence cod-ing for a peptide capable of mediating the protein import back into the endosymbiont. This is yet another "lucky coincidence," but it is not as unlikely as it might seem because the import peptides are typ-ically simple, repetitive sequences that could evolve by sheer chance (Neupert and Herrmann, 2007). Once there is a nuclear gene for a protein that functions in the mitochondrion, the functionally redun-dant mitochondrial genes can be lost without any deleterious effect. This redundancy *creates another ratchet that puts the mitochondria squarely on the path of reductive evolution,* given that the nuclear genome is constantly exposed to the DNA from lysed endosymbionts, resulting in multiple "trials" for the transfer of each endosymbiont gene to the nuclear genome. The end result is that the great majority of the proteins that function in the mitochondria are encoded in the nuclear genome, with only those genes that have to be expressed inside the mitochondrion for its proper functioning (see the discus-sion earlier in this chapter) remaining in the organellar genome. The same pattern holds for other endosymbionts, particularly plastids. Of course, reductive evolution also involves the irreversible loss of many endosymbiont genes that are made redundant without even recruit-ing a host protein, but rather because the function itself becomes irrelevant for the endosymbiont or because metabolites from the host, such as nucleotides and amino acids, are imported into the endosymbiont, obviating the need for the respective metabolic path-ways. Endosymbionts with a substantially reduced genome outcom-pete those with larger genomes simply because of their faster genome replication and division, so the genome reduction is fixed during evolution.

Assigning a single overarching cause to any major evolutionary transition is an inevitable oversimplification and an epistemological fallacy because causes are, after all, human constructions (see Appendix A). Nevertheless, I believe that the coherence among many key eukaryotic innovations that appear to be interpretable as response to endosymbiosis, particularly to the onslaught of endosymbiont-derived mobile elements, is too striking to be dismissed as wishful thinking alone. Such a scenario, even if not falsifiable in its entirety (see Appendix A), does include specific falsifiable predictions and stimulates experimentation. Indeed, after Bill Martin and I proposed the specific version of the symbiogenesis scenario in which the central stage belongs to Group II introns (Martin and Koonin, 2006a), it survived two rather stringent falsification tests. The discovery of transcription-translation coupling in Archaea is one of these (French, et al., 2007). An even more meaningful test is the solution of the structure of a Group II intron that left no reasonable doubt about the origin of spliceosomal introns from prokaryotic retroelements; this relationship was still considered tenuous at the time the hypothesis was formulated (Toor, et al., 2008). A complementary line of experimental study would target unusual bacteria such as Planctomycetes that possess intracellular compartments enclosing the chromosome (Fuerst, 2005). Certainly, these organisms are prokaryotes by every criterion. Moreover, comparative genomic analysis shows that they do not encode homologs of the protein subunits of the nuclear pore complex (Mans, et al., 2004). The prediction of the present model is that, although Planctomycetes and some related bacteria possess a "nucleus-like" compartment, they maintain the transcription-translation coupling typical of prokaryotes, that is, functional ribosomes enter the compartment and initiate translation of the mRNA before their transcription is complete, or else the nascent mRNA molecules are cotranscriptionally extruded through holes in the compartment walls.

If, on the contrary, experiments show that translation is decoupled from transcription, such a result will seriously challenge the model. The king of all falsifications would be the discovery of an extant archezoan, a free-living eukaryote without any traces of ever harboring an MLO, but possessing all other eukaryotic cellular signatures. (A parasite with a dramatically reduced genome would not make a compelling case.) The chances that an archezoan is discovered one day decrease with each finding of yet another group of protists that have no regular mitochondria but harbor a MLO.

The remarkable story of eukaryotic introns

The "genes in pieces" (exon-intron) architecture of the protein-coding (and some RNA-coding) genes in eukaryotes is a truly astonishing feature (we might not always see it that way, only because we are so used to the concept of splicing, given that the discovery is more than 30 years old at the time of writing). Why would genes be interrupted by multiple noncoding sequences, most of which have no demonstrable function and are excised from the transcript by an elaborate molecular machine (evolved solely for this purpose) only to be destroyed? This almost defies imagination. When the introns were discovered in 1977, Walter Gilbert quickly came up with the enticing "introns early" hypothesis that formed the basis of the so-called "exon theory of genes" (Gilbert, 1978). In essence, Gilbert proposed that introns accompanied life from the earliest stages of its evolution and played a key role in the evolution of protein-coding genes by allowing joining of short sequences encoding primordial peptides via recombination of adjacent noncoding sequences. The formulation of this idea was followed by more than 20 years of attempts to validate the existence of primordial introns by analysis of various features of extant ones (de Souza, et al., 1998). We will not review this effort here. Suffice it to say that no convincing evidence has ever been found. Of course, it does not help the introns-early case that no prokaryotes possess a spliceosome or spliceosomal-type introns, although Gilbert and his colleagues argued that this is a result of evolutionary "streamlining." The strongest argument against "introns early" probably is the demonstration of the ancestral relationship between bacterial self-splicing introns and the spliceosomal introns. This finding implies that, even if there were introns at the earliest stages of the evolution of life (we return to this issue in Chapters 10 and 11), these introns were completely different from modern ones, and the latter cannot carry any "memory" of the primordial evolution. The spliceo-somal introns and the entire splicing system are thus a purely eukary-otic feature, one of those that define the "eukaryote state."

So why do so many introns interrupt eukaryotic genes? The only reasonable answer seems to be that they are there because their ancestors invaded eukaryotic genes during eukaryogenesis or soon afterward, and mechanisms to efficiently remove them from primary transcripts evolved and ensured survival of the organismal lineage with the strange genes in pieces. After that, the selective pressure to eliminate introns in many lineages of eukaryotes was insufficiently strong to get rid of most of them, although this is precisely what hap-pened in other lineages that evolved under stronger purifying selec-tion (see Chapter 8). This is certainly not to deny functional importance to introns altogether: Some of them are known to con-tribute to expression regulation (Le Hir, et al., 2003), whereas others even contain nested genes (Assis, et al., 2008). Moreover, introns pro-vide for the possibility of alternative splicing, a key mechanism for the creation of structural and functional diversity of proteins in multicel-lular eukaryotes (see Chapter 8). On the whole, however, the persist-ence of introns seems to depend largely on the strength of purifying selection against them. The population-genetic aspects of intron loss and gain are considered in Chapter 8; here I briefly discuss the results of comparative-genomic reconstructions of intron evolution and additional ideas on the nature of the genomes of the earliest eukaryotes related to the earlier scenario of eukaryogenesis.

Eukaryotes widely differ in their characteristic intron density: Many protists and unicellular fungi contain only a few introns in the entire genome, whereas animals, plants, and some other protists are intron-rich, with several introns interrupting the coding sequences of most genes (Jeffares, et al., 2006). Notably, the positions of a large fraction of introns are conserved between orthologous genes from distant organisms, including plants and animals (Rogozin, et al., 2003). Comparative-genomic reconstructions that take into account the conserved and variable intron positions lead to a counterintuitive conclusion that the genes of LECA were almost as intron-rich as those of modern mammals, and a large fraction of the LECA introns persist to this day in the same positions (see Figure 7-8; Csuros, et al.,

2011). Strange as this conclusion might seem, it is becoming increasingly unshakeable, as more genomes are available for analysis with more sophisticated reconstruction methods. The implication of this finding that is formally supported by the results of the reconstructions is that the subsequent evolution involved primarily intron loss occurring along most of the eukaryote branches, with only a few explosive gain episodes that seem to be linked to the emergence of new major branches such as plants or animals (see Figure 7-8). The spurt of intron gain at the base of the Plantae supergroup might have been caused by a new wave of Group II introns coming from the cyanobacterial symbiont. The source of intron gain at the base of the animal branch remains enigmatic and might even suggest a role for a hidden endosymbiosis in the origin of animals.

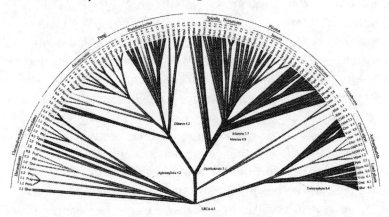

Figure 7-8   Reconstruction of intron gain and loss during the evolution of eukaryotes and the ancestral intron densities. The reconstruction was performed using the Markov Chain Monte Carlo method (Csuros, et al., 2011). The intron densities (introns per kilobase) for the extant forms and the inferred densities for key ancestral forms are shown. The thickness of the black shade is roughly proportional to the intron density. The human lineage is identified by a circle. Three supergroups of eukaryotes (Chromalveolata, Unikonta, and Plantae) and the major groups within each, for which complete genome sequences and accordingly data on intron location are available, are indicated.[1] Adapted from Csuros, et al., 2011. This Open Access article is licensed under Creative Commons Attribution License.

So the LECA seems to have accumulated introns to densities close to those in the most intron-rich among the extant genomes. What about intron dynamics during the stem phase, between eukaryogenesis and LECA? A simple estimate shows that, if intron invasion occurred "instantaneously," the protoeukaryote genome would consist mostly (up to 80%) of introns, given the large and uniform size of Group II introns (around 2.5 Kb; Koonin, 2009b). Most likely, this is an oversimplification. The process of intron accumulation probably was more gradual and was accompanied by shrinking of the inserted introns. Nevertheless, introns seem to have played a key role from the start of the evolution of eukaryotes, in agreement with the eukaryogenesis model discussed.

The three domains of life: Beyond the Woeseian tree

The symbiogenesis scenario of eukaryogenesis leads to an overhaul of the three-domain tree of life championed by Woese and colleagues, even apart from the consequences of the extensive HGT among prokaryotes discussed in Chapters 5 and 6. The Woeseian tree implies the archezoan scenario, endosymbiosis being considered a relatively late event in the history of the eukaryote domain that is irrelevant for the three-domain classification of life (see Figure 7-9A). By contrast, the symbiogenesis scenario posits that acquisition of the primary endosymbiont literally gave rise to the eukaryote domain and, in the process, contributed a large fraction (possibly the majority) of the genes to the evolving eukaryotic genome. Under this scenario, fusion of organisms from the two primary domains gave rise to the third domain; the resulting graph then is *not* a tree (see Figure 7-9B). An important corollary to which we return in Chapter 11 is that, when considering the origin of cells, we have to care about only the two prokaryotic domains, archaea and bacteria.

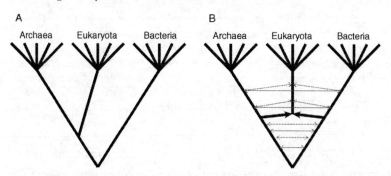

Figure 7-9   The three domains of life revisited: (A) The traditional three-domain tree following Woese; (B) The cyclic graph of the three domains according to the symbiogenesis scenario of eukaryogenesis and interdomain HGT.

Synopsis and perspective

Of the three domains of life, eukaryotes possess by far the most complex, strikingly elaborate cellular organization that for some might even summon the specter of "irreducible complexity" (Kurland, et al., 2006) because for most of the signature functional systems of the eukaryotic cells, we can detect no evolutionary intermediates. It seems natural to view eukaryogenesis as one of the grand challenges of evolutionary biology, one that we are better equipped to attack head-on than the even more fundamental problems of the origin of cells and the ultimate origin of life (see Chapters 11 and 12). Comparative genomics has so far neither solved the enigma of eukaryogenesis nor offered a definitive picture of the primary radiation of the major eukaryote lineages. Nevertheless, phylogenomic analysis has yielded many insights into the origin and earliest stages of the evolution of eukaryotes. Thus, phylogenomics has clarified the evolutionary links between the eukaryote kingdoms and led to the delineation of five or six supergroups. The relationships between the supergroups and the root position in the tree of eukaryotes remain extremely difficult to decipher, probably owing to compressed cladogenesis during the primary radiation of the major eukaryote branches. The expanding sampling of genomes from diverse branches of life is far from being a trivial pursuit; on the contrary, comparative analysis of diverse genomes continues to yield unexpected biological insights. Many more certainly should be expected.

In a perfect congruence, ultrastructural, functional, and comparative genomic data show that eukaryotes are archaeo-bacterial chimeras. Furthermore, genes of apparent bacterial origin are in a numerical excess of "archaeal" genes. Yet, strikingly, comparative analysis of the expanding collection of archaeal genomes increasingly shows that many key cellular systems of eukaryotes exist, in a primitive form, in archaea. The scatter of these systems among different archaeal lineages, along with the phylogenies of conserved proteins, suggests that the archaeal ancestor of eukaryotes belonged to a deep, possibly extinct archaeal branch with a highly complex genome and diverse cellular functionalities. The recent discovery of the potential direct ancestor of the ubiquitin system in a new archaeal genome suggests that we might currently under-appreciate the extent to which many signature functional systems of eukaryotes could have been preformed during the evolution of Archaea. This and other similar findings give credence to the "combinatorial scenario" for the origin of eukaryotes under which these preformed systems were transiently combined in the archaeal host of the primary endosymbiont. In contrast, the endomembrane systems of eukaryotes—particularly the nucleus, with its elaborate nuclear pore complex—are not found in archaea and seem to have been assembled at least partly from bacterial ancestral components. *It appears significant that eukaryotes inherited evolved, elaborate systems from Archaea (with the obvious exception of the mitochondrion), whereas the abundant bacterial molecular componentry was mostly inherited piecemeal and recombined to form novel molecular machines. This contrast seems to be a reflection of the asymmetry between the host and the endosymbiont: Despite all the drastic innovations that accompanied eukaryogenesis, many cellular systems of the archaeal host have survived and changed only in an evolutionary manner, through duplications and additional embellishments.*

Taken together, these findings seem to be best compatible with a symbiogenesis scenario for the origin of eukaryotes. In this scenario, eukaryogenesis was triggered by the endosymbiosis of an α-proteobacterium with an ancestral archaeon, with the endomembrane system and particularly the nucleus evolving as defense against intron invasion. Moreover, other key innovations of the eukaryotic cell, such as the nonsense mediated decay of aberrant transcripts and the remarkable proliferation of the ubiquitin-dependent system for degradation of aberrant proteins, seem to find a logical explanation as additional lines of defense against the same invasion. In less direct ways, the defense hypothesis may contribute to the understanding of the evolution of other major eukaryotic features, such as the disappearance of operons and the transition from circular to linear chromosomes. All in all, we now seem to have a rather coherent, although certainly still sketchy, narrative on eukaryogenesis. To conclude this chapter, I want to stress that, regardless of numerous details that remain unclear, the story of eukaryogenesis is an ideal exhibit for the main theme of this book: the interplay between chance and necessity in the evolution of life. Indeed, the capture of the proto-mitochondrial endosymbiont undoubtedly was a key event in eukaryogenesis, and the partners in the symbiosis were "chosen" by chance. Nevertheless, the symbiosis seems to have triggered a complex chain of events with many necessary elements, if only to provide for the survival of the chimeric organism, given our knowledge that, on this planet, eukaryotes did survive and reached incredible complexity and diversity.

Recommended further reading

Doolittle, W. F. (1998) "You Are What You Eat: A Gene Transfer Ratchet Could Account for Bacterial Genes in Eukaryotic Nuclear Genomes." *Trends in Genetics* 14: 307–311.

Apparently, the first description of the ratchet of gene transfer from endosymbionts to the host, albeit in the context of the archezoan scenario.

Embley, T. M., and W. Martin. (2006) "Eukaryotic Evolution, Changes, and Challenges." *Nature* 440: 623–630.

An insightful analytical review of the various scenarios of eukaryogenesis in the light of the realization that all known eukaryotes possess mitochondria or MLOs.

Koonin, E. V. (2010) "The Origin and Early Evolution of Eukaryotes in the Light of Phylogenomics." *Genome Biology* 11: 209.

An overview of relationships between eukaryote supergroups, the nature of LECA, and eukaryogenesis, with an emphasis on a highly complex LECA.

Koonin, E. V. (2006) "The Origin of Introns and Their Role in Eukaryogenesis: A Compromise Solution to the Introns-Early Versus Introns-Late Debate?" *Biology Direct* 1: 22.

An extension of the defense scenario of eukaryogenesis presenting the single chain of causation for the origin of various eukaryote-specific functional systems triggered by intron invasion.

Kurland, C. G., L. J. Collins, and D. Penny. (2006) "Genomics and the Irreducible Nature of Eukaryote Cells." *Science* 312: 1,011–1,014.

A spirited argument against the endosymbiosis scenario of eukaryogenesis and for the primordial origin of eukaryotic complexity.

Lane, N., and W. Martin. (2010) "The Energetics of Genome Complexity." *Nature* 467: 929–934.

A hypothesis on the inevitability of endosymbiosis, which is construed to be the only path to efficient bioenergetics required for the evolution of large, complex cells.

Martin, W., and E. V. Koonin. (2006) "Introns and the Origin of Nucleus-Cytosol Compartmentalization." *Nature* 440: 41–45.

The hypothesis on the defense against intron invasion as the selective factor behind the origin of the nucleus.

Martin, W., and M. Müller. (1998) "The Hydrogen Hypothesis for the First Eukaryote." *Nature* 392: 37–41.

A crucial hypothesis on the metabolic cooperation as the selective factor favoring the mutualistic relationship between the host and the endosymbiont.

Martin, W., T. Dagan, E. V. Koonin, J. L. Dipippo, J. P. Gogarten, and J. A. Lake. (2007) "The Evolution of Eukaryotes." *Science* 316: 542–543.

A refutation of the argument of Kurland, et al.

Zimmer, C. (2009) "Origins. On the Origin of Eukaryotes." *Science* 325: 666–668.

A popular discussion of different scenarios of eukaryogenesis.

8

The non-adaptive null hypothesis of genome evolution and origins of biological complexity

Evolutionary entropy and complexity

Few buzzwords in the last two decades have been as popular and at the same time defined in as many different, often contradictory, and sometimes misleading ways as *complexity*.[1] All this celebrity status notwithstanding, the notion of complexity obviously captures a common, fundamentally important phenomenon that permeates all of biology and also reaches beyond the biological realm. Unlike many scientific terms, *complexity* has an undeniable vernacular meaning. As in the case of pornography, we know it when we see it. Everyone recognizes that a mammal or a bird is more complex than a worm, whereas the worm is more complex than any unicellular organism. Intuitively, there is the additional connotation of "more advanced" or "closer to perfection" associated with greater complexity.

At a level beyond sheer intuition, what does it mean that a mammal is more complex than an amoeba? This is an important question if we strive to develop some sort of a satisfactory answer to the notorious "Why?" question: Why are there elephants and redwoods around us (even if increasingly fewer of these) instead of only bacteria and archaea with complements of genes that are necessary and sufficient for supporting the functioning of a minimal cell? In other words, what are the factors behind the emergence of complexity during evolution? In Chapter 7, we discussed evolutionary scenarios that attempt to explain *how* the strikingly complex organization of the eukaryotic cell (compared to the cells of prokaryotes) could have emerged. In this chapter, we face the "why" question directly, and the answers are going to be unexpected and perhaps disconcerting to some.

Defining *organizational*—or, when it comes to biology, organismal—complexity in a precise manner is inherently difficult. The attempts in this direction take account of different numbers of distinct parts in the compared systems. For instance, vertebrates have a greater number of tissues and cell types than worms, and this naturally translates into the statement that vertebrates possess a greater organismal complexity (Bonner, 2004). More important for the present discussion, eukaryotic cells have far more intracellular organelles than the cells of prokaryotes (which typically have no real organelles at all), a difference that surely reflects a greater complexity of the eukaryotic cell organization. Similarly, one could, in principle, measure the number of interactions between components or the numbers of connections in signal transduction networks, and compare the complexity of organisms or cells using the resulting numbers. All these definitions of complexity appear to miss "something" that we intuitively perceive as essential to complex organization. In any case, quantitative comparisons of organismal complexity do not seem to be of much use in actual research. Genomic complexity is defined more naturally and can be explored further. Indeed, at the end of the day, genome sequences are long strings of digital symbols (letters), and for this class of objects, formal, operational definitions of complexity are well established. Probably the best known and most intuitively plausible of these definitions is Kolmogorov complexity, which is related to Shannon information and to Boltzmann's classical statistical definition of entropy. Kolmogorov complexity is simply the length of the shortest string of symbols in which the given sequence (a genome) can be encoded. Obviously, the least complex sequence is a homopolymer (such as polyA), for which the length of the message is just one letter and the complexity (information content) is 2 bits (given four nucleotides). The most complex sequence is a completely random polymer with equal frequencies of all 4 nucleotides (or 20 amino acids, if we adopt this definition for amino acid sequences) in each position. The classical Shannon formula for the entropy (information content) of a nucleotide sequence of length L (see Figure 8-1A) can be written as follows:

$$H(L)= \sum_{i=1}^{L} f_i \log f_i$$

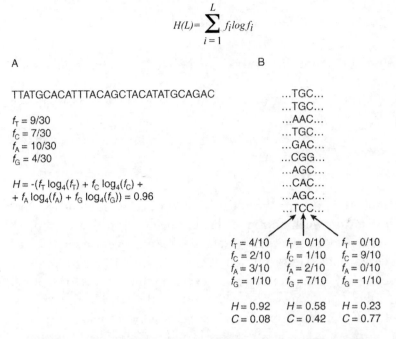

A

TTATGCACATTTACAGCTACATATGCAGAC

$f_T = 9/30$
$f_C = 7/30$
$f_A = 10/30$
$f_G = 4/30$

$H = -(f_T \log_4(f_T) + f_C \log_4(f_C) +$
$+ f_A \log_4(f_A) + f_G \log_4(f_G)) = 0.96$

B

...TGC...
...TGC...
...AAC...
...TGC...
...GAC...
...CGG...
...AGC...
...CAC...
...AGC...
...TCC...

$f_T = 4/10$	$f_T = 0/10$	$f_T = 0/10$
$f_C = 2/10$	$f_C = 1/10$	$f_C = 9/10$
$f_A = 3/10$	$f_A = 2/10$	$f_A = 0/10$
$f_G = 1/10$	$f_G = 7/10$	$f_G = 1/10$

| $H = 0.92$ | $H = 0.58$ | $H = 0.23$ |
| $C = 0.08$ | $C = 0.42$ | $C = 0.77$ |

Figure 8-1   Information content and complexity of a single sequence (A) and an alignment of homologous sequences (B). f denotes frequencies of nucleotides in a sequence (A) or an alignment column (B).

Here, f_i is the frequency of the letter i ($i= A, T, G, C$); hereafter, the base of the logarithm is assumed to be equal to the size of the alphabet (4 in the case of nucleotide sequences, and 20 for amino acid sequences).[2] Defined this way, information (entropy) tells us very little about the meaningful information content or complexity of a genomic sequence. The highest complexity (entropy or information content) obviously does not at all suggest that a sequence is complex in any biologically relevant sense. A completely random sequence actually is most likely to be meaningless, whereas a homopolymer sequence would have a limited biological relevance. However, a nearly random high-entropy sequence can well be functional, as can a low-entropy sequence—there is just no way to tell. A biologically meaningful definition of complexity is required, and one has been proposed by Chris Adami (Adami, 2002) and somewhat differently interpreted by myself (Koonin, 2004). Under this new definition, entropy and complexity are calculated for an alignment of orthologous sequences rather than a single sequence:

$$H(L)=\sum_{i=1}^{L} H_i = \sum_{i=1}^{L} \sum_{j} f_{ij} \log f_{ij}$$

Here, $H(L)$ is the total entropy of the alignment of n sequences of the length L, H_i is the per-site entropy, and f_{ij} are the frequencies of each nucleotide (j = A, T, G, C) in site i.[3] Clearly, for a fully conserved site, $H(i) = 0$, whereas, for a completely random site, $H(i) = 1$. Note that this definition of entropy directly conforms to the famous statistical definition of Boltzmann:

$$H = k ln W$$

Here, W is the number of microstates that corresponds to the macrostate for which entropy is being calculated so that entropy is zero for a completely ordered state and maximum for a completely disordered state. Thus, the definition of *evolutionary entropy* of a genome, $H(L)$, introduced by the previous formula seems to be physically valid; thus, it makes sense to reserve the term to denote this quantity. Evolutionary entropy also makes perfect biological sense: Low-entropy sites are most conserved and, by inference, most functionally important. It stands to reason that these sites carry more information on the functioning and evolution of the organisms in question—and on the interactions between the organisms and the environment as originally posited by Adami—than high-entropy (poorly conserved, relatively unimportant) sites. The quantity that has the meaning of *"biological (evolutionary) complexity"* of a genome can be defined as follows:

$$C(N)=N-\sum_{i=1}^{k} H(L_i)$$

Then, "biological (evolutionary) information density" can be defined as:

$$D(N)=C(N)/N = (N -\sum_{i=1}^{k} H(L_i))/N = 1 -\sum_{i=1}^{k} H(L_i)/N$$

Here, N is the total length (number of nucleotides) of a genome, L_i is the length of a genomic segment subject to measurable selection (typically, a gene), k is the number of such segments in the genome, and $H(L_i)$ is the evolutionary entropy for the segment L calculated using the previous formula.

The exact values of H are not easy to calculate for entire genomes because the distribution of evolutionary constraints is never known precisely (see Chapter 3). Furthermore, there is a degree of arbitrariness in the choice of orthologs to be included in the alignment for the calculation. However, these details are not important if we want only a reasonable ballpark estimate. Indeed, the fraction of sites under selection across the genome has been estimated with reasonable precision for some model organisms such as humans and *Drosophila* (see Chapter 3). For others, particularly prokaryotes and unicellular eukaryotes, the fraction of coding nucleotides plus the estimated fraction of regulatory sites can be taken as a reasonable approximation; for sites under selection, $H(i) = 0.5$ can be taken for the mean entropy value.

Comparing the estimates of $H(N)$, $C(N)$, and $D(N)$ for genomes of different life forms reveals a major paradox. The total biological complexity $C(N)$ monotonically increases with the genome size, particularly in multicellular eukaryotes compared to prokaryotes, but the entropy $H(N)$ increases dramatically faster; as a result, the evolutionary information density $D(N)$ sharply drops (see Figure 8-2). Thus, the organisms that are habitually perceived as the most complex (for example, humans) turn out to possess *"entropic"* genomes with low or even extremely low information density, whereas organisms that we traditionally think of as primitive, such as bacteria, have *"informational"* genomes in which information is tightly packed and information density is high. This paradox does not tell us much new, compared to what has already been said in Chapter 3 about the organization of different genomes. Nevertheless, it is instructive to formalize the notion of biological complexity and to express it in terms steeped in the concept of entropy, obviously one of the key concepts of physics. This formal examination of complexity shows that "something is rotten in the state of Denmark": The genomes of the organisms that we consider, for good reasons, to be most complex and most "advanced" (perhaps a less defensible idea) carry much more entropy and, hence, have a much lower biological information density than the genomes of the simplest cellular forms. To rephrase this paradox in a more provocative way, the genomes of unicellular organisms (especially prokaryotes) appear incomparably "better designed" than the genomes of plants or particularly animals.

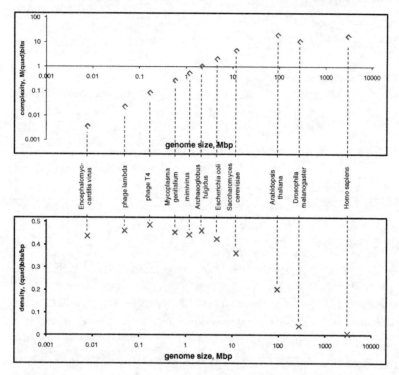

Figure 8-2   The dependence of evolutionary complexity (C) and biological information density (D) on genome size. The points are crude estimates obtained using the formulas given in this chapter, under the assumption of H(i) = 0.5 for nonsynonymous sites in protein-coding regions and H(i) = 1 for other sites. The plot is in double logarithmic size.

The complexity paradox seems to imply that the sophisticated features that are present in the genomes of "higher" organisms (large families of paralogous genes, complicated regulation of gene expression, alternative splicing, and much more) probably have not evolved as straightforward adaptations or "improvements." Explaining the advent of these embellishments is a big challenge to evolutionary biology; an answer (or possibly *the answer*) came in the form of a new theory of the evolution of complexity proposed by Michael Lynch in 2003 (Lynch and Conery, 2003).

Effective population size as the general gauge of evolutionary constraints: The non-adaptive theory of genome evolution

As discussed in the previous section, the most complex organisms on Earth have "high-entropy" genomes that appear to be extremely inefficient and "poorly designed." It takes a huge leap of faith to believe that adaptive evolution leads to such results. Informally, the motivation for a new theory of the evolution of genomic complexity can be presented as follows. Genomes of complex organisms contain a variety of features that are essential for their organizational complexity but appear to be useless and, hence, at least slightly deleterious at the time of their appearance. The most prominent of such features in the genomes of multicellular eukaryotes include introns that ensure the possibility of alternative splicing, which occurs in the great majority of mammalian genes and constitutes the principal basis of proteome diversity (Blencowe, 2006; Wang, et al., 2008), and duplicated genes, which are the major source of evolutionary innovation and diversification in eukaryotes (Lespinet, et al., 2002; Lynch and Conery, 2000). These genomes also carry numerous selfish elements and other DNA that is not subject to selection and, to the best of our understanding, is "junk." The persistence of all these sequences in the complex genomes is naturally attributed to weak (inefficient) purifying selection and, conversely, a major role of drift in the evolution of these organisms.

Under population genetic theory, the effectiveness of purifying as well as positive selection is proportional to the effective population size (Ne) of a given organism, assuming a uniform mutation rate. Only mutations for which $|s| >> 1/Ne$ (where s is the selection coefficient—in other words, the fitness differential between the wild type and the respective mutant) can be efficiently fixed (positive selection) or eliminated (purifying selection) during evolution. *Conversely, mutations with $|s| << 1/Ne$ are "invisible" to selection. This simple dependence is possibly the primary determinant of the constraints that affect diverse aspects of genome and phenome evolution*, particularly the fixation of the embellishments that are associated with the genomes of complex organisms (Lynch, 2007b, 2007c; Lynch and Conery, 2003). Indeed, differences in Ne seem to underlie the qualitative difference between the genome architectures of the unicellular and multicellular organisms described previously. Substantial genome expansion appears to be attainable only in organisms with small populations and the attendant weak selection. Unfortunately, the effective population sizes are not easy to estimate, although the available rough estimates vary within a huge range, from around 10^9 in bacteria to 10^5 or less in animals (see Box 8-1). Estimates that are more readily available from the level of genomic polymorphism and, as we shall see in the next section, could be even more relevant for understanding the evolution of genomic complexity, are for the product $N_e u$, where u is the per-site mutation rate. The $N_e u$ values vary by approximately two orders of magnitude, from about 0.001 in bacteria to about 0.1 in vertebrates (see Box 8-1). The power of selection is predicted to differ accordingly. The result is that, in prokaryotes, with their typically large populations, even very slightly deleterious mutations, with s values on the order of 10^{-8}, are efficiently eliminated; by contrast, in small populations of multicellular eukaryotes, only mutations with relatively large s, on the order of 10^{-4}, which entail substantial effects on fitness, can be wiped out by purifying selection. As we shall see in the next sections, this difference is crucial as a determinant of the course of evolution because the s values for the principal "embellishments" of complex genomes, such as introns, lie within that range. Thus, they are mostly eliminated by purifying selection in organisms with large N_e but not in organisms with small N_e. *Evolutionary conservation of any genomic element does not automatically imply that the conserved element is constrained by purifying selection owing to its functional importance; somewhat paradoxically, it might reflect weak purifying selection that is insufficient to eliminate non-adaptive ancestral features* (Koonin and Wolf, 2010b).

Certainly, Ne is not constant throughout the evolutionary history of a lineage. On the contrary, large fluctuations almost necessarily occur, leading to population bottlenecks (intervals of low Ne) during which evolution depends almost entirely on drift so that numerous slightly and even moderately deleterious genomic changes can be fixed, providing raw material for subsequent evolution. It is important to realize that even populations with large Ne can fix slightly deleterious mutations via draft/hitchhiking (see Chapter 2) and, moreover, carry a large cache of neutral and weakly deleterious mutations that never go to fixation, but may persist in the population as polymorphisms for a long time. Some of these persistent nonfixed mutations may go to fixation rapidly when selective pressure changes and a mutation becomes beneficial, or when a new mutation creates a beneficial combination with one of the persistent polymorphisms.

Box 8-1: Population characteristics and features of genome organization in diverse cellular life forms

Organisms	$\sim N_e$	$\sim N_e u$	Typical Genome Size Range (Mb)	Gene Density (Gene/ Kb)	Information Density (Bit/bp)	Intron Density (Introns/ Kbp)
Bacteria, archaea	10^8–10^{10}	0.01–1	0.5–10	~1	~0.4	0 (N/A)
Unicellular eukaryotes	~10^7	0.01–0.1	5–30	~0.5	~0.3	0–2
Annual plants	~10^6	~0.01	0.1–1×10^3	~0.1–0.2	~0.2	5–6
Invertebrates	~10^6	~0.01	0.1–1×10^3	~0.1–0.2	~0.1	2–7
Trees	~10^4	~10^{-4}	~1×10^3	~0.01	~0.01	5–6
Vertebrates	~10^4	~10^{-4}	~0.5–5×10^3	~0.001	<0.001	5-8

The data on N_e and $N_e u$ comes from Lynch, 2006; the data on intron density comes from Csuros, et al., 2011; for the intron density in prokaryotes, "not applicable" is indicated, given the absence of the spliceosome.

This simple (and presented here in a deliberately oversimplified form) theory steeped in population genetics *provides the null hypothesis for genome evolution* (Koonin, 2004). In the next sections, we consider this theory in some more detail and, most importantly, see how it holds against comparative genomic data.

Gene architecture in eukaryotes: A showcase for the non-adaptive theory of genome evolution

Evolution of the exon-intron gene structure in eukaryotes (see also Chapter 7) is an excellent case in point for the non-adaptive population genetic paradigm that allows one to better grasp the theory and its predictions. Before we discuss the specifics of gene architecture evolution from this perspective, it is necessary to understand the connection between the selection coefficient s and the load of deleterious mutations imposed by an added genomic element (Koonin, 2009b; Lynch, 2007b, 2007c). Each element added to a genome magnifies its vulnerability to mutational inactivation and so "encourages" the elimination of that element from the population. When an embellishment requires the conservation of n nucleotides for the respective gene to maintain its functionality, this requirement obviously creates the room for n deleterious mutations to occur, so the mutational disadvantage is $s = nu$. The recognition and efficient excision of each intron by the spliceosome requires the involvement of approximately 25 to 30 nucleotides within the intron and adjacent exons surrounding the donor and acceptor splice junctions. Then, the condition for the fixation of an intron in a population is $N_e u <<1/n$ or $N_e u << 0.04$.

Comparing the $N_e u$ values and intron densities in Box 8-1, we immediately see the excellent correspondence between the theory and the observations. The vertebrates with their low $N_e u$ values are obviously well below the threshold. Indeed, vertebrate genes have the highest known intron densities. Moreover, evolution of vertebrates seems to have involved very little intron turnover, in agreement with the theoretical prediction that the strength of purifying selection in these organisms is insufficient to eliminate introns. Invertebrates and plants are slightly below the threshold and have intermediate intron densities. In a sharp contrast, most unicellular eukaryotes are above the threshold, even if not by a great margin, and show a precipitous decline in intron densities (see Box 8-1).

The positions of many introns are conserved in orthologous genes of animals and plants (see Chapter 7); thus, most of these introns represent the heritage of the LECA. However, conservation of intron positions appears to be due to the weak purifying selection that precludes efficient elimination of introns in organisms with small Ne, not to constraints on intron positions, per se. A more detailed analysis of introns and exon-intron junctions leads to further observations that might seem inexplicable at first glance but appear to be in an excellent agreement with the predictions of the theory (Irimia, et al., 2007). Strikingly, all introns in intron-poor genomes of unicellular eukaryotes are short, with nearly uniform, apparently tightly controlled lengths and conserved, optimized splice signals at exon-intron junctions. By contrast, in intron-rich genomes, particularly in vertebrates, introns are often long and bounded by relatively weak, suboptimal splice signals. Further analysis of the evolution of splice junctions suggests that the splice signals in intron-rich genomes still evolved under selection for splicing signal optimization, but this selection was too weak to offset the stochastic deviation from the consensus sequences—in perfect agreement with the population genetic theory (Irimia, et al., 2009).

As Chapter 7 pointed out, evolutionary reconstructions strongly suggest that LECA already had a high intron density, and most of the subsequent evolution of eukaryotic genomes involved primarily loss of introns that could be either moderate, as in most animal and plant lineages, or dramatic, as in most unicellular eukaryotes (Carmel, et al., 2007; Csuros, et al., 2011). The episodes of intron gain appear to have been few and far between and were associated with the emergence of major new groups of organisms such as animals. The implications of this pattern under the non-adaptive population-genetic theory of genome evolution are intriguing. In principle, at least, it appears possible to reconstruct the population dynamics throughout the history of all eukaryotic lineages from the extant and inferred ancestral intron densities. Although the available data is insufficient for a detailed reconstruction, examination of the numbers in Figure 7-8 already leads to interesting conclusions. Given that vertebrates have only slightly greater intron densities than the LECA, that vertebrates and plants share numerous intron positions, and that substantial intron regain in the same positions is extremely unlikely, there were apparently no intron-poor intermediates along the entire evolutionary path from LECA to vertebrates. In other words, *our lineage has never gone through a phase of large effective population size and the ensuing intense selection during the entire course of the evolution of eukaryotes*. To a somewhat lesser extent, this also pertains to the path from LECA to plants. Moreover, episodes of massive intron gain almost certainly were associated with population bottlenecks. This makes perfect sense from the general perspective on the major evolutionary transitions, such as the emergence of animals, because an event of this magnitude required a variety of embellishments, including extensive gene duplication and accumulation of novel regulatory elements, which would be possible only in a drift-dominated evolutionary regime.

Perhaps the most striking conclusion pertains to the pre-LECA (stem) phase of evolution and the genome architecture of the early ancestral eukaryotes that antedated the LECA. An estimate based on the assumption of an instantaneous invasion of Group II introns from the endosymbiont into the host genome (see Chapter 7) suggests a bottleneck so severe (with an Ne of about 1,000, if not less) that it would be hardly compatible with survival for purely stochastic reasons (Koonin, 2009b). Thus, one is compelled to postulate some degree of gradual spread of the invading introns in the host genome. Nevertheless, even this less disruptive invasion scenario implies a very long and very thin bottleneck on the path from the original host of the endosymbiont to LECA (see Figure 8-3). Such a bottleneck is likely to be the only possible passage to the emergence of the eukaryotic cell organization, considering the numerous associated duplications and other innovations.

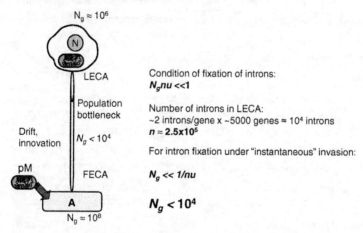

Figure 8-3   A reconstruction of the population dynamics during eukaryogenesis: an extreme bottleneck enabling eukaryogenesis. Ng = effective number of genes/loci; n = number of nucleotides required for intron (self-)splicing (about 25/intron), the target size for deleterious mutations; u = mutation rate/nucleotide/generation (~0.5x10^{-9}); A = archaeon, the presumed host of the proto-mitochondrial endosymbiont (pM); N = nucleus; FECA = First Eukaryotic Common Ancestor, the chimeric cell formed immediately upon endosymbiosis.

All these conclusions are rather straightforward implications of the non-adaptive population-genetic theory of genome evolution that, combined with the comparative genomic results, seems to give us an otherwise unimaginable window on the evolutionary past.

From junk to function: The importance of relaxed purifying selection for the evolution of complexity

What was the driving factor(s) behind the evolution of genomic (and perhaps the associated organismal) complexity? The non-adaptive population-genetic theory (Lynch, 2007c; Lynch and Conery, 2003) implies an astonishing answer: *The necessary and possibly sufficient condition for the emergence of complexity was the inefficient purifying selection in populations with small Ne.* The inefficient selection provided for the fixation of slightly deleterious features that, in larger populations, would have been eliminated, and for accumulation of junk, some of which was then recruited for diverse functions.

Let us rewrite the fixation condition from the preceding section as follows:

$$n \ll 1/N_e u$$

This simple inequality puts the limit on the size of the deleterious mutation target that remains invisible to purifying selection—or, in other words, the maximum number of nucleotides required for the function of a new genomic element that is likely to be fixed.

Estimates using $N_e u$ values from Box 8-1 reveal major differences between organisms: For example, in vertebrates, as many as 250 constrained nucleotides pass under the radar of purifying selection, whereas in prokaryotes, the fixation of any constrained sequences longer than about 10 nucleotides is unlikely.

These theoretical considerations imply that any substantial increase in genomic complexity is possible only in the regime of relaxed purifying selection. Consider three of the major staples of the genomic complexity in vertebrates that also provide for the complexity of the molecular phenome and, for all we know, tissue differentiation and other aspects of organismal complexity:

1. Alternative splicing that generates most of the protein diversity in these organisms

2. Combinatorial regulation of transcription, whereby genes are equipped with arrays of transcription factor–binding sites. Diverse combinations of transcription factors bind to these sites to provide elaborate regulation of expression (Venters and Pugh, 2009).

3. The huge noncoding RNome that includes the relatively well-understood microRNAs, a variety of other partially characterized small RNAs, the more mysterious long noncoding RNAs, and an enormous amount of the RNA "dark matter" (Amaral, et al., 2008).

Looking at each of these remarkable phenomena in some more detail, we discern unmistakable imprints of non-adaptive evolution under relaxed purifying selection in each of them.

As the preceding section pointed out, intron-rich genomes possess "weak" splicing signals, likely simply because the power of purifying selection in the respective populations is insufficient to exert a tight control on these nucleotide sequences. Put another way, occasional aberrant transcripts yielded by the relatively sloppy splicing in intron-rich organisms are not deleterious enough to be eliminated by purifying selection, given the low *Ne*. This tolerance of sloppy intron excision provides the niche for the evolution of alternative splicing. More precisely, *the sloppy splicing is alternative splicing*. Because the evolving small populations could not get rid of it, they "learned" how to utilize some of the alternative (initially aberrant) transcripts for a variety of functional roles. These roles often have to do with the fact that alternative proteins are modifications of the "normal" proteins and, therefore, may be apt to function as modified versions of the original protein or else as dominant negative regulators. Under the logic of evolution, *alternative splicing is analogous to HGT in prokaryotes, in that both are profitable alternatives to gene duplication, whereby a modified activity is gained in one step as opposed to an extended period of evolution.* Considering the reconstruction in Figure 7-8, one would conjecture that there was a lot of erratic splicing in LECA and, accordingly, a great diversity of transcripts, but little, if any, functional alternative splicing. The subsequent evolution of different branches of eukaryotes seems to have proceeded according to two opposite scenarios:

1. Elimination of the majority of introns and tightening of the splice junctions around the remaining introns so that no appreciable amounts of aberrant transcripts are produced

2. Retention of about the same level of erratic splicing as in LECA (given roughly the same intron density), accompanied by the evolution of functional alternative splicing—that is, recruitment of many, but certainly not all, and probably not the majority—of the aberrant transcripts for the production of functional alternative protein forms

Most of the unicellular eukaryote lineages that evolved in the direction of large Ne and efficient purifying selection took the first route; the second scenario applies to animals and plants that have never reached large effective population size and have had to cope with the ancestral erratic splicing. There seems to be no third way: Either develop the means to eliminate aberrant transcripts or utilize them, or die.

Transcription factor–binding sites in eukaryotes consist of about eight to ten nucleotides, so the cost of adding one site is $s \approx 10u$, or about 10^{-7}, taking the characteristic vertebrate u value (Lynch, 2007c). Thus, the genomes of complex multicellular eukaryotes seem to be virtually free to accumulate transcription factor–binding sites, allowing the emergence of complex site cassettes. Unicellular eukaryotes possess limited opportunities for evolution in this direction; for prokaryotes, this path to innovation seems to be blocked by purifying selection.

The noncoding RNome of vertebrates might be the utmost manifestation of genomic complexity. The protein-coding exons account for about 1.5% of the mammalian genome sequence, whereas exons corresponding to noncoding RNAs are estimated to occupy more than 4% of the genome—about 80% of the coding potential of the genome is dedicated to RNA molecules that are not translated into proteins (Eddy, 2002). This is a dramatic contrast to the coding repertoires of prokaryotes and even unicellular eukaryotes, in which noncoding RNAs constitute only a small fraction. Even more strikingly, a number of recent reports show that a large fraction—probably around 60%—of the mammalian genome is transcribed at a detectable rate (Lindberg and Lundeberg, 2010; Mendes Soares and Valcarcel,

2006). The nature of this "dark matter" is far from clear. Sometimes expression is taken to imply functional relevance of the transcribed regions of the genome. However, given the lack of any appreciable evolutionary conservation for most of these transcribed sequences and the relative ease of the emergence of spurious (weak) transcription initiation sites in random DNA sequences, it appears almost certain that most of the dark matter is transcriptional noise. However, this spuriously transcribed part of the genome and "junk" DNA in general comprise a vast reservoir for the generation of new microRNAs and other noncoding RNAs with regulatory and structural functions, many of which are poorly conserved during evolution and evolve with a high turnover rate. The discovery of the vast expanse of the animal RNome shows that the complex genomes of multicellular organisms and the simple genomes of unicellular life forms are qualitatively different. This difference is naturally interpretable within the framework of the non-adaptive population genetic theory of genome evolution. Under this theory, evolution of life forms with small Ne and the ensuing weak purifying selection results in the accumulation of large amounts of intronic and intergenic junk DNA, some segments of which are at times recruited for various functions. The scale of the transformation of the expression landscape of the genome that seems to be caused primarily by simple population genetic factors is striking and seems to be commensurate with the intuitively grasped difference in complexity (most obviously, in size) between a mammal and a protist. Recalling the discussion of sequence evolution in Chapters 3 and 4, the extensive set of nonfunctional transcripts constitutes the nearly neutral space that is open for the evolution of complexity in multicellular organisms. Such nearly neutral space inevitably emerges in the course of evolution of organisms with low Ne for purely entropic reasons.

Although the extent of recruitment is rather small compared to the total amount of the noncoding (junk) DNA, it is huge compared to the total size of the protein-coding sequences. Given the population bottleneck that most likely accompanied eukaryogenesis (see Figure 8-3), it seems likely that significant amounts of junk DNA evolved at a very early stage in the history of eukaryotes and might have already been present in the LECA—and so was the extensive spurious transcription. Afterward, one would envisage "symmetry breaking" mimicking the bifurcation that we described when discussing the history of introns: Lineages that evolved large Ne have tightened their genome by eliminating most of the junk DNA. In contrast, lineages that never evolved to have large Ne have "compensated" by gradually recruiting increasing parts of the (former) junk as functional RNAs (see Figure 8-4).

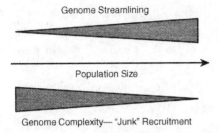

Genome Streamlining

Population Size

Genome Complexity— "Junk" Recruitment

Figure 8-4   The streamlining and "junk recruitment" routes of genome evolution.

Continuing along similar lines, the non-adaptive theory suggests a straightforward explanation for the switch from simple Jacob-Monod type regulation of transcription to the complex regulation strategy that eukaryotes employ. Instead of using just one binding site for a single regulator of an operon (or, in rare cases, a few sites), as is the case with prokaryotes, transcription of most eukaryotic genes is regulated in the so-called combinatorial mode, whereby multiple transcription factors interact with multiple, often numerous sites upstream of a gene (Ravasi, et al., 2010). In prokaryotes, transcription factor–binding sites contain enough information to ensure accurate recognition of a unique site in the relatively small genome sequence. By contrast, a single eukaryotic site typically carries too little information to provide for precise recognition (in other words, the genome contains many sites with an equal or even greater affinity to the cognate transcription factor; Wunderlich and Mirny, 2009). This inadequacy of single binding sites in eukaryotes is the result of weak purifying selection that cannot maintain many precisely conserved sites per genome (see the discussion of intron evolution earlier in this chapter) and also cannot resist genome expansion, thereby inflating the search space for transcription factors. Hence, the combinatorial model could be the only solution to the efficient regulation problem. The evolution of this mode of regulation is facilitated by genome expansion, particularly the relatively high rate of short tandem duplications. The evolution of the complex gene expression regulation that is a hallmark of eukaryotes and a necessary condition for the evolution of complex multicellular forms seems to be a showcase for the junk-to-function evolution that is underpinned by weak selection. As with other aspects of the evolution of complexity, selection here acts to prevent entropic collapse, not directly to "improve" regulation.

The evolution of advanced adaptations in small populations with weak selection might appear paradoxical, and perhaps for good reason: Evolving such complex features seems to require efficient positive selection that is possible only in populations with a large Ne. This is admittedly a difficult problem. The solution, however counterintuitive, seems to require "weak anthropic reasoning" (see Chapter 12 and Appendix B): Species in which these complex features have not been fixed, primarily via random drift and constructive neutral evolution (see the discussion later in this chapter), simply had no chance to survive.

Genome streamlining as the principal route of evolution and complexity as a genomic syndrome

We are intuitively inclined to assume that evolution proceeds from simple to complex forms. As Darwin wrote in the concluding chapter 14 of *Origin*, "[F]rom so simple a beginning endless forms most beautiful and most wonderful have been, and are being, evolved" (Darwin, 1859). Certainly, this intuitive notion makes sense (and presents a formidable problem) when it comes to the origin of the first life forms (we turn to this problem in Chapter 12). However, was gradually increasing complexity the prevailing trend throughout the history of most lineages over the entire course of the evolution of life? Both population genetic theory and comparative genomic reconstructions suggest that this might not be so. For a clear illustration, we once again turn to Figure 7-8. The emergence of the two branches of multicellular eukaryotes seems to have been accompanied by a moderate increase in intron density, which suggests a population bottleneck that is associated with a (sometimes dramatic) increase in overall genome entropy (the H value, from the first section of this chapter). The increase in entropy provides the neutral space that is required for the subsequent increase of the overall biological complexity (high C values). Recall that, in these cases, the biological information density drops (low D values): These lineages evolve in the "entropic regime." However, even among plants and animals there are major lineages such as insects, in which evolution involved *genome streamlining, or a decrease of the evolutionary entropy of the genomes*. This was associated with a less precipitous drop in the overall complexity and an increase in the biological information density. Turning to the majority of the branches in the eukaryote tree (see Figures 7-2 and 7-8) that include unicellular forms, we see the unequivocal pattern of genome streamlining: The genome entropy dramatically drops and the overall complexity also decreases, albeit less dramatically, whereas the information density sharply increases.

It is still too early to tell how general, in the overall context of the evolution of life, is the trend of genome streamlining that we derive from the reconstruction in Figure 7-8, because the taxonomic density of sequenced genomes from diverse branches of life is still insufficient. Nevertheless, the results of the limited reconstructions available suggest that the outlined picture could be general enough. For instance, the reconstruction of the common ancestor of the extant archaea suggests that the genome of the ancestral form was at least as complex (in terms of the overall complexity, C, because it is difficult to directly reconstruct entropy and, hence, information density) as the typical modern members of the group (Csuros and Miklos, 2009). Moreover, a clear trend emerges in the reconstruction results themselves: The estimated complexity of ancestral forms is revised upward with the increasing number of genomes used for the reconstruction and with the refinement of the employed maximum likelihood models. Qualitatively similar results have been obtained in the reconstructions of the gene set of the LECA (see Chapter 7): Even deliberately conservative approaches applied to a limited set of genomes suggest that the LECA was at least as complex as a typical extant unicellular eukaryote (Koonin, 2010a).

Given these indications from ancestral genome reconstructions and within the framework of the non-adaptive population genetic theory of genome evolution, it is tempting to propose a general model of the evolution of genome entropy and complexity. Under this model, evolution typically occurs in a punctuated manner, through stages of high entropy associated with population bottlenecks that subsequently evolve in one of the two distinct modes (see Figure 8-5):

1. Low entropy (high biological information density) states associated with high Ne, under the streamlining scenario
2. High entropy (low biological information density) states associated with low Ne, under the "recruitment scenario"

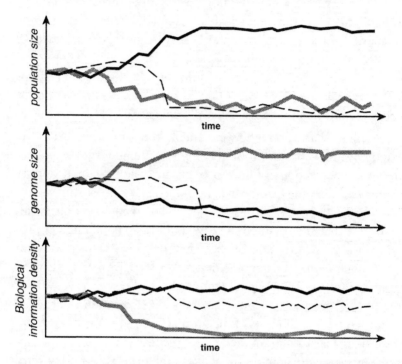

Figure 8-5   A general model of the dynamics of the effective population size, genome size, and biological information density under the non-adaptive theory. Each panel shows three routes of genome evolution: solid line = genome streamlining (free-living autotrophic bacteria and archaea, some unicellular eukaryotes); gray line = junk recruitment and genome complexification (eukaryotes, particularly multicellular forms); broken line = the ratchet of genome degradation (parasites and symbionts, particularly intracellular forms).

This pattern of evolution recurs throughout the history of life. The high-entropy bottlenecks correspond to the emergence of new major groups, whereas the subsequent radiation of the lineages within these groups typically involves "symmetry breaking" between these two scenarios. The correspondence between this model and the compressed cladogenesis model discussed in Chapter 6 should be obvious. Importantly, the episodes of sudden entropy increase are few and far between, whereas most of the history of life passed in the "normal evolution" regime between these episodes. During the phases of "normal evolution," genome streamlining that involves shrinking of the genome under the strong purifying selection in populations with a large Ne seems to be more common than the limited complexification seen in the groups of organisms that we traditionally view as complex, certainly including our own mammalian lineage.

Genome streamlining is a regime that is readily demonstrable in so-called *in vitro* Darwinian evolution experiments. Sol Spiegelman and colleagues performed probably the best-known series of such experiments in the 1960s (Mills, et al., 1973; Spiegelman, 1971). They placed a small amount of bacteriophage RNA in a test tube containing the replicase (the phage enzyme responsible for the genome replication), the nucleotide substrates, and the required ions, and allowed it to replicate for a short time. Part of the content then was transferred to another tube containing the same mix, and the procedure was repeated. Under these conditions, the only selective pressure on the phage RNA is to replicate as fast as possible, and the results of the evolution in this regimen were as drastic as it gets: After some 70 passages, the size of the RNA dropped from about 3,500 nucleotides to about 400 nucleotides, the minimal molecule that the polymerase could efficiently replicate.

Beyond the null hypothesis: Limitations of the population genetic perspective on genome evolution

After reading the preceding sections of this chapter, one is bound to question the validity of a grand explanation of the course of evolution by a single overarching factor. These misgivings are fully justified. Let me emphasize once again that the strongest claim of the population genetic theory of genome evolution is that non-adaptive, N_e-driven evolution could be an *appropriate null hypothesis*. Its fundamental importance notwithstanding, Ne determines the course of evolution only on a coarse-grained scale. The actual evolutionary trajectories are determined—and constrained—by specific biological contexts. For example, in an extensive survey of the selective constraints in the evolution of prokaryotes conducted by my colleagues and myself, we failed to detect a negative correlation between the strength of purifying selection and genome size that is predicted by the straightforward population genetic perspective (Novichkov, et al., 2009). On the contrary, larger genomes tend to evolve under stronger constraints than smaller genomes, even when only free-living microbes are analyzed. The implication is that the lifestyle of an organism could be a critical determinant of genome evolution that favors, in particular, gene acquisition via HGT in variable environments, more or less independent of the N_e.

Genomics provides many other indications of the limited power of the population genetic theory of genome evolution and specifically the genome streamlining concept. Streamlined genomes are expected to be found in organisms that are abundant (reach high N_e values) in more or less constant environments and, accordingly, should be subject to strong purifying selection. These genomes appear to be characterized not so much by their small size, given the insurmountable constraints associated with particular lifestyles (for example, autotrophic prokaryotes cannot shed genes beyond the lower limit of about 1,300 genes), as by their extreme compactness and (virtual) lack of pseudogenes and integrated selfish elements. All such elements are supposed to be rapidly wiped out by the intense purifying selection, which is so powerful that even short intergenic regions contract to the bare-minimum length required for regulatory functions. The most abundant known organism, the marine photosynthetic bacterium *Pelagibacter ubique,* seems to perfectly fit this prediction, having no detectable pseudogenes or mobile elements, very few paralogs, and extremely short intergenic regions. However, comparative genomics of the numerous strains of *Prochlorococcus,* another group of extremely abundant marine photoautotrophs that belong to Cyanobacteria, reveals features that do not seem to be compatible with streamlining—namely, genomic islands containing a variety of phage-related genes (Novichkov, et al., 2009).

More generally, the interactions between cellular life forms and selfish mobile elements substantially modify the genome structure compared to the predictions of the population genetic theory. The relationships between hosts and selfish elements (parasites) are often described as an "arms race" (more on this in Chapter 10). These interactions can be adequately described only by taking into account the distinct population dynamics of both the hosts and the parasites. The host-parasite conflict leads to an equilibrium that cannot be derived from the population dynamics of the host alone, so even some of the apparently most streamlined genomes harbor a substantial number of selfish elements.

Genome streamlining and genome shrinking are not the same. Bacterial parasites and intracellular symbionts, as well as the only known archaeal parasite, *Nanoarchaeum equitans*, have the smallest genomes among prokaryotes, but these are not streamlined genomes. Instead, these organisms appear to undergo neutral genome degradation. Indeed, although some of these genomes are extremely small, because in parasites and symbionts many genes become dispensable, they tend to contain considerable numbers of pseudogenes. In some cases, they also sustain propagation of selfish elements. Well-characterized examples are *Rickettsia, Wolbachia,* pathogenic *Mycobacteria,* and some lactobacilli (Frank, et al., 2002; Lawrence, et al., 2001). Parasites and symbionts do not typically reach large Ne values. Nevertheless, they gradually lose genes that become dispensable via a ratchet-type mechanism (a gene once lost is extremely unlikely to be regained, especially considering the lifestyles of these organisms) that is buttressed by a deletion bias in the mutation process (Mira, et al., 2001) and by the curtailment of HGT (see Chapter 5). Another key prediction of the population genetic theory does hold for these organisms: They typically have high Kn/Ks values, indicative of a weak purifying selection pressure. This is expected, given their small Ne values. It seems, therefore, that, for certain lifestyles, different predictions of the theory may be decoupled.

Darwin's eye, irreducible complexity, exaptation, and constructive neutral evolution

In the previous sections, we discussed different facets and driving forces of genomic complexity. Organismal complexity, on the other hand, was considered only as a consequence of the genomic trends. Traditionally, it was organismal complexity that bothered and fascinated biologists—both in our day and in Darwin's time, and even earlier. A detailed discussion of this problem is beyond the scope of this book, but some notes on the major concepts that have been developed to rationalize phenotypic complexity are due.

Darwin perceived the evolution of complex organs as a formidable problem, but he also believed that the problem should be solvable within the framework of his theory. As already touched upon in Chapter 2, the essence of the difficulty is the apparent irreducibility of complexity: "What good is half an eye?" In other words, how could a complex organ that consists of multiple parts evolve by natural selection if individual parts have no known functions?[4] Facing this difficulty, Darwin remained firmly convinced in the power of natural selection, as epitomized in the famous passage on the evolution of the eye:

> Reason tells me, that if numerous gradations from a simple and imperfect eye to one complex and perfect can be shown to exist, each grade being useful to its possessor, as is certainly the case; if further, the eye ever varies and the variations be inherited, as is likewise certainly the case and if such variations should be useful to any animal under changing conditions of life, then the difficulty of believing that a perfect and complex eye could be formed by natural selection, though insuperable by our imagination, should not be considered as subversive of the theory. (Darwin, 1859)

Darwin's narrative proposes one of the possible conceptual solutions to the problem of the evolution of organizational complexity. We might characterize Darwin's idea as "the nonobvious intermediate hypothesis": Although one cannot immediately perceive probable evolutionary intermediates from the structure and functions of an evolved complex structure, such intermediates actually existed; more often than not, at least some of their features can be inferred through comparative study (comparative anatomy in Darwin's time, comparative cytology and biochemistry in the twentieth century, and, additionally, comparative genomics in our day). The idea is definitely relevant and fruitful, and seems to apply particularly to the eye and other complex animal organs. However, the Darwinian explanation seems to be less obviously applicable to complex molecular structures, as we saw in Chapter 7 with regard to the elaborate supramolecular structures of eukaryotic cells.

The second major route to complex organization is exaptation, the simple but powerful concept of Stephen Jay Gould and Richard Lewontin (see Chapter 2): Molecules or complexes that evolved under selection for a particular function can be and apparently often are recruited (exapted) for a different although often mechanistically related function (Gould, 1997a). We have seen many cases of indisputable exaptation when discussing fundamental novelties that emerged during eukaryogenesis (see Chapter 7), for example the nuclear pore complex. Exaptation is often supplemented by fortuitous recombination of pre-existing molecules or devices, especially at the times during evolution when recombination is stimulated, as it almost certainly was at eukaryogenesis, with the flow of genetic material from the symbiont to the host. On rare occasions, random combinations of preexisting devices yield new functions that might resolve outstanding problems and are fixed by selection.

The third seminal idea that arguably complements the non-adaptive population genetic theory of genome evolution and might indicate the most general path to organizational complexity is the model of constructive neutral evolution (CNE), which Arlin Stoltzfus proposed in 1999 (Stoltzfus, 1999). The essence of CNE is the emergence of dependence between fortuitously interacting molecules that makes the interaction indispensable and thus leads to the evolution of organizational complexity. The CNE is a ratchet-like process, similar to many other evolutionary phenomena discussed in this book: *A dependence once evolved is effectively irreversible*. A perfect example of CNE seems to be the evolution of the spliceosome in eukaryotes (see Chapter 7). Under the CNE model, a fortuitous split of some Group II introns invading the host genome at an early stage of eukaryogenesis would yield the ancestral snRNAs (the active moiety of the spliceosome) and allow for the deterioration of the self-splicing terminal structures of all introns. At a parallel or subsequent stage, fortuitous interaction of RNA-binding proteins, particularly the archaeal Sm protein, with the intron RNA would allow the deterioration of the intron-encoded RT. Clearly, these changes that create dependence between components of the evolving spliceosome are irreversible, hence the ratchet and the fixation of the evolving complex organization. To quote the recent generalization of this concept by Michael Gray and coworkers (Gray, et al., 2010), the complexity emerging through CNE might be not so much irreducible as it is "irremediable."

A direct parallel to the CNE model is the subfunctionalization scenario of the evolution of gene duplications proposed by Lynch and colleagues (Lynch and Katju, 2004). Under this scenario, gene duplications can be fixed without direct adaptation because, after duplication, the newly emerged paralogs are free to differentially accumulate mutations that eliminate, in each of the paralogs, some of the multiple functions of the ancestral gene. Once that happens, both paralogs become indispensable—yet another ratchet mechanism of constructive neutral evolution. The observations of the mostly symmetrical relaxation of purifying selection in paralogs immediately after the duplication are compatible with the subfunctionalization model (Kondrashov, et al., 2002).

Synopsis and perspective: The non-adaptive evolutionary paradigm and reappraisal of the concept of evolutionary success

The emergence and evolution of complexity at the levels of the genotype and the phenotype, and the relationship between the two, is a central (if not *the* central) problem in biology. Even leaving aside for now the problem of the actual origin of the very substantial complexity associated with the cellular level of organization (see Chapter 11), one cannot help wondering why the evolution of life didn't stop at the stage of the simplest autotrophic prokaryotes, with 1,000 to 1,500 genes. Why instead did evolution continue, to produce complex prokaryotes possessing more than 10,000 genes and, far more strikingly, eukaryotes, with their huge, elaborately regulated genomes; multiple tissue types; and even ability to develop mathematical theories of evolution?

The traditional thinking on these problems explicitly or implicitly focused on complexity as a sublime manifestation of adaptation and the power of natural selection. Accordingly, the more complex organisms traditionally are considered more advanced, more successful, and, in a sense, more important than simpler creatures. Gould notably proposed a very different, stochastic perspective on the evolution of complexity that he described using the metaphor of a drunkard's walk outside a bar[5]: Even if a person moves completely randomly after having a few too many drinks, given enough time, he or she will eventually end up quite far from the bar door for example, in the ditch across the road (Gould, 1997b). Ditto for the evolution of complexity: Given enough time, evolution starting "from so simple a beginning" should be expected to reach high complexity by purely stochastic processes. This view of complexity is entirely reasonable but is far too general to pass as a satisfactory theory.

As soon as comparison of the genomes of simple (prokaryotes) and complex (animals and plants) life forms became possible, researchers realized that there was something strange about these genomes, something hardly compatible with the idea of steadily increasing genome complexity in parallel with the growing organismal complexity. Indeed, the genomes of multicellular eukaryotes might be more complex than those of prokaryotes and even unicellular eukaryotes, but these complex genomes also appear awfully disordered and full of mobile elements and other junk; they represent high-entropy states, as emphasized by the estimates in this chapter. Conceptual thinking on this paradox of comparative genomics led to the theory of non-adaptive genome evolution, which is mostly coached in standard formulas of population genetics. However, the simple apparatus notwithstanding, this theory turned the existing ideas on the nature of genome evolution upside down. Under the non-adaptive theory, the evolution of genome complexity is not an adaptation, per se, but rather a consequence of the initial increase in entropy caused by weak purifying selection and the conversely increased power of drift, which are characteristic of population bottlenecks. Paradoxical as this might be, the increase in genome entropy that is the necessary condition for the subsequent complexification can be legitimately viewed as a "genomic syndrome," the inability of organisms with small effective population size to cope with the spread of selfish elements and other entropy-increasing processes. Of course, evolution of complexity is a complex process itself, and the evolution of recruited sequences involves many apparent adaptations. However, the original entropic push is a maladaptation that the population initially is not equipped to overcome. In part, the subsequent functional adaptation of the originally neutral sequences offsets the burden of the increased genomic entropy—in other words, it allows organisms to survive the expansion of their own genomes.

Reconstructions of the history of genomes and cells, viewed within the framework of the non-adaptive genome evolution paradigm, led to a rather shocking realization. It turns out that much— probably most—of the history of life is not a history of "progressive" evolution toward increasing complexity.[6] Instead, numerous evolving lineages followed the path of *genome streamlining*, in which the genome entropy and overall biological complexity of the genome drop, often substantially, whereas biological information density increases. Some other lineages, such as our own, followed the route of *junk recruitment* (for the regulatory and structural roles, as is the case with the animal RNome), which led to a marked increase of the overall complexity but only a slight decrease in entropy. Thus, in these lineages, the biological information density shows only a modest increase, compared to the high-entropy state associated with bottlenecks during transitional epochs.

The models of constructive neutral evolution and subfunctionalization of paralogs complement the non-adaptive theory of genome evolution by providing compelling scenarios of non-adaptive evolution of complexity at the level of molecular phenotypes. From a broader perspective, these theoretical developments that are compatible with the empirical data of comparative genomics complete the overhaul of evolutionary biology started by the neutral theory of molecular evolution. The neutral theory showed that the majority of mutations that are fixed during evolution are effectively neutral, thus establishing neutrality as the appropriate *null hypothesis* for all molecular evolutionary studies. The new developments do the same for genome and molecular phenome evolution. Clearly, the null hypothesis is not expected to be a complete description of any process, let alone such a complicated, multifaceted process as the evolution of life. As we have seen, on many occasions, some of the predictions of the non-adaptive theory fail due to additional, overriding constraints stemming from the specific features of organisms' lifestyle. And of course, positive selection and the adaptations it causes are crucial aspects of evolution. However, it appears that these factors are manifest locally against the global background of more fundamental processes, such as the pressure of purifying selection that is determined by the effective population size and evolutionary ratchets that may lead to non-adaptive emergence of complexity.

To finish this chapter, a few words on the notions of evolutionary success and "progress" are due. The idea of "progress" might be considered thoroughly discredited in the narrow, anthropomorphic sense of the word, but increasing complexity is still commonly perceived as a feature of "advanced" life forms and a major evolutionary trend. The opposite perspective that probably was most eloquently presented by Gould in several of his books associates evolutionary success of a group exclusively with its abundance in the biosphere and ability to thrive in diverse niches. Under the non-adaptive theory, it is natural to link "success" to a large effective population size. From that perspective, evolution of complexity has nothing to do with the success of a group and is instead instigated by a failure (population bottleneck) at some stage and continued inability to evolve large populations subject to efficient selection. The truly successful and efficient are simple and streamlined.

Recommended further reading

Adami, C. (2002) "What Is Complexity?" *Bioessays* 24: 1,085–1,094.

A conceptual article that quantitatively defines "physical complexity" of genomes (related to the evolutionary complexity discussed in this chapter).

Gray, M. W., J. Lukes, J. M. Archibald, P. J. Keeling, and W. F. Doolittle. (2010) "Cell Biology. Irremediable Complexity?" *Science* 330: 920–921.

An updated, nontechnical discussion of the constructive neutral evolution hypothesis. Worth quoting for the crystal-clear summary: "Many of the cell's macromolecular machines appear gratuitously complex, comprising more components than their basic functions seem to demand. How can we make sense of this complexity in the light of evolution? One possibility is a neutral ratchet-like process described more than a decade ago, subsequently called constructive neutral evolution. This model provides an explanatory counterpoint to the selectionist or adaptationist views that pervade molecular biology."

Koonin, E. V. (2004) "A Non-adaptationist Perspective on Evolution of Genomic Complexity or the Continued Dethroning of Man." *Cell Cycle* 3: 280–285.

A discussion of the meaning of evolutionary complexity, in light of Lynch's nonadaptive theory and the general implications of the new ideas on the evolution of complexity.

Lynch, M. (2007) "The Frailty of Adaptive Hypotheses for the Origins of Organismal Complexity." *Proceedings of the National Academy of Sciences USA*. 104 Supplement 1: 8,597–8,604.

A concise discussion of a new, pluralistic perspective on organismal evolution, according to which natural selection is but one of the important factors of evolution. To quote: "[T]he origins of many aspects of biological diversity, from gene-structural embellishments to novelties at the phenotypic level, have roots in non-adaptive processes, with the population-genetic environment imposing strong directionality on the paths that are open to evolutionary exploitation."

Lynch, M. (2007) "The Origins of Genome Architecture." Sunderland, MA: Sinauer Associates.

A seminal book on the nonadaptive theory of genomic complexity evolution and its various implications (for a detailed discussion, see Chapter 8).

Lynch, M. (2006) "Streamlining and Simplification of Microbial Genome Architecture." *Annual Review of Microbiology* 60: 327–349.

Application of the non-adaptive theory of complexity evolution to the evolution of prokaryote genomes, with the emphasis on the role of genome streamlining under strong purifying selection in populations with a large N_e.

Lynch, M., and J. S. Conery. (2003) "The Origins of Genome Complexity." *Science* 302: 1,401–1,404.

Key article that presents the non-adaptive theory of genome complexity evolution and describes simple population genetic estimates underpinning the theory.

Novichkov, P.S., Y. I. Wolf, I. Dubchak, and E. V. Koonin. (2009) "Trends in Prokaryotic Evolution Revealed by Comparison of Closely Related Bacterial and Archaeal Genomes." *Journal of Bacteriology* 191: 65–73.

A comparative study of the selection regimes in different evolutionary lineages of prokaryotes, demonstrating the limitations of the straightforward population genetic approach.

Stoltzfus, A. (1999) "On the Possibility of Constructive Neutral Evolution." *Journal of Molecular Evolution* 49: 169–181.

Important conceptual article that introduces the possibility of non-adaptive evolution of complexity and the ratchet of "constructive neutral evolution."

9

The Darwinian, Lamarckian, and Wrightean modalities of evolution, robustness, evolvability, and the creative role of noise in evolution

The drama of Lamarckism

As the preface to this book pointed out, one of Darwin's key achievements was demonstrating the essential interaction between chance and necessity in the evolution of life. According to Darwin, most of the heritable variation is random, and the directionality of evolution is brought about entirely by natural selection that governs the fixation or elimination of the random mutations (Darwin, 1859). As we have repeatedly discussed, randomness also substantially contributes to the fixation stage through the drift and draft routes, which are critically dependent on population dynamics (see Chapter 8). However, Darwin allowed a significant, albeit subsidiary role for a fundamentally different type of variation, the so-called Lamarckian inheritance.

Lamarckian inheritance refers to nonrandomly acquired phenotypic changes, particularly those that are directly affected by the use of organs and are accordingly assumed to be adaptive (beneficial for the organism). The controversial French naturalist Jean-Bapteste Lamarck believed that directed changes are inheritable and constitute the basis of evolution. Lamarck was the author of the first coherent theory of the evolution of life, which he presented in his *Philosophie Zoologique;* "inheritance of acquired (adaptive) characters" played a key role in this theory (Lamarck, 1809). As repeatedly emphasized here and in contrast to Lamarck, Darwin assigned a greater importance to random, undirected change that, in his theory, provided the bulk material for natural selection. However, in the later editions of *The Origin of Species*, Darwin assigned increasingly greater weight to the Lamarckian mechanism of evolution, apparently out of concern that random variation and natural selection might be insufficiently powerful to fuel the evolutionary process in its entirety (Darwin, 1872).

The "inheritance of acquired (adaptive) characters" remains a fundamental problem whose relevance goes far beyond the dramatic and intriguing history of biology in the nineteenth and twentieth centuries. The interaction between random and directed genome change (if the latter exists at all) is central to the main theme of this book. Here we go to the very heart of the conundrum of chance and necessity by considering the element of biologically relevant nonrandomness that might exist even at the first stage of the evolutionary process, when variation is generated. More specifically, the key question about the Lamarckian mechanism of inheritance and evolution is this: Can environmental factors cause adaptive evolution of the genome directly, without recourse to the circuitous path of natural selection?

Lamarckian inheritance was supposedly discredited beyond repair by the notorious experiments of Weismann with tail-less rats.[1] The demise of this type of inheritance was further cemented by the bizarre and tragic episode of Paul Kammerer's supposedly fraudulent experiments with midwife toad coloration, which led to his suicide.[2] In the twentieth century, "Lamarckism" gained an extremely bad reputation when the Lysenkoist pseudoscience in the Soviet Union appropriated it.[3] Lately, however, several lines of research seem to converge to indicate that mechanisms that in various degrees meet the criteria of Lamarckian inheritance could be important contributors to the evolution process (Koonin and Wolf, 2009b).

The classical Lamarckian scheme involves inheritance of specific, adaptive phenotypic characters that an individual acquires during its life span. In this narrow sense, the Lamarckian scenario does appear untenable because of the apparent nonexistence of mechanisms for direct reverse engineering of acquired characters into the genome. However, this irreversibility of the genetic information flow that Francis Crick formulated and that became known as the Central

Dogma of molecular biology (Crick, 1970) strictly applies only to the flow of information between nucleic acids and proteins. The irreversibility emerges at the stage of matching amino acids to the cognate tRNAs that determines the incorporation of amino acids into nascent proteins in response to the cognate codons in the respective mRNAs. There is no path back to the genome from any changes that might occur in a protein sequence. However, the case of nucleic acids, particularly RNA, is different: RNA biogenesis includes no irreversible step that would render impossible the transfer of information back to the genome. This distinction is important to keep in mind when we address different classes of genomic changes, some of which can be triggered by environmental factors.

In this chapter, I discuss the putative Lamarckian and quasi-Lamarckian mechanisms of evolution, along with other important phenomena, such as evolution of evolvability, fidelity of biological information transmission, and the role of noise in evolution. These are related to the same key question: Is there any evolutionary logic to the mutational processes that generate genome variation, or are these processes governed by chance alone?

The Lamarckian, Darwinian, and Wrightean modes of evolution and the criteria for identifying Lamarckian inheritance

Before turning to the wide range of phenomena that seem to display all or some features of the mechanism of evolution that is associated with the name of Lamarck, it is necessary to define the Lamarckian paradigm and the criteria that an evolutionary process must satisfy to be considered Lamarckian. In doing so, I do not dwell on the differences between Lamarck's original views and the numerous subsequent (mis)representations; instead, I try to distill the essence of what is commonly known as inheritance of acquired characters and Lamarckian evolution.

Lamarck's concept of heredity, which is one of the two cornerstones of his evolutionary synthesis (Gould, 2002), stands on two principles that Lamarck promoted to the status of fundamental laws in *Philosophie Zoologique* and other works:

1. Use and disuse of organs
2. Inheritance of acquired characters

Lamarck directly linked the "use and disuse" clause to effects of the environment on the "habits" of an organism and, through those habits, on the "shape and nature" of body parts. Of course, he considered these environment-effected adaptive changes to be heritable. Wrote Lamarck, "[N]ature shows us in innumerable...instances the power of environment over habit and of habit over the shape, arrangement and proportions of the parts of animals." So Lamarck's idea of heredity is based on the threefold causal chain: environment–habit–form. Lamarck insisted on the essentiality of change in habits as an intermediary between the environment and (inheritable) change of organismal form:

> Whatever the environment may do, it does not work any direct modification whatever in the shape and organization of animals. But great alterations in the environment of animals lead to great alterations in their needs, and these alterations in their needs necessarily lead to others in their activities. Now if the new needs become permanent, the animals then adopt new habits that last as long as the needs that evoked them. (Lamarck, 1809)

Lamarck was by no means alone in his belief in the inheritance of acquired characters: It appeared to be the folk wisdom of his day. However, he was more specific than others in spelling out the causal chain of heredity and, more importantly, he made this scheme the foundation of his far more original concept of evolution.

The second foundation of Lamarck's evolutionary synthesis was his belief in the innate tendency of evolving organisms toward increasing organizational complexity—or, simply, progress—which, in Lamarck's view, shaped biological evolution along with heredity as he understood it. Although Lamarck often used the phrase *pouvoir de la vie* to denote this fundamental tendency, his idea was completely materialistic, even mechanistic, as he attributed the trend toward progress to the motion of fluids in the animal body. Those fluids, he thought, would carve channels and cavities in soft tissues, and gradually lead to the evolution of increasing organizational complexity. For a good measure, to explain why simply organized life forms persisted despite the purported progressive course of evolution, Lamarck maintained that spontaneous generation was a constant source of primitive organisms. The ideas of spontaneous generation and the innate tendency toward progress are hopelessly obsolete. However, it remains a widespread belief that evolution leads to increasing complexity through numerous successive adaptations. (Of course, nowadays, scientists who continue to advocate the existence of such a trend would not describe it as an "innate tendency.") As discussed in Chapter 8, this is plainly not the case: There is no overall trend toward an increasing complexity of life forms over the course of evolution, even if the maximum observable complexity increases for stochastic reasons. In this chapter, we address the more relevant and interesting problem of Lamarckian package, the inheritance of acquired characters and its contribution to the evolutionary process.

In terms compatible with modern genetics, Lamarck's scheme (see Figure 9-1) posits the following:

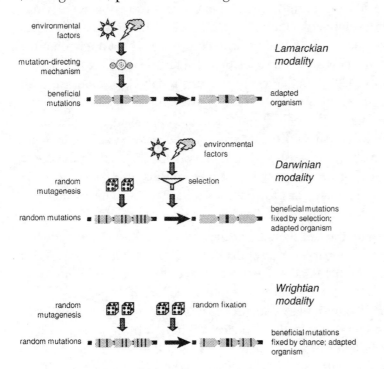

Figure 9-1   The Darwinian, Lamarckian, and Wrightian modalities of evolution. Adapted from Koonin and Wolf, 2009b.

1. Environmental factors cause genomic (heritable) changes.
2. The induced changes (mutations) are targeted to a specific gene (or genes).
3. The induced changes provide adaptation to the original causative factor.

Obviously, the adaptive reaction to a specific environmental factor has to be mediated by a molecular mechanism that channels the genomic change toward the relevant gene(s). The distinction from the Darwinian route of evolution is straightforward: In the latter, the environment is not the agency causing adaptive changes to occur, but rather the source of extraneous selective pressure that promotes the fixation of those random changes that are adaptive under the given conditions (see Figure 9-1). The Darwinian scheme is simpler and less demanding than the Lamarckian scheme because Darwin required no specialized mechanisms to direct mutations to the relevant genomic loci and restrict the changes to the specific mutations that provide the requisite adaptation. Conversely, the Lamarckian mode of evolution would be more efficient and faster than the Darwinian mode. Indeed, there is a steep price to be paid for using the Darwinian mechanism instead of the Lamarckian one: Numerous mutations that emerge in genomes are strongly deleterious, so their carriers are eliminated; others are nearly neutral and are sometimes fixed through drift or draft but make no immediate contribution to adaptive evolution. So the Lamarckian scheme would be really helpful for evolving organisms—if only it was feasible. The difficulty of discovering or even conceiving of mechanisms of directed adaptive change in genomes has for decades relegated the Lamarckian scenario of evolution to the garbage pile of science history.

Despite the substantial mechanistic differences and notwithstanding the apparent "wastefulness" of the Darwinian modality that contrasts the potential efficiency of Lamarckian evolution (however, see the discussion later in this chapter), the Darwinian and Lamarckian schemes are similar: Both are essentially adaptive in the final outcome and, in that regard, are radically different from random drift or draft. The latter processes may be denoted the "Wrightian modality of evolution," after Sewall Wright, one of the founding fathers of population genetics and the originator of the key concept of random genetic drift (see Figure 9-1 and Chapter 2). In the next sections, I discuss the recent studies on several phenomena that seem to call for a resurrection of a version of the Lamarckian scenario as an important contribution to the genome and organism evolution.

Lamarckian and quasi-Lamarckian phenomena in evolution

The CRISPR-Cas system of antivirus immunity in prokaryotes: The showcase of a bona fide Lamarckian mechanism

A system of antivirus defense/adaptive immunity in archaea and bacteria that has been recently characterized in a series of sometimes serendipitous discoveries seems to function via a straightforward Lamarckian mechanism. This system is known as CRISPR-Cas (or simply CRISPR, for brevity); CRISPR stands for Clustered Regularly Interspaced Short Palindromic Repeats, and Cas stands for CRISPR-associated proteins (products of *cas* genes) (Deveau, et al., 2010; Karginov and Hannon, 2010; Koonin and Makarova, 2009; van der Oost, et al., 2009). The CRISPR repeats are interspersed, in the sense that they contain short unique spacers embedded within palindromic repeat units. Archaeal and bacterial genomes contain cassettes of multiple CRIPSR units—in many cases, more than one cassette per genome. Although the CRISPR repeats were discovered as early as the 1980s, years before the first complete bacterial genome was sequenced, only much later was it realized that CRISPR cassettes are almost always adjacent in genomes to an array of *cas* genes. The *cas* genes are predicted to encode a variety of enzymes involved in nucleic acid metabolism, including several nucleases, a helicase, and a polymerase.[4] Serendipitously, some of the unique spacers in the CRISPR cassettes have been shown to be identical to fragments of bacteriophage and plasmid genes, so the hypothesis was proposed that the CRISPR system utilized the phage-derived sequences as guide molecules to destroy phage mRNAs analogously to the eukaryotic RNA interference (RNAi) (Makarova, et al., 2006). Although most of the mechanistic details remain to be elucidated, the principal predictions of this hypothesis have been validated: The presence of a spacer sequence precisely complementary to a region of a phage genome is essential for resistance, the guide RNAs containing the

CRISPR spacers form complexes with multiple Cas proteins and are employed to abrogate the infection, and new spacers conferring resistance to cognate phages can be acquired. Quite interestingly, it appears that some CRISPR systems target viral mRNA as postulated by the original hypothesis, whereas others destroy viral DNA itself (Barrangou, et al., 2007; Brouns, et al., 2008; Hale, et al., 2009; Marraffini and Sontheimer, 2008).

The mechanism of heredity and genome evolution embodied in the CRISPR-Cas system seems to be bona fide Lamarckian (see Figure 9-2):

- An environmental cue (a selfish genetic element, such as a virus) is employed to directly modify the genome.
- The resulting modification (unique, element-specific spacer) directly affects the same cue that caused the modification.
- The modification is clearly adaptive and is inherited by the progeny of the cell that encountered the selfish element.

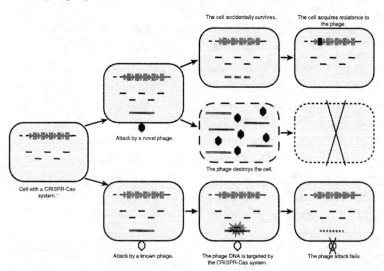

Figure 9-2   The CRISPR-Cas system and its mechanism of action: the showcase of Lamarckian evolution. Adapted from Koonin and Wolf, 2009b.

The CRISPR-mediated heredity appears to be short-lived: Even closely related bacterial and archaeal genomes do not carry the same inserts. The implication is that, as soon as a bacterium or archaeon ceases to encounter a particular agent (virus), the cognate spacer rapidly deteriorates. Indeed, the inserts hardly could be evolutionarily stable in the absence of strong selective pressure because a single mutation renders them useless. Moreover, much like the animal adaptive immune system, the CRISPR system on rare occasions seems to display autoimmunity: Spacers identical to fragments of regular host genes are inserted into the CRISPR cassettes and presumably impair the expression of the cognate genes (Stern, et al., 2010). Despite the ephemeral nature of the CRISPR heredity, its Lamarckian character is undeniable: Adaptive evolution of organisms occurs directly in response to an environmental factor and the result is specific adaptation (resistance) to that particular factor.

Other (quasi) Lamarckian systems functioning on the CRISPR principle

It is interesting and instructive to compare the hereditary and evolutionary features of the CRISPR system with those of eukaryotic RNA interference (RNAi) and, more specifically, small interfering (si)RNA and PIWI-interacting (pi)RNA, the defense systems of eukaryotes that are generally functionally analogous to CRISPR. To begin, let us recall a remarkable and rather enigmatic fact: The protein machinery of eukaryotic RNAi is unrelated to the Cas proteins; instead, the protein components of this quintessential eukaryotic machinery have been assembled from prokaryotic domains that originally were involved in other functions (see Chapter 7; Shabalina and Koonin, 2008). The apparent absence of orthologs for any of the Cas proteins in eukaryotes seems to suggest that this system is somehow excluded by selection from the eukaryotic domain of life, although the nature of the underlying selective pressure is obscure. The only clue might be the generic cause of the deterioration of operons that we discussed in Chapter 7: The operons disappear via a recombinational ratchet, and genes that require a particularly tight coordination of expression or are deleterious out of the operonic context are eliminated by purifying selection.

Unlike CRISPR-Cas, the RNAi systems do not function via a straightforward Lamarckian mechanism. Nevertheless, they display clear "Lamarckian-like" features. The siRNA system (a distinct branch of RNAi) "learns" from an external agent (a virus) by generating siRNAs complementary to viral genes (Kim, et al., 2009), a process that certainly resembles the CRISPR mechanism but also is reminiscent, at least metaphorically, of the Lamarckian "change of habits." Moreover, the system has a degree of memory because, in many organisms, siRNAs are amplified, and the resistance to the cognate virus can persist for several generations (Ding, 2010). Such persistence of siRNA is one of the manifestations of the increasingly recognized RNA-mediated inheritance, sometimes called paramutation (Hollick, 2010). The key difference from CRISPR is that (as far as currently known) siRNAs are not incorporated into the genome, so *Lamarckian-type epigenetic inheritance* but not bona fide genetic inheritance seems to be involved.

However, even this distinction is blurred in the case of transposon-derived piRNAs, the most abundant small RNAs in animals that form rapidly expanding genomic clusters that provide a defense against transposable elements in the germ line (Bourc'his and Voinnet, 2010). In the case of piRNA, as with the CRISPR, fragments of mobile element genomes are integrated into the host genome, where they rapidly proliferate, apparently under the pressure of selection for efficient defense (Assis and Kondrashov, 2009). This system seems to meet all the criteria for the inheritance of acquired characters and the Lamarckian mode of evolution. It is particularly remarkable that the sequestered germline, a crucial innovation of multicellular eukaryotes that seems to hamper some forms of (quasi) Lamarckian inheritance, such as those associated with HGT (see the discussion later in this chapter), itself evolved a specific version of a Lamarckian-like mechanism.

A series of notable recent findings in both plants and animals indicates that eukaryotes employ reverse transcription to integrate DNA copies of RNA virus genomes into the host chromosomes and might use these integrated sequences to produce siRNAs or proteins that confer immunity to the cognate viruses (Feschotte, 2010; Horie, et al., 2010; Koonin, 2010c). These mechanisms remain to be corroborated by more extensive research, but at face value, they seem to be analogous to the CRISPR and, hence, are Lamarckian.

Horizontal gene transfer: A major Lamarckian component

One of the key novelties brought about by comparative genomics is the demonstration of the ubiquity and high frequency of HGT among

prokaryotes, and a considerable level of HGT in unicellular eukaryotes as well (see Chapters 5 and 7). Prokaryotes readily obtain DNA
from the environment, with phages and plasmids serving as vectors,
or without any vectors, through the transformation mechanism,
which is mediated by membrane pumps specialized in DNA inges-
tion. The absorbed DNA often integrates into prokaryotic chromosomes and can be fixed in a population if the transferred genetic
material confers even a slight selective advantage onto the recipient,
or even neutrally. The HGT phenomenon shows obvious Lamarckian
features: DNA is acquired from the environment, and naturally the
likelihood of acquiring a gene that is abundant in the given habitat is
much greater than the likelihood of obtaining a rare gene. The second component of the Lamarckian scheme, the direct adaptive value
of the acquired character, is not manifest in all fixed HGT events but
is relevant and common enough.

Perhaps the most straightforward and familiar case in point is
the evolution of antibiotic resistance (Martinez, 2008; Wright,
2007). When a sensitive bacterium enters an environment where an
antibiotic is present, the only chance for the newcomer to survive is
to acquire a resistance gene(s) by HGT, typically via a plasmid. This
common (and extremely practically important) phenomenon
appears to be a clear-cut case of Lamarckian inheritance. Indeed, a
trait—in this case, the activity of the transferred gene that mediates
antibiotic resistance—is acquired under a direct influence of the
environment and is obviously advantageous—often essential, in this
particular habitat. A similar pattern exists among photosynthetic
genes in the ocean: The genes for bacteriorhodopsin, the protein
central to light-driven bioenergetics (proton-motive force) in
halophilic archaea and also in numerous bacteria, as well as genes
for photosystems I and II involved in chlorophyll-dependent photosynthesis, seem to spread via HGT at a high rate, often through
bacteriophage vehicles (Alperovitch-Lavy, et al., 2011; Falkowski,
et al., 2008; Sullivan, et al., 2006). These genes confer a major
selective advantage on to the recipient organisms and so are fixed at
a high frequency.

More generally, any instance of HGT in which the acquired gene
provides an advantage to the recipient, in terms of reproduction in
the given environment (that is specifically conducive to the transfer of

the gene in question), seems to meet the Lamarckian criteria. Comparative-genomic studies indicate that HGT is the principal mode of bacterial adaptation to the environment through the extension of metabolic and signaling networks that integrate new, horizontally acquired genes and, hence, incorporate new capabilities within pre-existing frameworks (Maslov, et al., 2009). Quantitatively, in prokaryotes, HGT, with its Lamarckian component, appears to be a far more important route of adaptation than gene duplication (Pal, et al., 2005).

A provocative indication that HGT might be an adaptive phenomenon is the already mentioned discovery of the GTAs. As pointed out in Chapter 5, GTAs are derivatives of defective bacteriophages that package apparently random sets of bacterial genes and transfer them within bacterial and archaeal populations. Striking observations on gene transfer in marine bacterial communities indicate that GTAs are quite promiscuous with respect to the bacteria they infect and seem to provide for very high rates of HGT (McDaniel, et al., 2010). The properties of GTAs remain to be investigated in detail, but it is a distinct possibility that these agents are dedicated vehicles of HGT that evolved under the selective pressure to enhance gene transfer. If that is the case, one would have to conclude that HGT itself is partly an adaptive phenomenon (see also the discussion of the HGT optimization hypothesis in Chapter 5).

All in all, there seems to be no escape from the conclusion that some of the most important routes of genome evolution—at least, in prokaryotes—are (quasi) Lamarckian.

Stress-induced mutagenesis and activation of mobile elements: Quasi-Lamarckian phenomena

Darwin emphasized the evolutionary importance of random, undirected variation, whereas the Lamarckian modality of evolution centers on directed variation that is specifically caused by environmental factors. The real evolution defies such oppositions. A crucial case in point is the complex of diverse phenomena that collectively can be denoted stress-induced mutagenesis, one major facet of which is activation of mobile elements. A phenomenon of this class was first described by Barbara McClintock, who demonstrated (in a series of classic experiments that eventually won the Nobel Prize) the activation of "gene jumping" in plants under stress and the importance of this stress-induced mobility of distinct "controlling elements" for the emergence of resistance phenotypes (McClintock, 1984).

The later, also famous and controversial experiment of John Cairns and coworkers on the reversion of mutations in the *lac* operon induced by lactose brought the Lamarckian mechanism of evolution to the fore in a dramatic fashion (Brisson, 2003; Cairns, et al., 1988; Rosenberg, 2001). Cairns and colleagues discovered a strong enhancement of frameshift mutation reversion in the *lac* operon in the presence of lactose and boldly speculated that the classical Lamarckian mechanism of evolution was responsible for the observed effect— that is, that lactose directly and specifically caused mutations in the *lac* operon. Subsequent, more thorough investigations, including the work of Patricia Foster and Cairns himself, showed that this was not the case: Stress such as starvation did induce mutations, but not in specific loci (Foster, 2000). The mutations underlying the reversion of the *lac-* phenotype and other similar phenotypes have been shown to be strictly stress induced (*lac-* cells plated on a medium with lactose as the only carbon source experience starvation stress) rather than emerging from the pool of pre-existing rare, spontaneous mutations.

Stress-induced mutagenesis—specifically, the mutagenic repair pathway in *Escherichia coli*, known as SOS repair—was discovered long before the experiments of Cairns. Moreover, Miroslav Radman (Radman, 1975) and Harrison Echols (Echols, 1981) have independently proposed the seminal idea that this mutagenic form of repair could be an adaptive antistress response mechanism rather than a simple malfunctioning of the repair systems. The two decades of subsequent research validated this striking conjecture beyond reasonable doubt. Several lines of compelling evidence support the adaptive character of error-prone DNA repair (Foster, 2007; Galhardo, et al., 2007; Rosenberg, 2001). The activity of the SOS pathway and the other mutagenic repair mechanisms in bacteria is elaborately regulated, in particular, through the switch from high-fidelity to error-prone double-strand break repair affected by the dedicated RNA polymerase sigma-factor, RpoS, apparently to reach the optimal mutation rate. Most importantly, stress-induced mutations produced by error-prone repair processes, although not targeted to specific genes, are not randomly scattered in the genome, either. On the contrary, these mutations are clustered around double-stranded DNA breaks that are caused by various stress factors and attract the error-prone repair machinery. This "sloppiness" of the repair machinery might have evolved as a distinct adaptive mechanism that allows coordinated evolution of clustered, functionally linked genes (a central feature of genome architecture in prokaryotes) in rare cells where beneficial mutations emerge, while limiting the damage to other parts of the genome. Stress-induced mutagenesis, particularly retrotransposon mobilization, has been demonstrated also in yeast and in animals, suggesting that this route of adaptive evolution is universal to cellular life forms.

At least in bacteria, stress-induced mutagenesis is not rare or exotic, but is an extremely widespread process. Among hundreds of investigated natural isolates of *E. coli*, more than 80% showed induced mutagenesis in aged colonies, and the excess of stress-induced mutations over constitutive mutations varied by several orders of magnitude (Bjedov, et al., 2003).

Strikingly, stress-induced and apparently adaptive genome instability is also central to the progression of cancer. It is well known that tumors develop (evolve) under conditions of perpetual hypoxic stress, which induces extensive genome rearrangement and mutation. These stress-induced changes are the basis for the survival of mutants that are capable of uncontrolled growth in spite of the stress. Despite the differences in the specific mechanisms of mutagenic repair and its regulation, animal (including human) malignant tumors are, in principle, not so different from bacterial populations evolving under stress.

Adaptive evolution resulting from stress-induced mutagenesis is not a strictly Lamarckian phenomenon because the stress does not cause mutations directly and specifically in genes that are responsible for stress resistance. Instead, organisms evolved mechanisms that, in response to stress, induce nonspecific mutagenesis. However, this process appears to be fine-tuned to minimize the damage from deleterious mutations in those rare genomes that carry a beneficial mutation. This type of mechanism is best defined as quasi-Lamarckian. Indeed, in the case of stress-induced mutagenesis, consider the following:

1. Environmental conditions trigger mutations.
2. The induced mutations lead to adaptation to the stress factor(s) that triggered mutagenesis.
3. Mutagenic repair is subject to elaborate regulation, which leaves no doubt regarding the adaptive nature of this process.

A direct link exists between the Lamarckian aspects of stress-induced mutagenesis and HGT through the phenomenon of antibiotic-induced HGT of resistance determinants. Many antibiotics induce the SOS response, which, in turn, leads to the mobilization of integrating conjugative elements that serve as vehicles for the HGT of antibiotic resistance genes (Barriss, et al., 2009). The analogy to GTAs is obvious and fully relevant. Here we observe convergence of different mechanisms of genome change in the Lamarckian modality.

The continuum of Darwinian and Lamarckian mechanisms of evolution

In the preceding sections, we discussed a considerable variety of phenomena. Some seem to strictly meet the Lamarckian criteria, whereas others qualify as quasi-Lamarckian (see Box 9-1). The crucial difference between Darwinian and Lamarckian mechanisms of evolution is that the former relies upon random, undirected variation, whereas the latter is based on variation directly caused by an environmental cue and resulting in a specific response to that cue (see Figure 9-1). Neither Lamarck nor Darwin was aware of the mechanisms of emergence and fixation of heritable variation, so it was relatively easy for them to entertain the idea that phenotypic variation directly translates into heritable (what we now consider genetic or genomic) changes. However, the strict Lamarckian scenario is extremely demanding, in that a molecular mechanism must exist for the effect of a phenotypic change to be precisely channeled into the corresponding modification of the genome (mutation). There seem to be no general mechanisms for such reverse genome engineering, and it is not unreasonable to surmise *that such mechanisms are kept under a tight control by selection against genome destabilization.* Furthermore, the transfer of information from proteins to nucleic acids would be extremely difficult physico-chemically—conceivably, this difficulty reflects the separation between template and catalytic biomolecules that emerged at the earliest stages of the evolution of life (see Chapter 12). The Central Dogma of molecular biology (Crick, 1970), which states that *there is no information flow from protein to nucleic acids,* is a partial embodiment of this separation. However, in principle, the reverse flow of specific information from the phenotype—or the environment viewed as an extended phenotype—to the genome is not impossible, considering the wide spread of reverse transcription and DNA transposition. Highly sophisticated mechanisms are required for this bona fide Lamarckian scenario to work; in two remarkable cases, the CRISPR-Cas and the piRNA systems (described earlier in this chapter), such mechanisms have been discovered.

Box 9-1: Lamarckian and quasi-Lamarckian phenomena

Phenomenon	Biological role/function	Phyletic spread	Lamarckian criteria		
			Genomic changes caused by environmental factor	Changes are specific to relevant genomic loci	Changes provide adaptation to the causative factor
Bona Fide Lamarckian					
CRISPR-Cas	Defense against viruses and other mobile elements	Most of the Archaea and many bacteria	Yes	Yes	Yes
piRNA	Defense against transposable elements in germline	Animals	Yes	Yes	Yes
HGT (specific cases)	Adaptation to new environment, stress response, resistance	Archaea, bacteria, unicellular eukaryotes	Yes	Yes	Yes

Phenomenon	Biological role/function	Phyletic spread	Lamarckian criteria		
			Genomic changes caused by environmental factor	Changes are specific to relevant genomic loci	Changes provide adaptation to the causative factor
Quasi-Lamarckian					
HGT (general phenomenon)	Diverse innovations	Archaea, bacteria, unicellular eukaryotes	Yes	No	Yes/no
Stress-induced mutagenesis	Stress response/ resistance, adaptation to new conditions	Ubiquitous	Yes	No or partially	Yes (but general evolvability enhanced as well)

The existence of additional bona fide Lamarckian systems is imaginable and even likely as suggested, particularly, by the discovery of virus-specific sequences that potentially confer resistance to the cognate viruses, in plant and animal genomes (see Chapter 10). However, these mechanisms hardly constitute the mainstream of genome evolution, perhaps owing to the aforementioned selection against excessive genomic instability. In contrast, the mechanisms denoted quasi-Lamarckian in the preceding sections are ubiquitous. Conceptually, these mechanisms seem to be no less remarkable—and no less sophisticated—than the genuine Lamarckian scenario: The quasi-Lamarckian processes translate random mutations into specific, adaptive responses to environmental cues.

The theme of powerful, often adverse effects of the environment on organisms seems to be common to different facets of the (quasi) Lamarckian mode of evolution described here, for the CRISPR-Cas system, stress-induced mutagenesis and other phenomena (see Box 9-1). This association is most likely not spurious: It seems entirely logical that strong (extraordinary) signals from the environment trigger (quasi) Lamarckian processes, whereas relatively weak ("business as usual") signals translate into the Darwinian modality of evolution (see Figure 9-3).

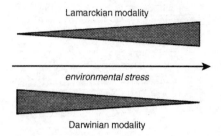

Figure 9-3   Environmental pressure and the transition from the Darwinian to the (quasi) Lamarckian modality. Adapted from (Koonin and Wolf, 2009b).

In a discussion of the evolutionary impact of HGT, Anthony Poole suggested that the Lamarckian aspect of HGT becomes illusory when "a gene's view" of evolution is adopted (Poole, 2009). Indeed, it appears that the Lamarckian modality is associated primarily, if not exclusively, with the organismal level of complexity and does not apply to the most fundamental level of evolution that involves genes, independently evolving portions of genes (such as those encoding distinct protein domains), and mobile elements (see Chapter 6). Thus, Lamarckian evolution seems to be an "emergent phenomenon." This is perhaps not surprising, considering the need for complex mechanisms for integrating new material into the genome, to realize the Lamarckian scheme.

Generally, the comparison between the Darwinian and Lamarckian scenarios suggests that *evolution is a continuum of processes, from entirely random to intrinsically adaptive ones*, that are exquisitely orchestrated to ensure a specific response to a particular challenge. The critical realization suggested by many recent advances referred to in this chapter is that genomic variation itself is a far more complex phenomenon than previously imagined and is regulated at multiple levels to provide adaptive reactions to changes in the environment. Eliminating the conflict between the Lamarckian and Darwinian evolutionary scenarios, far from being of purely historical significance, affects our fundamental views on the role and place of chance in evolution. This then seems to be a veritable, if underappreciated, paradigm shift in modern biology.

Fidelity of information transmission in biological systems and its (non)adaptive evolution

Evolution of life is fully based on digital information transmission processes—across generations via genome replication and from the genome to the effector molecules (RNA and proteins)—as discussed in Chapter 2. No information transmission channel is error-free, as first formally asserted by Claude Shannon, who founded information theory by connecting information transmission with the laws of thermodynamics. As noted in Chapter 2, the fidelity of genome replication cannot be lower than a certain minimum; conversely, the mutation rate cannot exceed a certain threshold, to avoid the mutational meltdown of the population. Obviously, the mutation rate cannot be too low, either, to allow at least minimum evolvability (the potential for evolution—see the next section for discussion). Less clear is whether this lower limit is a practical consideration in real biological systems. Thus, a fundamental question is this: How does selection control the mutation rate (if at all)? More specifically, does purifying selection simply keep the mutation rate below the meltdown threshold, or is there, at least in some organisms and perhaps in particular situations, selection for a sufficiently high mutation rate to provide raw material for evolution?

Selection for a sufficient fidelity of replication (and more generally all processes of information transmission) is one of the central aspects in all evolution. This is immediately obvious from the enormous diversity, complexity, and multilayer organization of repair systems that exist in all cellular life forms (Aravind, et al., 1999; Friedberg, et al., 2005). In prokaryotes, up to 10% of the coding capacity of the genomes may be dedicated to repair system components that act at all stages of DNA replication and also eliminate various mutational lesions that occur outside the replication process. Conversely, there is a class of replicators that possess (virtually) no repair mechanisms, the RNA viruses. Indeed, these viruses show extremely high nucleotide misincorporation and overall mutation rates, a phenomenon well known due to the medical importance of the rapid evolution of influenza viruses or HIV (Holmes, 2009). These viruses appear to evolve not too far from the mutational meltdown threshold (Drake and Holland, 1999). All RNA viruses possess small genomes (less than 30Kb), which is partly a consequence of the physical fragility of long RNA molecules but is also related to the lack of repair mechanisms. (One might argue that complex repair system in such viruses could not evolve because they would not be advantageous, given the intrinsic genome instability.) Actually, the RNA viruses with the largest genomes (the animal nidoviruses) appear to possess a distinct, although simple, repair system (Eckerle, et al., 2007). Furthermore, the discovery of RNA demethylases in the genomes of a variety of plant RNA viruses implies that even these simplest genomes may evolve repair when the virus propagates under conditions of increased environmental stress (Aravind and Koonin, 2001; van den Born, et al., 2008).

Within the framework of the mutational meltdown concept, a natural idea is that the mutation rate per nucleotide should be inversely proportional to the genome size of an organism so that the *number of mutations per genome per generation remains roughly constant.* Jan Drake first made this conjecture explicit, so it is often called the "Drake hypothesis" (Drake, 1991). The Drake hypothesis seems to hold quite well for viruses and prokaryotes. However, unexpectedly (at least, at first glance), an updated survey undertaken by Michael Lynch reveals the opposite dependence in eukaryotes: The per-nucleotide mutation rate is positively correlated with the genome size (Lynch, 2010). Following up on the non-adaptive theory of the evolution of complexity, Lynch showed that a limited increase in the mutation rate would not be "visible" to purifying selection in small populations that are typical of multicellular eukaryotes and so could not be coped with in the course of evolution of these organisms. Hence, a "semi-adaptive" hypothesis on the evolution of mutation rates emerges: There is selective pressure to lower the mutation rate below the meltdown threshold and somewhat beyond because the evolving population becomes more robust, but not to *minimize* the mutation rate. Under this hypothesis, there is no selection for evolvability, as such, that would prevent the mutation rate from falling below any minimal value; the mutation rate simply remains relatively high for purely stochastic reasons (see also the next sections in this chapter).

However, the situation is not so simple, as illustrated by the results of long-term experiments on evolving *E. coli* populations undertaken by Richard Lenski and colleagues. These experiments show that intense selection for adaptation of the bacterial population to a new environment often involves the emergence of mutator alleles (that is, bacteria that have a high mutation rate due to impairment of one of the repair enzymes) that outcompete the ancestral bacteria with low mutation rates (Sniegowski, et al., 1997). More precisely, it appears that mutator alleles reach a high frequency in the population and even fixation by hitchhiking with the adaptive mutations the mutator causes. However, when the external selective pressure is removed, mutators become disadvantageous and are selected against (Denamur and Matic, 2006). These findings lead to a crucial generalization: *Depending on a variety of factors, such as the environmental stress and effective population size, selection for either low or high mutation rate can and does occur.*

The error rate of transcription is much higher than that of replication, and the error rate of translation is much higher yet (see Figure 9-4). Although the experimental measurements of the amino acid misincorporation rate during translation are scarce and limited to a few model systems, it is clear that the fidelity of translation is almost shockingly low. Indeed, the rate of non-cognate amino acid incorporation is 10^{-4} to 10^{-5}—interestingly, close to the replication error rate in RNA viruses. Thus, about 20% of the protein molecules synthesized in any cell contain at least one wrong amino acid (Drummond and Wilke, 2009). The consequences of errors of transcription and translation, sometimes aptly called *phenotypic mutations*, obviously are less dramatic than the consequences of genetic mutations, for the very reason that phenotypic mutations are generally not inherited (notable exceptions exist, such as reverse transcription followed by the incorporation of a DNA copy of a mistranscribed RNA into the genome; Burger, et al., 2006). Given the relatively short lifetime of any RNA or protein molecule, no phenotypic mutation can have a major fitness effect on its own, so it is not surprising that much greater error rates are tolerated for phenotypic mutations than for genetic mutations. However, it is equally obvious that excessively high rates of phenotypic mutation are incompatible with life. Thus, as is the case with DNA repair systems, multiple mechanisms for keeping transcription and translation errors in check certainly exist. Proofreading activity of DNA-dependent RNA polymerases has been detected and shown to decrease the error rate by orders of magnitude (Alic, et al., 2007; Sydow and Cramer, 2009). Moreover, still poorly characterized processes of post-transcriptional repair of methylation damage in RNA have been discovered as well (Begley and Samson, 2003; Falnes, 2005). Probably the best understood of the mechanisms that control the rate of phenotypic mutations is proofreading by aaRS, in which aminoacyl-tRNAs charged with non-cognate amino acids are hydrolyzed and recycled (Hussain, et al., 2010; Ling, et al., 2007). The aaRS proofreading is complemented by the downstream ribosomal proofreading, in which the ribosome rejects non-cognate tRNAs (Blanchard, et al., 2004; Daviter, et al., 2006). However, a large increase of the translation accuracy seems to clash with the requirement of a high rate of protein production. A substantially increased translation fidelity can be readily achieved by mutating specific positions in rRNA or ribosomal proteins, but these mutations are deleterious, apparently, because of slow translation (Dong and Kurland, 1995; Johansson, et al., 2008).

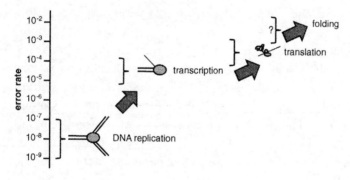

Figure 9-4   The characteristic error rates at different stages of biological information transmission.

The experiments with the accurate but slow ribosomal mutants suggest that the trade-off between the speed and accuracy of translation is linked to limits on the mechanistic capabilities of the translation system and might be hard to overcome by mutations in the components of this machinery. Hence, other types of adaptations seem to have been selected to limit translational errors and their deleterious consequences. It is well established that "high-status" genes (those that are highly expressed and evolve slowly) have a stronger codon bias than "low-status" genes. The optimal codons for which the high-status genes are enriched provide for a lower mistranslation rate, as well as a higher rate of translation, thus partially overcoming the aforementioned trade-off (Drummond and Wilke, 2009, 2008). The difference in codon bias between high-status and low-status genes is explained by the cost of selection; because of it, significant selection for optimal codons can occur only in high-status genes.

The principal deleterious effect of mistranslation is thought to be protein misfolding (Drummond and Wilke, 2009, 2008), although amino acid misincorporation at catalytic sites certainly could be an additional factor. As already discussed in Chapter 4, selection for robustness to misfolding is a major aspect of protein evolution—possibly even its main driving force. It is less clear whether misfolding of native sequences or mistranslation-induced misfolding is most important. Regardless, although protein folding is not usually viewed as an information-transmission process, this is what it actually is. Indeed, folding involves the *flow of information from the one-dimensional amino acid sequence to the three-dimensional protein structure.* Exactly the same applies to structural RNAs. The rate of misfolding is hard to determine experimentally, and this has not been done for large sets of proteins or RNAs. If the trend of error rates depicted in Figure 9-4—the farther from the genome, the less accurate an information transmission step is—is any indication, the error rate of folding is expected to be even greater than the error rate of translation. This prediction also stands to reason, given the enormous complexity of the folding process and the vast number of (mis)folding pathways that are, in principle, available to a folding protein or RNA molecule (Bowman, et al., 2011; Pande, et al., 1998). Given the high complexity of the folding landscape, the seminal discovery (made originally by Christian Anfinsen and subsequently confirmed in numerous experiments) that proteins can spontaneously fold into the native conformation came as a considerable surprise (Anfinsen, 1973).

Almost 50 years after Anfinsen's discovery, it remains a matter of debate whether spontaneously folding proteins find the global or a local free energy minimum. What has become clear is that only small proteins fold spontaneously; the majority of proteins require special molecular devices, namely other proteins known as chaperones, to fold into the native structure. Chaperones function in a remarkable manner: The chaperone molecules form a "cage" (known also as the Anfinsen cage) that isolates the folding protein from the cytoplasm and partially unfolds it, thus facilitating the search for the native conformation (Ellis, 2003). Most of the chaperones are abundant, highly conserved, high-status proteins. Originally, some of the chaperones were discovered as "heat shock proteins;" that is, proteins that are strongly upregulated at elevated temperature (and, as shown later, other stress conditions) and counteract protein misfolding, which is enhanced by the stress (Vabulas, et al., 2010). Although not as well characterized, protein chaperones mediate RNA folding as well (Russell, 2008; Woodson, 2010). On the whole, the control over protein (and probably RNA) folding undoubtedly is a major function in all cells.

Beyond the chaperone-type devices, all cells dedicate a versatile repertoire of molecular machines to controlled degradation of proteins, particularly misfolded ones, and RNA. Similarly to molecular chaperones, these machines—the proteasome, in the case of proteins and the exosome (degradosome in bacteria), in the case of RNA—are ubiquitous in the three domains of life, highly abundant in most cells, and subject to regulation under stress (Hartung and Hopfner, 2009; Volker and Lupas, 2002; see also Chapter 7). Furthermore, these machines, along with additional back-up systems of regulated proteolysis, are major intracellular consumers of energy (ATP). Bacteria additionally possess a highly conserved system of the so-called trans-translation that releases stalled ribosomes from aberrant mRNAs on which translation fails to terminate properly and targets such mRNAs and their (also aberrant) protein products for degradation (Keiler, 2008).

As we discussed in great detail in Chapter 7, eukaryotes possess an important information-processing step that effectively has no counterpart in prokaryotes: splicing of primary transcripts. The accompanying quality-control system apparently has evolved concomitantly with eukaryogenesis (see Chapter 7): the nonsense-mediated decay (NMD) machinery which recognizes and destroys aberrant mRNAs that contain in-frame stop codons in exons other than the last, 3'-terminal exon of the coding sequence (Behm-Ansmant, et al., 2007; Stalder and Muhlemann, 2008).

Thus, the control of the error rate and its effects in biological processes of information transmission is one of the key aspects of evolution. For reasons that we understand only partially (at best), the error rates do not seem to drop very far below the highest acceptable value: the mutational meltdown threshold and the corresponding error catastrophe threshold of phenotypic mutations that is not well characterized but may be presumed to exist. In the case of the mutation rates, the simple non-adaptive population genetic theory seems to explain the observed values with reasonable accuracy (Lynch, 2010). Similar reasoning has been applied to phenotypic mutations (Burger, et al., 2006), but in this case, the solution seems less clear. There is a tension between the relatively high error rates of transcription, translation, splicing, and most likely folding, and the extreme elaboration of damage-control devices such as proteasomes, exosomes, the NMD system, and others. The evolution of these multiple echelons of damage-control systems implies that the *deleterious effects of phenotypic mutations at the rates they occur in reproducing cells are non-negligible, but the cost of selection for increased fidelity would be unsustainable, so the alternative routes of damage control evolution have been taken repeatedly.*

In general, it seems that the *fight against entropy is one of the crucial aspects of evolution.* The selection for entropy control and decrease is universal and is distributed down the line of information transmission, from replication to protein and RNA folding and sorting. The antientropic evolution partly lowers the mutation/error rates themselves and partly acts at the level of damage control. Evolutionary experiments indicate that selection for increased mutation rates does occur. Nevertheless, it remains uncertain whether such selection for *increased* noise is widespread or whether the residual levels of noise that the antientropic mechanisms cannot remove are sufficient to provide the variance required for evolution. We discuss this key problem in the following sections.

Noise in biological systems and its creative role in evolution

No information channel can be free of noise (see Chapters 4 and 8, and the preceding section). This fundamental, thermodynamically determined aspect of information transmission makes evolution possible through the intrinsic non-negligible error rate of replication, even in the absence of selection for increased variability (see Chapter 2). As noticed in Chapter 4, the effectively neutral mutations that accumulate for purely entropic reasons comprise nearly neutral networks that act as reservoirs of evolution, including positive selection, especially in changing environments. Let us emphasize once again that not only fixed mutations, but also polymorphisms that are particularly numerous in large populations contribute to evolutionarily relevant neutral networks. Moreover, we have seen repeatedly that increased mutation rate can be beneficial—and selected for—as is the case for stress-induced mutagenesis and mutator alleles.

Here I want to concentrate on phenotypic mutations and their possible role in evolution. The obvious default view seems to be that phenotypic noise is of no consequence for evolution (so far as its deleterious effects are under control), for the obvious reason that phenotypic mutations are not inherited. However, this could be a myopic perspective, thanks to the so-called look-ahead effect of phenotypic mutations. As we repeatedly see throughout this book, a key problem in evolution is to traverse the rugged evolutionary landscape, moving from local fitness peaks or plateaus to higher altitudes on other peaks or plateaus. This path often goes through crevices, some of them deep. To cross these crevices, fixation of two or more mutations is needed. Obviously, this is a nontrivial task, given that the first mutation leads to decreased fitness. In principle, such feats are achievable via drift, but the chances of fixation of multiple mutations are exceedingly low when the first required mutation is deleterious; they drop to zero when the first mutation is lethal. Here is where phenotypic mutations can come to rescue. When two mutations are required to acquire a trait, phenotypic mutations can provide a low level of the second mutation once the first mutation has occurred. If the first mutation is deleterious or even lethal, whereas the two mutations together are beneficial, phenotypic mutations might rescue the organism and yield a beneficial phenotype (reach the elusive high place on the fitness landscape), albeit at a low frequency. Mathematical modeling shows that this look-ahead effect is a realistic possibility within a reasonable range of effective population sizes and selective coefficients (Whitehead, et al., 2008).

How common and important is the look-ahead effect? We do not know, but if it was important and had an adaptive value, one would predict that stress should induce mistranslation in a controlled fashion, much as it induces mutagenesis (see the discussion of stress-induced mutagenesis earlier in this chapter). Remarkably, this is indeed the case, as shown by recent studies in which oxidative stress led to a dramatically increased mistranslation rate, at least in part through the impairment of the proofreading function of the aaRS (Ling and Soll, 2010; Netzer, et al., 2009). Whether stress-induced phenotypic mutations are regulated and controlled as described for stress-induced genetic mutagenesis remains to be investigated. The finding that the misincorporation of methionine, the amino acid that protects proteins from oxidative damage, is specifically enhanced under oxidative stress (Netzer, et al., 2009) suggests that controlling mechanisms for phenotypic mutations might exist, so stress-induced phenotypic mutagenesis could be viewed as an adaptive strategy.

The evolutionary significance of phenotypic mutations is a potentially broad and important subject that recent investigations have barely touched. However, it seems virtually certain that the phenotypic noise (entropy) is not evolutionarily indifferent. Instead, beyond the obvious selective pressure to keep it in check, noise might be a constructive factor in a variety of evolutionary processes. Phenotypic mutations are part of a more general, pervasive trend that we touched upon in Chapter 8. The central idea is that noise at all levels and in all pathways of biological information transmission—typical genetic and phenotypic mutations, spurious transcription, errors of splicing, protein misfolding—while being a burden on the evolving organisms, also enhances their evolutionary potential. The straightforward but crucial corollary is that *populations with a low intensity of purifying selection that are unable to eliminate or substantially reduce noise possess the highest evolutionary potential.*

Evolution of evolvability, robustness of biological systems, and the feasibility of evolutionary foresight

Evolvability, the potential of biological systems for evolutionary change, is one of the many important concepts that were introduced by Richard Dawkins (Dawkins, 2006). In the first decade of the twenty-first century, it has become a fashionable term and a subject of intense debate (Brookfield, 2009; Kirschner and Gerhart, 1998; Masel and Trotter, 2010; Radman, et al., 1999). As such, evolvability is unproblematic as an intrinsic property of replicating systems (see the EPR principle introduced in Chapter 2). As discussed earlier in this chapter, evolution of any organism (with some reservations, in the case of RNA viruses) involves selection for decreasing the rates of both genomic and phenotypic mutations. However, this selection is relatively inefficient for reasons that receive a population-genetic explanation in the case of genomic mutations but are less clear in the case of phenotypic mutation, given the existence of elaborate systems for eliminating aberrant products. Thus, evolvability is clearly subject to selection, in the straightforward sense of keeping noise in check.

The null hypothesis on evolvability, then, is that the residual noise that purifying selection cannot eliminate is sufficient (and, indeed, necessary) to provide the variance that is the raw material for evolution. This "constructive" noise, then, is a non-adaptive by-product of evolution. The bone of contention is the intriguing possibility that evolvability evolves in a nontrivial way—that is, that increased variation under certain circumstances and/or in certain genomic loci could be a selectable trait. The idea of "adaptive evolvability" is anathema to many biologists because it reverberates with "evolutionary foresight" or temporal nonlocality of evolution, whereas the generally accepted dogma is that evolution is strictly local ("myopic"). However, this general belief notwithstanding, the findings presented in this chapter clearly indicate that certain forms of evolvability are evolvable and could be adaptive, and that, in a general sense, evolution is capable of foresight.

Let us briefly summarize the evidence in support of the adaptive evolution of evolvability. Perhaps the strongest evidence is the extensive set of experimental data on the systems of stress-induced, error-prone repair; effectively, these are mechanisms for stress-induced mutagenesis. These systems, such as the SOS repair/mutagenesis machinery in bacteria, do not seem to possess mechanisms to specifically direct mutations to genes that are involved in coping with a particular form of stress. Nevertheless, they certainly promote survival by increasing the overall mutation rate and so increasing the chance of adaptation. The elaborate regulation of these systems and their high prevalence in microbes that inhabit changing environments leave no doubt that stress-induced mutagenesis is an adaptive phenomenon and constitutes a generic form of evolutionary foresight— or, more precisely, *evolutionary extrapolation*. The evolutionary process cannot possibly "know" what is about to come, but in stress-prone environments, organisms that evolve the capacity to transiently increase the mutation rate enjoy an increased chance of survival. This effect is further enhanced by the mutation clustering mediated by the recruitment of the stress-induced repair enzymes to the lesions in DNA (Galhardo, et al., 2007; Rosenberg, 2001). In addition, evolutionary experiments indicate that under stress elevated mutation rates caused by mutations in repair genes can provide a selective advantage to the organisms that carry mutator alleles. Speaking anthropomorphically for a moment, evolution cannot predict what is actually going to happen, but can extrapolate from the difficult past that bad things will necessarily happen sooner or later, so the only chance to survive is to be prepared and start mutating rapidly once the challenge arrives. It is a risky strategy, but apparently the only evolvable one. Whether an analogous strategy is realized at the level of phenotypic mutations is less clear, but the possibility appears realistic, given the findings on the induction of translation errors, possibly of a specific kind, under stress.

A striking piece of evidence in support of the evolution of evolvability is provided by the GTAs, the specialized virus-like agents of HGT in bacteria and Archaea (see Chapter 4 and Chapter 10). Considering that HGT is the dominant evolutionary process in prokaryotes, the existence of dedicated devices that, to the best of our understanding, evolved solely under the selective pressure to enhance HGT shows that mechanisms of evolution themselves evolve and, in some cases, may be viewed as adaptations.

Another crucial aspect of the evolvability problem has to do with the robustness of biological networks and the so-called capacitation phenomenon. Certain proteins, the best characterized of which is the molecular chaperone HSP90 (HSP is the generic acronym for Heat Shock Protein, a protein induced by high-temperature stress), possess the properties of *evolutionary capacitors*, or mediators of the effects of genetic and phenotypic variation (Masel and Siegal, 2009). Inactivation of HSP90 leads to the appearance of numerous mutant phenotypes by unmasking hidden variation: Proteins with amino acid replacements that have no phenotypic effect in the presence of HSP90 misfold in its absence (Rutherford, et al., 2007). It turns out that capacitation is quite a general phenomenon: In yeast, about 300 genes (more than 5% of the total) behave as capacitors. The shared property of capacitors is that they are hubs of interaction networks, so presumably the disruption of the interactions between a capacitor and other proteins unleashes the hidden variation (Levy and Siegal, 2008). The effect of capacitation is likely to be important at the level of phenotypic mutations as well. The flip side of the capacitation coin is the possibility that HSP90 and perhaps other capacitors also have the ability to act as *potentiators* of genotypic and phenotypic mutations by allowing mutant proteins that would otherwise misfold to fold correctly and exert phenotypic effect.

Perhaps the most striking known mechanism of evolvability is presented by fungal prion proteins (Halfmann and Lindquist, 2010; Masel and Bergman, 2003). The extensively characterized yeast prion (PSI+) is a translation termination (polypeptide chain release) factor that has the ability to spontaneously convert to the prion form, which then nucleates the self-perpetuating formation of amyloid-like protein aggregates. The prion aggregates sequester the release factor and stimulate frequent readthrough of stop codons. The result is the appearance of numerous extended variants of proteins and the unmasking of hidden variability in 3'-UTR sequences that normally evolve under weak purifying selection. The prion formation is strongly stimulated by various stresses (Tyedmers, et al., 2008), and the prion state is heritable, a remarkable form of protein-based inheritance.[5] Notably, the mechanism of prion action increases phenotypic variance through a combined effect of phenotypic mutations (stop codon readthrough during translation) and unmasking of hidden genetic variation. The increased survival of prion-carrying strains under stress indeed has been demonstrated (Tyedmers, et al., 2008). It is still unclear how common the prion-mediated potentiation of evolvability is. What is most remarkable is that the prion property of release factors does not appear to have any other function and thus probably evolved specifically under the pressure to promote evolvability.

Capacitors appear to be bona fide *regulators of evolution.* On one hand, these genes provide robustness to biological systems and dampen the effect of mutations. Somewhat paradoxically, however, capacitation also promotes evolvability through the potentiation effect, by allowing evolving organisms to increase the size of nearly neutral networks and, hence, the potential for adaptive evolution (Wagner, 2008b).

The connection between robustness and evolvability is a key aspect of evolution in general. This link seems to be a "systemic property" in which robustness both protects evolving systems from the deleterious effects of variation and increases their evolutionary potential (Kaneko, 2007). Quantitative analysis of population genetic models shows that robustness can either increase or decrease evolvability, depending on the population dynamics and the structure of the fitness landscape. In particular, it has been demonstrated that adaptive evolution accelerates with the increased size of the neutral network (robustness) as long as phenotype accessibility remains constrained within the given fitness landscape (Draghi, et al., 2010). Within the framework of the fitness landscape concept, the phenomenon of selection for robustness and evolvability became known as *survival of the flattest*. Simulations of evolution with digital organisms have shown that, at high mutation rates, genotypes with relatively low replication rates but large nearly neutral networks—that is, those occupying relatively low but flat areas of the fitness surface—outcompeted genotypes that replicated faster but occupied high, steep peaks (Wilke, et al., 2001). By contrast, at low mutation rates, evolution occurs by *survival of the fittest*—that is, the genotypes that occupy the tallest peak—and intermediate regimes of evolution have been discovered in which both the fittest and the flattest survive (Beardmore, et al., 2011). Thus, robustness and evolvability come at a trade-off with fitness. Evolution favors one or the other or both (thus, promoting diversity) depending on the conditions.

Synopsis and perspective

In this chapter, we discussed a diverse gamut of data, models, and hypotheses that are united by a common thread: They drive evolutionary biology away from the important but simplistic triad of heredity–variance–selection that is at the core of Modern Synthesis. Even the more realistic conceptual framework of today's evolutionary biology that includes prominent contributions from drift, draft, recombination, and HGT is substantially incomplete. The studies reviewed in this chapter reveal more complex, unexpected contributions both on the side of chance/randomness and on the side of adaptive, even directional processes (see Figure 9-5).

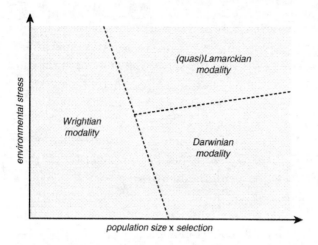

Figure 9-5   The structure of the evolutionary process: a multifactorial view.

Speaking of chance, entropy (noise) at all levels of biological information transmission can be a constructive factor of evolution, in large part because of the robustness of biological networks. To what extent this robustness is an evolved, adaptive property, as opposed to being an intrinsic property of the networks, is a deep, interesting question that remains to be thoroughly investigated. Importantly, although no one so far has discovered a direct path from phenotypic mutations to the genome, phenotypic noise is also a potentially important factor of evolution, thanks to the look-ahead effect and also special mechanisms of evolvability enhancement that act through phenotypic mutations, as is the case of fungal prions.

In a different plane, the numerous phenomena of epigenetic inheritance, such as those involving RNAi (see earlier), as well as the better-studied ones based on heritable DNA methylation patterns, are important mechanisms of evolution (Johnson and Tricker, 2010; Richards, 2006). In part, the epigenetic phenomena (which we do not have the opportunity to discuss here in detail) play the same role as the look-ahead affect of phenotypic mutations: They create a buffer of plasticity that gives populations a chance to cross deep valleys in the rugged fitness landscape.

As far as "necessity" goes, a close examination of various widespread processes that contribute to the generation of genomic variation shows that evolution does not entirely rely on stochastic mutation. Instead, generation of variation is often controlled via elaborate molecular machinery that instigates adaptive responses to environmental challenges of various degrees of specificity. Genome evolution appears to span the entire spectrum of scenarios, from the purely Darwinian, based on random variation, to the *bona fide* Lamarckian, in which a specific mechanism of response to a cue is fixed in an evolving population through a distinct modification of the genome. In a broad sense, all these routes of genomic variation reflect the interaction between the evolving population and the environment in which the active role belongs either to selection alone (pure Darwinian scenario) or to directed variation that itself might become the target of selection (Lamarckian scenario).

Mechanisms of evolution are subject to selection and evolve themselves: *Evolvability is evolvable.* Many evolutionary biologists might be uneasy about this statement because it could be read as acceptance of "evolutionary foresight." These concerns notwithstanding, the extensive studies on stress-induced mutagenesis and the emerging realization of the potential key role of specialized devices, GTAs, in horizontal gene transfer leave no doubt that the evolutionary potential of organisms is itself subject to selection and evolves. Evolution of evolvability is directly observable in laboratory experiments with evolving bacterial populations. I will say it again: *Evolution has the ability to extrapolate from repeated events of the past and to effectively predict generic aspects of the future.*

To conclude this chapter, it is worth emphasizing that the novel routes of evolution discussed here do not require any *unknown elementary mechanisms*. Thus, none of these previously underappreciated or overtly denied evolutionary phenomena runs afoul of central principles of molecular biology, particularly Crick's Central Dogma that proclaims the irreversibility of information transfer from nucleic acids to protein. For instance, the CRISPR system that seems to embody the Lamarckian scenario of evolution and so to violate a major taboo operates through a combination of molecular mechanisms that, in principle, are common and well known, even if the details could be unique to this system. These mechanisms include various complementary interactions between nucleic acids, integration of DNA fragments into specific loci in a genome, and recognition and cleavage of distinct RNA structures by enzymatic complexes—a unique machinery that the evolutionary process "tinkered" from generic components.

Recommended further reading

Draghi, J. A., T. L. Parsons, G. P. Wagner, and J. B. Plotkin. (2010) "Mutational Robustness Can Facilitate Adaptation." *Nature* 463 (7,279): 353–355.

This work shows "using a general population genetics model, that mutational robustness can either impede or facilitate adaptation, depending on the population size, the mutation rate, and the structure of the fitness landscape. In particular, neutral diversity in a robust population can accelerate adaptation as long as the number of phenotypes accessible to an individual by mutation is smaller than the total number of phenotypes in the fitness landscape."

Galhardo, R. S., P. J. Hastings, and S. M. Rosenberg. (2007) "Mutation As a Stress Response and the Regulation of Evolvability." *Critical Reviews in Biochemistry and Molecular Biology* 42: 399–435.

Stress-induced mutagenesis is interpreted as a bona fide system for the regulation of evolvability that facilitates adaptation and survival.

Koonin, E. V., and Y. I. Wolf. (2009). "Is Evolution Darwinian or/and Lamarckian?" *Biology Direct* 4: 42.

A discussion of the evolutionary phenomena that appear to involve Lamarckian or quasi-Lamarckian mechanisms.

Levy, S. F., and M. L. Siegal. (2008) "Network Hubs Buffer Environmental Variation in Saccharomyces Cerevisiae." *PLoS Biology* 6: e264.

A pioneering experimental study of capacitation showing that numerous network hubs possess the properties of evolutionary capacitors.

Lynch, M. (2010) "Evolution of the Mutation Rate." *Trends in Genetics* 26: 345–352.

A survey of experimentally determined mutation rates in the full spectrum of organisms, revealing the paradoxical dependence between the mutation rate and genome size in eukaryotes.

Marraffini, L. A., and E. J. Sontheimer. (2010) "CRISPR Interference: RNA-Directed Adaptive Immunity in Bacteria and Archaea." *Nature Reviews Genetics* 11: 181–190.

A review of the molecular mechanisms of the CRISPR-Cas system.

Masel J., and M. V. Trotter. (2010) "Robustness and Evolvability." Trends in Genetics 26: 406–414.

The title speaks for itself.

Rajon, E., and J. Masel. (2011) "Evolution of Molecular Error Rates and the Consequences for Evolvability." *Proceedings of the National Academy of Sciences USA* 108: 1,082–1,087.

Important study that distinguishes between local adaptations to decrease the effect of genomic and phenotypic mutations (evolution toward increased robustness), and global adaptations (evolution toward decreased mutation rate). Population genetic modeling shows that local adaptations are feasible only in large populations with intense selection, whereas small populations evolve global adaptations.

Wagner, A. (2008) "Neutralism and Selectionism: A Network-Based Reconciliation." *Nature Reviews Genetics* 9: 965–974.

An important paper describing the evolution of (nearly) neutral networks that comprise the reservoir of potentially adaptive variants.

Whitehead, D. J., C. O. Wilke, D. Vernazobres, and E. Bornberg-Bauer. (2008) "The Look-ahead Effect of Phenotypic Mutations." *Biology Direct* 3: 18.

A conceptually important modeling study demonstrating the potential evolutionary relevance of phenotypic mutations.

10

The Virus World and its evolution

Viruses have been discovered in a rather inconspicuous manner, as a kind of peculiar pathogen and possibly a special form of toxin that causes plant diseases such as tobacco mosaic. The agents of these diseases passed through fine filters that captured bacteria, so it has been correctly concluded that these agents are distinct from (typical) bacteria. Shortly thereafter, the first viruses infecting animals were discovered. These included the Rous Sarcoma Virus, the first known tumorigenic virus, and strange agents that seemed to devour bacteria, which were named bacteriophages but actually turned out to be bacterial viruses. Virology then enjoyed illustrious development in the twentieth century (Fields, et al., 2001), for two reasons: First, many viruses were important both medically and agriculturally. Second, viruses are the simplest genetic systems and, thus, became the favorite models, first for early molecular genetics (primarily through the work of the famous Phage Group, led by Max Delbruck [Cairns, et al., 1966]), and later for genomics.[1] However, genetics by the 1970s and genomics by the late 1990s had matured enough to productively work with cellular models. As a result, virology has left the central stage of research in fundamental biology (some cameo appearances notwithstanding).

The first decade of the new millennium was marked with a veritable renaissance of virus research, instigated by two series of discoveries. The first was the discovery of giant viruses, such as the mimivirus, which have particles and genomes of decidedly cellular proportions, thus blurring the superficial boundary between viruses and cells in terms of size (Raoult, et al., 2004; Van Etten, et al., 2010). The second, more spectacular development is metagenomics, which showed, to the extreme surprise of biologists, that viruses are the most abundant biological entities on Earth (Edwards and Rohwer, 2005). These advances stimulated a much broader interest in the evolution of viruses. I view these results as the discovery of a vast, ancient Virus World that was an integral and crucial part of life since its inception on Earth. All that time, the Virus World intensely interacted with cellular life forms, which evolved a huge variety of antivirus defense systems, but retained its identity and, in many respects, was key to the entire history of life. In this chapter, we discuss the Virus World, its evolution, and the arms race between viruses and cells that permeates all evolution.[2] I argue that viruses comprise one of the two "empires of life," the other one obviously represented by cellular organisms. Chapter 11 addresses the contribution of viruses to the origin and evolution of cells.

The extraordinary diversity and ubiquity of viruses

What is a virus?

Definitions in biology are difficult and never quite satisfactory. Nevertheless, before considering various aspects of the evolution of viruses in this chapter, we need to define viruses. Indeed, at a coarsegrained level, it is not difficult to provide such a general definition. Over the last century, the knowledge of viruses has progressed from the vague notions of their discoverers, Dmitri Ivanovsky and Martinus Beijerinck, to exquisite molecular detail. Here we very generally define viruses as follows: *obligate intracellular parasites or symbionts that possess their own genomes encoding information required for virus reproduction and, hence, a degree of autonomy from the host genetic system, but do not encode a complete translation system or a complete membrane apparatus.* This definition applies to any "truly" selfish genetic element: The key phrase here is *encoding information required for virus reproduction and, hence, possessing a degree of autonomy from the host genetic system.* Thus, regular genes and operons do not fit this definition, even though they may possess some selfish properties, because they encode no dedicated "devices" for their own reproduction. Within the vast space of biological entities that encode "something" that is necessary for their own reproduction but not the translation system or the membrane, this definition is all-encompassing.

I deliberately did not specify that the "replication device" has to be a protein, so viroids (plant pathogens with genomes of only about 300 nucleotides that hijack the host transcription machinery for their replication) definitely qualify. Nor did I specify that the viral genome should encode a capsid (that is, the protein scaffold of the virion). This could seem somewhat counterintuitive, given that viruses historically have been known largely as particles (virions), starting with the first successful crystallization of tobacco mosaic virus by Wendell Stanley in 1934. So notable is the capsid that Patrick Forterre and Didier Raoult recently defined viruses as "capsid-encoding organisms," as opposed to cellular life forms defined as "ribosome-encoding organisms" (Raoult and Forterre, 2008). This definition seems to be on the right track with respect to the separation of cells and viruses as the two major forms of life, but it is unnecessarily narrow and fails to objectively delineate the Virus World.

The capsid certainly is an extremely important and common feature. In this chapter, however, we describe clear evolutionary relationships, accompanied by similarities in genome architecture and replication cycles, between traditional, capsid-encoding viruses and "naked" selfish elements such as plasmids and various mobile elements. Under the present definition (even as any definition has its limitations), all these agents belong to the vast Virus World.

From time to time, discussions flare up on the notorious subject of whether viruses are "alive." The latest installment of this debate has attracted considerable attention (Moreira and Lopez-Garcia, 2009). In itself, the question is a purely semantic and accordingly unimportant one. The definition of a virus given here clearly indicates that viruses belong in the realm of biology; as discussed later in this chapter, comparative genomics reveals multiple connections between the genomes of viruses and cellular life forms. An unfortunate implication of the denial of the "alive" status to viruses is that viruses are of no substantial relevance to the evolution of cellular life forms. In this chapter and the next, we shall see that the opposite is true.

The diversity of the replication-expression strategies among viruses

All cellular life forms possess dsDNA genomes that are transcribed into mRNAs, which are translated into multiple proteins, as well as various noncoding RNAs. This uniformity of the genetic cycle among cellular life forms is in stark contrast to the diversity of the replication-expression cycles of viruses, some of which possess RNA genomes of different polarities, whereas others have ssDNA genomes (Baltimore, 1971; see Figure 10-1).[3] Some viruses and virus-like elements have also incorporated the transition from RNA to DNA as part of their regular cycle, through the combination of the activities of a virus-encoded reverse transcriptase (RT) and the host DNA–dependent RNA polymerase. Positive-strand RNA viruses have the distinction of implementing the simplest imaginable genetic cycle, whereas the reverse-transcribing elements provide the link from the RNA world to the DNA world. This plasticity of viral replication cycles may have deep evolutionary implications, as discussed in the next chapter.

When the diversity of virus replication-expression strategies became apparent, it was tempting to "play Mendeleev"—that is, to generate an exhaustive table of possible replication-expression cycles, populate it with the observed ones, and then try to predict which of the remaining cells will be filled through future discoveries and which might be "off limits" for some fundamental reasons. To my knowledge, the first such attempt was undertaken by my teacher in virology, Vadim Agol (Agol, 1974; originally published in an obscure Russian journal, where I read it). This was the article that, through its captivating elegance and the precious attempt (at least, that was how I felt about it at the time, between my freshman and sophomore years at the university) to use deep, if simple symmetry considerations in biology, induced me to study viruses in the first place. I have never regretted that decision; years later, I developed my own version of genome strategy classification (Koonin, 1991).[4] Beyond the Central Dogma, which stands firmly, there seems to be one *fundamental prohibition: ssDNA is never translated, so RNA is involved in the reproduction cycle of any genetic element*. Unlike the case of proteins according to the Central Dogma, there seems to be no straightforward chemical basis for this "ban" (actually, translation of ssDNA has been demonstrated experimentally [Hulen and Legault-Demare, 1975; McCarthy and Holland, 1965]). However, the only translation system known to us apparently has evolved specifically to make proteins on an RNA template (more on this in Chapter 12). This exception apart, all replication-expression cycles that are conceivable on the basis of RNA and DNA molecules seem to be realized in the Virus World, even if some exotic genome forms, such as a DNA-RNA hybrid, are rarely seen (see Figure 10-1).

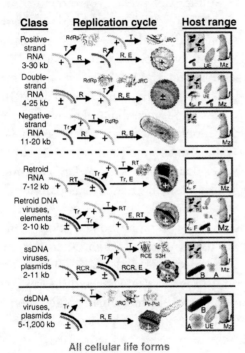

All cellular life forms

Figure 10-1   The diversity of the replication-expression strategies in viruses and virus-like elements. For each class of viruses and related elements, the approximate range of genome sizes is indicated (Kb, kilobases). A + denotes a positive strand (same polarity as mRNA) and a – denotes a negative strand. Tr = transcription; T = translation; R = replication; E = encapsidation; A = archaea; B = bacteria; F = fungi; Mz = Metazoa; P = plants; UE = unicellular eukaryotes. For each class of viruses (elements), typical structures of hallmark proteins and characteristic electron-microscopic images of viruses are shown. RdRp = RNA-dependent RNA polymerase; JRC = jelly roll capsid protein; RT = reverse transcriptase; RCRE = rolling-circle replication (initiating) endonuclease. The rightmost panel shows the host range; the size of the respective image and acronym is roughly proportionate to the abundance of the given virus class in the respective taxon. Adapted from Koonin, et al., 2006.

The range of genomic complexity, functional content, and diversity of genome architectures among viruses

Viruses with different genome strategies span a vast range of genome sizes: The genomes of the largest known virus, the mimivirus, and the smallest viruses (such as circoviruses) differ by three orders of magnitude. If we include viroids that encode no proteins but are bona fide selfish genetic elements and even pathogens, the range expands to almost four orders of magnitude (see Figure 10-1). Given that virus genomes in general show a wall-to-wall packing of protein-coding genes (see Chapter 3), the number of genes spans nearly the same range. The genome size strongly depends on the nature of the genome and the replication-expression cycle. Conspicuously, only dsDNA viruses reach large (by viral standards) genome sizes of greater than 35 Kb and (so far) up to 1.1 Mb (Van Etten, et al., 2010). All classes of RNA viruses, all retro-transcribing elements and all ssDNA viruses, possess small genomes that never exceed 35 Kb, and that only in one group of rather uncommon animal RNA viruses (coronaviruses and their relatives that comprise the order Nidovirales (Gorbalenya, et al., 2006). The underlying reason is apparent: The greater chemical stability and the regular structure of dsDNA are conducive to the information storage and replication functions, hence evolution of repair systems for dsDNA further widens the functional divide between dsDNA and other forms of nucleic acids. In addition to the diversity of replication-expression strategies and the vast size span, viral genomes assume all possible molecular configurations, including linear or circular DNA or RNA molecules and single or multiple genomic segments (chromosomes; see Figure 10-1).

The functional repertoires of viral genes differ dramatically depending on the expression cycle and, even more importantly, on the genomes size and genetic complexity. Small viral genomes encode almost exclusively proteins involved directly in genome replication, along with virion subunits. Often—and in all known RNA viruses and retro-transcribing elements—a virus encodes the polymerase that is involved in its own genome replication. This is easy to rationalize because the cellular hosts normally do not encode an RNA-dependent RNA polymerase or RT capable of replicating or reverse-transcribing long RNA molecules. The RNA-dependent RNA polymerases and RTs that are encoded in the genomes of cellular life forms and perform "normal" cellular functions, such as the telomerase and the RNA-dependent RNA polymerase involved in RNA interference in eukaryotes, only produce short oligonucleotides (see Chapter 7 and the discussion later in this chapter). By contrast, DNA viruses have the potential to recruit the host replication (and transcription) machinery and widely use this opportunity. Thus, many viruses of this type, particularly most of the identified viruses infecting Archaea, as well as numerous temperate bacteriophages (such as the classic model organism, the lambda phage), do not even encode their own replicative polymerase or any other proteins directly involved in replication. In these cases, DNA sequences that recognize and recruit the host replication machinery appear to be the primary determinants of the autonomous reproduction of the virus (see the definition above), although viral proteins play other important roles in virus reproduction, such as shutdown and reprogramming of the host gene expression and metabolism. By contrast, in addition to proteins that constitute the viral genome replication apparatus, viruses with the largest genomes encode a panoply of diverse proteins involved in repair processes, membrane trafficking, a variety of metabolic pathways, and, in some cases, even translation system components. Typically (and in all cases where translation is concerned), a virus would not encode a complete system or pathway, but only enzymes for one or two steps that complement or modify the corresponding functionalities in the host cell.

Viruses with different types of genomic nucleic acid show a nonuniform and nontrivial distribution across the host taxa. In particular, the extraordinary diversity of dsDNA bacteriophages and archaeal viruses is in stark contrast to the absence of bona fide dsDNA viruses in plants. Conversely, RNA viruses are extremely abundant and diverse in plants and animals, but are currently represented by only two compact families in bacteria and so far have not been detected in archaea (see Figure 10-1). In some cases, the biological underpinning of the virus host range is perfectly clear. For instance, in plants, large viruses would face severe problems with cell-to-cell spread because plasmodesmata (the channels between the walled plant cells) are impermeable to large particles or even large DNA molecules. For the most part, however, the causes of a particular distribution of viruses among hosts remain obscure. For example, it is difficult to tell why RNA viruses are so pervasive in plants and animals but not in prokaryotes; we return to this question later in this chapter from a different, evolutionary perspective.

Metagenomics of viruses, environmental virology, gene transfer agents, and the ubiquity of viruses

Viruses are ubiquitous companions of cellular life forms: Every cellular organism studied in reasonable detail appears to harbor its own viruses. In those organisms—such as nematodes—where bona fide viruses so far have not been discovered, numerous mobile elements are integrated in the genome.

Recently, the study of *viromes* (the totality of viruses detectable in a given habitat) has become a vibrant research field (Edwards and Rohwer, 2005; Kristensen, et al., 2010). The isolation of a virome is rather straightforward methodologically because virus particles (at least, the great majority, with the possible exception of the giant variety) pass through filters that are impenetrable for even the smallest cells. Thus, it is relatively easy to collect particles from the flow-through of such filters and analyze the content. The study of viromes brought big surprises. The first one has to do with the sheer concentration of virus particles. Shockingly, at least in marine environments, viruses (primarily bacteriophages) are the most abundant biological entities, with the total number of virus particles exceeding the number of cells by at least an order of magnitude. The comparison is not entirely fair because a single virus-infected cell can produce hundreds of virus particles, but nevertheless, these findings indicate that viruses are enormously abundant and active in the environment. Actually, viruses are now viewed as major geochemical agents because viral killing of microbial cells strongly affects sediment formation (Suttle, 2007, 2005). The second big surprise is the enormous genetic diversity of the viromes and the unexpected character of their genetic content. The gene composition of the DNA virome is dramatically different from that of known bacteriophages. The virome is dominated by rare and unique genes that have no homologs among the sequences in current databases. Bacteriophage-specific genes, although substantially enriched compared to microbiomes, are a tiny minority; most of the genes for which homologs are detectable seem to represent a random sampling of bacterial genes. Barring the possibility of heavy contamination, which is unlikely, given the rigorous protocols used to isolate viromes, one has to conclude that the viromes are dominated by entities other than conventional viruses (Kristensen, et al., 2010).

What is the nature of the dark matter that constitutes most of the viromes? We do not really know, but a plausible hypothesis is easy to come up with. Recall the Gene Transfer Agents (GTAs) that we already discussed in Chapters 5 and 9. The GTAs are a special kind of pseudoviruses (Lang and Beatty, 2007). They form virus (phage-like) particles that consist of proteins encoded in a defective prophage that resides in the respective bacterial or archaeal chromosome. However, the GTA particles do not contain the prophage DNA (so they are not bona fide bacteriophages), but instead encapsidate apparently random fragments of the bacterial chromosome. It is easy to speculate that *the dark matter of the viromes consists primarily of GTAs* (Kristensen, et al., 2010; and so the viromes are only "pseudoviral"). This simple hypothesis, which, of course, needs empirical validation, has far-reaching implications. Indeed, if virus (or virus-like) particles are the most common biological entities on Earth, and most of them are GTAs, then the inevitable logical conclusion is that GTAs dominate the biosphere. Combined with the presence of numerous, often "hidden" prophages and other mobile elements in bacterial and archaeal genomes (Cortez, et al., 2009) and the even greater abundance of (largely inactivated) selfish elements in the genomes of many eukaryotes (including humans), these findings suggest that the Virus World largely *"builds" the genomes of cellular life forms and so shapes the evolution of life in general*. The fundamental consequences of this conclusion remain to be thought through and further investigated; we return to them more than once in this chapter and the remaining chapters.

Although marine metagenomics is still a young field, it has already done wonders for our understanding of the Virus World, even beyond the striking findings on the genetic composition of viromes. One group of discoveries emerged from the analysis of the sequences produced by the Global Ocean Survey, the grand metagenomic initiative of J. Craig Venter (Yooseph, et al., 2007). It turns out that, in addition to myriad bacterial sequences (the main target of the project), the Global Ocean Survey database contains numerous sequences homologous to conserved genes of Nucleo-Cytoplasmic Large DNA Viruses (NCLDV; more on them later in this chapter) that infect eukaryotes. Regardless of the exact source of the viral DNA (giant viruses, infected picoeukaryotes that pass bacterial filters, or, most likely, both), the diversity of several NCLDV families has increased far beyond the range revealed by traditional virology (Monier, et al., 2008). The second group of notable findings came from the metagenomics of marine RNA viruses: Numerous RNA viruses infecting unicellular marine eukaryotes have been discovered and, quite unexpectedly, were found to all belong to only one super-family of viruses previously identified in animals and plants: the picorna-like viruses (Koonin, et al., 2008). Together these discoveries show that we are only scratching the surface of the Virus World; the true dimensions of this world might defeat the boldest imagination.

Virus evolution: Polyphyly versus monophyly and the hallmark genes

The previous sections introduced the Virus World and showed that it is commensurate in its scale with the world of cellular life forms—and probably quantitatively dominates the biosphere. Moreover, through the GTAs and a huge diversity of mobile elements, the Virus World exerts defining effects on the evolution of cells. So it is essential to look into the evolution of viruses if we strive for any deep understanding of the evolution of life in general.

Comparative genomics provides no evidence of a monophyletic origin of all viruses. By "monophyly" here, we mean *the origin from a common ancestral virus or a virus-like selfish element* (Koonin, et al., 2006). Many groups of viruses simply share no common genes, effectively ruling out any conventional notion of common origin. When applied to viruses, the notion of "common genes" is not a simple one: In the Virus World, commonality is not necessarily limited to clear-cut orthologous relationships between genes that are easily detectable through highly significant sequence similarity. Instead, as discussed in the next sections, distant homologous relationships among viral proteins and between viral proteins and their homologs from cellular life forms could convey more complex but important messages on the evolution of viruses. This complexity notwithstanding, cases of major virus groups abound that either share no homologous genes under any definition or have in common only distantly related domains with obviously distinct evolutionary trajectories. For example, most of the viruses of hyperthermophilic Crenarchaeota encompass no genes in common with any other viruses (Prangishvili, et al., 2006b), whereas RNA viruses share only extremely distant domains in their respective replication proteins with DNA viruses and plasmids that replicate via the rolling circle mechanism.

In sharp contrast, the monophyly of several large classes of viruses, including vast assemblages of RNA viruses and complex DNA viruses, can be demonstrated with confidence (see Box 10-1). Some of these monophyletic classes of viruses even cross the boundaries set by genome strategies: Thus, the monophyletic class of reverse-transcribing elements includes both RNA viruses and viruses, mobile elements, and plasmids with DNA genomes. The rolling circle replication class combines ssDNA and dsDNA viruses and plasmids. Furthermore, based on similarities in the structure of RNA replication complexes, along with the presence of homologous, even if distant, replication enzymes, a plausible hypothesis has been proposed that positive-strand RNA viruses, double-stranded RNA viruses, and retro-transcribing elements all have a common origin (Ahlquist, 2006). On the whole, however, the conclusion seems inevitable that *viruses comprise many distinct lines of descent* (Koonin, et al., 2006; see Box 10-1).

A brief natural history of viral genes

Sequence analysis revealed several categories of virus genes that markedly differ in their provenance (Koonin, et al., 2006). The optimal granularity of the classification could be debated, but at least five classes that can be assorted into three larger categories are clearly distinguishable.

Genes with readily detectable homologs in cellular life forms:

1. Genes with closely related homologs in cellular organisms (typically, the host of the given virus) present in a narrow group of viruses.

2. Genes that are conserved within a major group of viruses, or even several groups, and have relatively distant cellular homologs.

Box 10-1: The largest monophyletic classes of viruses and selfish genetic elements

Class of Viruses	Virus Groups	Hosts	Support for Monophyly
Positive-strand RNA viruses	Superfamily I: picorna-like Superfamily II: alpha-like Superfamily III: flavi-like The exact affinity of RNA bacteriophages within this class of viruses remains uncertain (possibly a fourth lineage)	Animals, plants, protists, bacteria (one family of bacteriophages)	Conserved RdRp JRC in most superfamily I viruses, and subsets of superfamilies II and III viruses Reconstructed ancestor with RdRp and JRC
Retro-transcribing viruses and mobile elements	Retroviruses, hepadnaviruses, caulimoviruses, badnaviruses LTR and non-LTR retroelements Retrons Group II self-splicing introns, the progenitors of eukaryotic spliceosomal introns	Animals, fungi, plants, protists, bacteria, archaea	Conserved RT
Small DNA viruses, plasmids, and transposons with rolling circle replication	Gemini-, circo-, parvo-, papo-vaviruses, phages (such as ΦX174), archaeal and bacterial plasmids, eukaryotic helitron transposons	Animals, plants, archaea, bacteria	Conserved RCRE, JRC, S3H (in eukaryotic viruses)

Class of Viruses	Virus Groups	Hosts	Support for Monophyly
Tailed bacteriophages (Caudovirales)	Families: Myoviridae (such as T4), Podoviridae (such as T7), Siphoviridae (such as λ)	Bacteria, euryarchaea	Complex, overlapping arrays of genes conserved in subsets of tailed phages Genes of all tailed phages thought to comprise a single pool
Nucleo-Cytoplas-mic Large DNA Viruses (NCLDV)	Poxviruses, asfarviruses, iridoviruses, phycodnaviruses, mimiviruses	Animals, algae, protists	Core set of 11 conserved genes, including JRC, S3H, and an FtsK-like packaging ATPase, found in all NCLDVs. Reconstructed ancestor with about 50 genes

Abbreviations: JRC = Jelly Roll Capsid protein; LTR = Long Terminal Repeat; RdRp = RNA-dependent RNA polymerase; RCRE = Rolling Circle Replication (initiation) Endonuclease; RT = Reverse Transcriptase; S3H = Superfamily 3 Helicase.

Virus-specific genes:

3. ORFans—genes without detectable homologs, except possibly in closely related viruses.

4. Virus-specific genes that are conserved in a (relatively) broad group of viruses but have no detectable homologs in cellular life forms.

Viral hallmark genes:

5. Genes shared by many diverse groups of viruses, with only distant homologs in cellular organisms and with strong indications of monophyly (common origin) of all viral members of the respective gene families. The phrase *viral hallmark genes* was coined to denote these genes that appear to be signatures of the "virus state."

The relative contributions of each of these classes of genes to the gene sets of different viruses depend on the viral genome size and genetic complexity, which differ by more than three orders of magnitude (see Figure 10-1). Viruses with small genomes, such as most of the RNA viruses, often have only a few genes, the majority of which belong to the hallmark class. By contrast, in viruses with large genomes, such as poxviruses, all five classes are broadly represented. To illustrate the diversity of viral genomic composition, Figure 10-2 shows the breakdown of the gene sets of three viruses with a small, intermediate-sized, and large genome, respectively, into the five classes of genes. Notably, moderate-sized and large genomes of bacteriophages and archaeal viruses are dominated by ORFans that often comprise more than 80% of the genes in these viruses. Rapidly evolving phage ORFans probably supply many, if not most, of the ORFans found in prokaryotic genomes (the lack of detectable sequence conservation notwithstanding); hence, they play a key role in the evolution of prokaryotes (Daubin and Ochman, 2004).

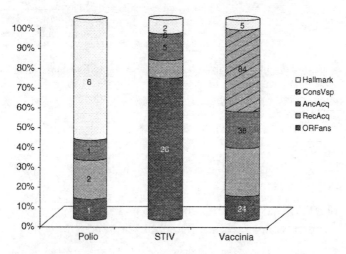

Figure 10-2 Decomposition of viral genes into five evolutionary classes: a virus with a small genome: poliovirus (7.4 Kb); a virus with an intermediate-sized genome: Sulfolobus turreted icosahedral virus (STIV); a virus with a large genome: vaccinia virus (195 Kb). ConsVsp = Conserved virus-specific genes; AncAcq = ancient acquisitions; RecAcq = recent acquisitions. The data comes from Koonin, et al., 2006.

The origins of the five classes of viral genes are likely quite different. The two classes of genes with readily detectable homologs in cellular life forms appear to represent, respectively, relatively recent (Class 1) and ancient (Class 2) acquisitions from the genomes of cellular hosts. Where virus-specific genes come from is a much harder and more intriguing question. One possibility is that these genes evolved from other viral and/or host genes, with a dramatic acceleration of evolution linked to the emergence of new, virus-specific functions, such that all traces of the ancestral relationships are obliterated. This notion is compatible with the fact that many (probably most) Class 4 genes (virus-specific genes conserved within a group of viruses) are virion components, a quintessential viral function. We postpone the discussion of other routes for the origin and evolution of virus-specific genes until after the discussion of the evolution of the virus hallmark genes. The hallmark genes that cross the barriers between extremely diverse virus lineages are of the greatest interest and relevance for understanding the evolution and the ultimate origins of viruses.

Viral hallmark genes: Beacons of the ancient Virus World

There are no traceable vertical relationships between large groups of viruses outside the major monophyletic classes listed in Box 10-1. However, a considerable number of genes that encode proteins with key roles in genome replication, expression, and encapsidation are shared by overlapping arrays of (otherwise) seemingly unrelated groups of viruses, although none of these genes is present in *all* viruses (see Box 10-2). Most of the virus hallmark genes have no highly conserved homologs in cellular life forms (except in easily recognizable proviruses or mobile elements), although distant homologs exist. The two genes that are most widely dispersed among viruses are the so-called jelly roll capsid protein and the superfamily 3 helicase. Each of these proteins crosses the boundary between RNA and DNA viruses and spans an astonishing range of virus groups, from some of the smallest positive-strand RNA viruses to the NCLDV, the class of viruses that includes the giant mimivirus (see Box 10-2). More specifically, the jelly roll capsid protein is the principal building block of the icosahedral (spherical) viral capsids, the most common form of the capsids that greatly differ in size but are quite similar in symmetry and overall shape across a huge range of viruses that employ all kinds of replication-expression strategies and infect hosts from all walks of cellular life. Similarly, the superfamily 3 helicase participates in the genome replication of a huge variety of RNA and DNA viruses.

Box 10-2: Proteins encoded by the most common virus hallmark genes

Protein	Function	Virus Groups	Homologs in Cellular Life Forms	Comments
Jelly roll capsid protein (JRC)	Main capsid subunit of icosahedral virions	Picornaviruses, comoviruses, carmoviruses, dsRNA phage, NCLDV, herpesviruses, adenoviruses, papovaviruses, parvoviruses, icosahedral DNA phages and archaeal viruses	Distinct jelly roll domains are seen in eukaryotic nucleoplasmins and in protein-protein interaction domains of certain enzymes.	Certain icosahedral viruses, such as ssRNA phages and alphaviruses, have unrelated capsid proteins. In poxviruses, the JRC is not a virion protein, but forms intermediate structures during virion morphogenesis.
Superfamily 3 helicase (S3H)	Initiation and elongation of genome replication	Picornaviruses, comoviruses, eukaryotic RCR viruses, NCLDV, baculoviruses, some phages (such as P4), and plasmids (particularly archaeal ones)	S3H is a distinct, deep-branching family of the AAA+ ATPase class.	Characteristic fusion with primase in DNA viruses and plasmids

Protein	Function	Virus Groups	Homologs in Cellular Life Forms	Comments
Archaeo-eukaryotic DNA primase	Initiation of genome replication	NCLDV, herpesviruses, baculoviruses, some phages	All viral primases appear to form a clade within the archaeo-eukaryotic primase family.	Characteristic fusion with S3H in most NCLDV, some phages, and archaeal plasmids
UL9-like super-family 2 helicase	Initiation and elongation of genome replication	Herpesviruses, some NCLDV, some phages	Viral UL9-like helicases form a dis-tinct branch in the vast superfamily of DNA and RNA heli-cases.	Fusion with primase in asfarviruses, mimiviruses
Rolling circle replication initiation endonuclease (RCRE)/origin-binding protein	Initiation of genome replication	Small eukaryotic DNA viruses (parvo-, gemini-, circo-, papova), phages, plasmids, and eukaryotic helitron transposons	No cellular RCRE or papovavirus-type origin-binding protein exists. However, these proteins have a derived form of the palm domain that is found in the majority of cellular DNA polymerases.	Papovaviruses have an inactivated form of RCRE that functions as an origin-binding protein.

Protein	Function	Virus Groups	Homologs in Cellular Life Forms	Comments
Packaging ATPase of the FtsK family	DNA packaging into the virion	NCLDV, adenoviruses, polyd-naviruses, some phages (such as P9 and M13), nematode transposons	This is a distinct clade in the FtsK/HerA superfamily of P-loop NTPases that includes DNA-pumping ATPases of bacteria and archaea.	
ATPase subunit of terminase	DNA pack-aging into the virion	Herpesviruses, tailed phages	The terminases comprise a derived family of P-loop NTPases that is distantly related to Superfamily I/II helicases and AAA+ ATPases.	

Protein	Function	Virus Groups	Homologs in Cellular Life Forms	Comments
RNA-dependent RNA polymerase (RdRp)/reverse transcriptase (RT)	Replication of RNA genomes	Positive-strand RNA viruses, dsRNA viruses, retro-transcribing viruses /elements, possibly, negative-strand RNA viruses	This is another major group of palm domains that are distinct from those in DNA polymerases.	The RdRps of dsRNA viruses are homologs of positive-strand RNA virus polymerases. The provenance of negative-strand RNA virus RdRp remains uncertain, although sequence motif and especially structural analysis suggest their derivation from positive-strand RNA virus RdRps.

Abbreviations: NCLDV = Nucleo-Cytoplasmic Large DNA Viruses.

Other proteins listed in Box 10-2 are not as common as JRC or S3H, but they still form multiple, unexpected connections between groups of viruses that otherwise appear to be unrelated. As a case in point, consider the rolling circle replication initiation endonuclease, which unites a great variety of small ssDNA and dsDNA replicons, including viruses, plasmids, and transposable elements that repro-duce in animals, plants, bacteria, and archaea. Extensive sequence analysis has shown that the DNA-binding domain of the replicative protein of polyoma and papilloma viruses (such as the T antigen of SV40) is a derived, inactivated form of the rolling circle replication initiation endonuclease (Iyer, et al., 2005). Thus, through this detailed analysis of one of the hallmark proteins, the well-known connection among a variety of small ssDNA-replicons (both viruses and plasmids) is extended to a group of similar-sized dsDNA-replicons. A similar expansion of the set of viral groups covered by a particular hallmark gene resulted from the detailed analyses of the ATPase that is responsible for the packaging of viral DNA into the capsid and the archaeo-eukaryotic primase involved in the initiation of DNA replication (Iyer, et al., 2005; Iyer, et al., 2004b; see Box 10-2).

The genome replication of positive-strand RNA viruses, dsRNA viruses, negative-strand RNA viruses, and retro-transcribing viruses/elements is catalyzed by another class of viral hallmark enzymes, the RNA-dependent RNA polymerases and RT. The positive-strand RNA virus polymerases and the RT form a distinct monophyletic group within the vast class of the so-called Palm-domains that are characteristic of numerous polymerases (Iyer, et al., 2005; Koonin, et al., 2008). The RdRps of dsRNA viruses and negative-strand RNA viruses are likely to be highly diverged derivatives of the same polymerase domain (Delarue, et al., 1990; Gorbalenya, et al., 2002; Koonin, et al., 1989). This viral hallmark gene might bring us right back to the earliest stages of the evolution of life, the RNA World (see Chapters 11 and 12 for much more detail)—and to the beginnings of the Virus World. The Palm-domain is likely to be the primordial polymerase protein that replaced the ribozyme polymerases of the (hypothetical) RNA world. This conjecture is supported not only by the wide spread of this domain in modern life forms, but also by the structural and, by inference, evolutionary link between the Palm-domain and the RNA recognition motif (RRM) domain, an ancient RNA-binding domain that might have initially facilitated the replication of ribozymes (Aravind, et al., 2002). The RNA-dependent RNA polymerase and the RTs are excluded from the regular replication cycles of cellular life forms, although most eukaryotic genomes, especially those of animals and plants, encompass numerous copies of RT-encoding retroelements; prokaryotes have some such elements as well (see also Chapters 5 and 7). These elements, however, are selfish and, from the evolutionary standpoint, belong to the Virus World. Perhaps the most notable incursion of an RT into the cellular domain is the catalytic subunit of the eukaryotic telomerase, the essential enzyme that is involved in the replication of chromosome ends.[5] Of course, it should not be forgotten that all eukaryotic introns evolved from prokaryotic retroelements (see

Chapter 7). Remarkably, the only other known RNA-dependent RNA polymerase that is unrelated to the Palm-domain containing polymerases and is a component of the eukaryotic RNAi system (see Chapter 7) also appears to be of viral origin (Iyer, et al., 2003).

The list of viral hallmark genes given in Box 10-2 is conservative. Most likely, other genes merit the hallmark status as well, but clear evidence is hard to find. Sequencing of additional viral genomes, combined with comprehensive comparative analysis, might reveal additional genes that, despite a relatively limited spread among viral lineages, will qualify as hallmark. Indeed, this could be the case for many, if not most, Class 4 genes, the viral genes that are conserved in large groups of viruses but not in cellular life forms.

The combination of features of viral hallmark proteins is highly unusual and demands an evolutionary explanation. Indeed, the hallmark genes are, without exception, responsible for essential, central aspects of the viral life cycles, including genome replication, virion formation, and packaging of the genome DNA into the virion (see Box 10-2). These genes span extremely diverse classes of viruses that often possess different reproduction strategies and differ by up to three orders of magnitude in genome size. Finally, all viral hallmark genes have remote homologs in cellular life forms (see Box 10-2), but the viral versions appear to be monophyletic.

Two straightforward hypotheses on the origins of the hallmark viral proteins offer contrasting evolutionary scenarios to account for their existence and spread (Koonin, et al., 2006).

1. The hallmark genes are the heritage of a Last Universal Common Ancestor of Viruses (LUCAV). This scenario implies that, despite all evidence to the contrary (see earlier), all extant viruses are genuinely monophyletic, although their subsequent evolution involved massive gene loss in some lineages as well as extensive acquisition of new genes from the hosts in others.

2. By contrast, under the hypothesis of polyphyletic origin of viruses, HGT could explain the spread of the hallmark genes across the range of virus groups.

Upon closer inspection, none of these hypotheses seems to be a viable general explanation for the existence and distribution of the viral hallmark genes. Indeed, the relatively small number and the mosaic spread of the hallmark genes (see Box 10-2) do not seem to be conducive to the LUCAV notion, although it is apparent that a great number of diverse viruses, if not all of them, share some common history. Conversely, the extremely distant (although still discernible) similarity between the hallmark proteins from diverse virus groups with dramatically different replication strategies is poorly compatible with an HGT scenario.

Later in this chapter, I outline a scenario of virus origin and evolution that does not involve a (traditionally interpreted) LUCAV, but integrates aspects of the common origin and HGT hypotheses and is naturally linked to specific models of the evolution of cells. The simplest explanation for the fact that the hallmark proteins involved in viral replication and virion formation are present in a broad variety of viruses but apparently not in any cellular life forms is that the latter never had these genes in the first place. Instead, the most plausible scenario posits that *the hallmark genes antedate cells and descend directly from a primordial gene pool*. Conceivably, in such a primordial pool, selection would act primarily on functions directly involved in replication, which is compatible with the properties of the majority of the hallmark genes (see Box 10-2). Given the spread of the hallmark genes among numerous groups of dramatically different viruses, a crucial corollary is that the major classes of viruses themselves derive from the precellular stage of evolution. This corollary is the key point of the *ancient Virus World* concept. The crucial feature of the Virus World *is the uninterrupted flow of genetic information through an enormous variety of selfish elements, from the precellular stage of evolution to this day*.

The conflicting concepts of virus origin and evolution

Before we discuss the full scope of the emerging concept of the origin of viruses from the precellular gene pool, we need to briefly examine the existing hypotheses on virus origin and evolution. Traditionally, these ideas have revolved around three themes (see Figure 10-3):

1. Origin of viruses from primordial genetic elements
2. Degeneration of unicellular parasites to the virus state

3. "Escaped genes" scenario, which derives viruses from genes of cellular organisms that have run away from the cell genome and switched to the selfish mode of reproduction

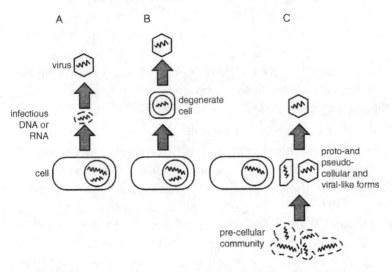

Figure 10-3   The three competing hypotheses on the origin of viruses: (A) The escaped gene scenario; (B) The cell degeneration scenario; (C) The primordial gene pool scenario.

The "primordial" hypothesis was somewhat in vogue in the very early days of virology, and it is remarkable that Felix d'Herelle, the discoverer of bacteriophages and one of the founders of virology, proposed as early as 1922 that phages might be evolutionary precursors of cells (D'Herelle, 1922). A few years later, in 1928, J. B. S. Haldane propounded this hypothesis in his classic essay on the origin of life (Haldane, 1928; we return to Haldane's prescient thoughts in Chapter 11). However, once it became clear that all viruses are obligate intracellular parasites, the primordial hypothesis was habitually dismissed on the strength of the simple and, on the surface, unbeatable argument that intracellular parasites could not possibly antedate full-fledged cells. On the contrary, the high prevalence of host-derived genes (as opposed to virus-specific genes) in many viruses (particularly those with large genomes) might be construed as support for the "escaped genes" or even the "cell degeneration" hypotheses. In the heyday of molecular biology, when the fundamental distinctions between viruses and cells were clearly realized so that the origin of viruses from cells (however degenerate) was considered highly unlikely, the escaped genes hypothesis seemed to have been, more or less by exclusion, the default concept of the origin of viruses (Luria and Darnell, 1967). However, more recently, the discovery of giant viruses, and especially the fact that these viruses possess some quintessential "cellular" genes, such as those for multiple translation system components, has led to a resurrection of the cell degeneration hypothesis (Claverie, 2006). Indeed, purely in terms of the genome size and genetic complexity, the discovery of the giant viruses obliterates the border between viruses and cellular life forms.

All these arguments notwithstanding, the existence of the virus hallmark genes seems to effectively falsify (or at least call into serious doubt) both the cell degeneration and the escaped-genes concepts of viral evolution. With regard to the cell degeneration hypothesis, let us consider the NCLDV (Koonin and Yutin, 2010), the class of large viruses to which the cell degeneration concept might be most readily applicable and indeed has been applied, in the wake of the discovery of the giant mimivirus (see Box 10-1). Among the nine signature genes that (almost) all NCLDVs share, three crucial ones (jelly roll capsid protein, superfamily 3 helicase, and packaging ATPase) are virus hallmark genes. Even the simplest ancestral NCLDV would not be functional without these genes. However, cellular derivation of this ancestral NCLDV would have to invoke decidedly nonparsimonious, *ad hoc* scenarios, such as concerted loss of all hallmark genes from all known cellular life forms or their derivation from an extinct major lineage of cell evolution. The same logic essentially refutes the escaped genes concept, inasmuch as the hallmark genes have never had a cellular "home" to escape from. Again, to save "escaped genes," an extinct cellular domain would have to be postulated from which the hallmark genes could escape.

Thus, the most parsimonious scenario for the evolution of viruses seems to be the precellular Virus World. It appears most likely that the major classes of viruses—at least, all the strategies of genome replication and expression—evolved already in the precellular era. It might be inappropriate to call the putative primordial selfish elements "viruses," given the absence of cells on the stage of their evolution. However, calling them "virus-like" agents or something similar cannot change the fact that there are no traces of cellular origins of viruses and does not in any sense invalidate the ancient Virus World hypothesis.

Two caveats must be kept in mind. First, when considering the three scenarios for the origin of viruses, we speak of viruses (or viral genomes) as *independently evolving genetic elements*. Many—in some viruses, perhaps most—viral genes might be of cellular origin (see Figure 10-2) but viruses as (quasi) autonomous entities do not seem to have cellular roots. Second, although for any particular lineage of viruses the scenarios are mutually exclusive, different groups of viruses in principle might have different origins. In any case, we so far do not have any strong evidence that the escaped genes or cell degeneration scenarios are the best explanation for the origin of any known viruses.

The continuity of the Virus World and connections with the world of cellular organisms

The analysis of viral hallmark genes seems to suggest their origin in the primordial gene pool. The existence of this primordial pool appears to be a logical inevitability, regardless of which specific scenario for the early stages of evolution is adopted. We address these questions in earnest in Chapters 11 and 12. Here I want to emphasize a different and remarkable aspect of virus evolution, the apparent continuity of the Virus World from the precellular era to this day and into any foreseeable future. Indeed, if the hallmark genes (i) originate from the primordial gene pool, (ii) have only very distant homologs among genes of cellular life forms and apparently have never been parts of cellular genomes, and (iii) are essential for the reproduction of viruses that harbor them, then the conclusion is inescapable that these genes have been passed from virus to virus (or virus-like element) throughout the entire 4-billion-year span of the evolution of life. This being the case, we must conclude that *viral genomes, although not monophyletic in the traditional sense, evolved by mixing and matching of genes in a giant genetic network that is the Virus World*. Numerous genes from cellular life forms also penetrate this network, primarily through genomes of large viruses such as the NCLDV and large bacteriophages that have acquired numerous genes from their hosts at different stages of evolution. However, most of the acquired genes are not essential for viral genome replication

and expression, per se (excluding some cases of possible nonorthologous displacement of hallmark genes); typically, these genes are
involved in virus-host interactions. Thus, all exchange of genes with
the hosts notwithstanding, it seems that viruses always evolve from
other viruses, even if, in many cases, the paths of evolution are circuitous. We are in a position to formulate a principle that is symmetrical to the dictum of Rudolf Virchow that captures the
understanding that cells propagate by division but never, to our
knowledge, emerge *de novo*: *Omnis cellula e cellula* (all cells from
[other] cells). Applied to the Virus World, this principle becomes:
Omnis virus e virus.

Certainly, there are viruses and various virus-like agents that
stretch this continuity principle. The most conspicuous examples
could be the numerous, diverse viruses of hyperthermophilic
Crenarchaeota (Prangishvili, et al., 2006a, 2006b) and the group of
polydnaviruses that infect insects. Some of these viruses share no
genes with any other viruses; the crenarchaeal viruses generally have
few genes with recognizable provenance, whereas polydnaviruses
contain a variety of genes derived from the host (Dupuy, et al., 2006).
However, even among these strange viruses, some retain a hallmark
gene or two; moreover, polydnaviruses show clear signs of inactivation
in these surviving hallmark genes. Apparently, polydnaviruses evolved
from different groups of full-fledged insect viruses, the ascoviruses
and the nudiviruses, through a route that closely resembles the evolution of the GTAs (Bezier, et al., 2009; Bigot, et al., 2008). A nudivirus
genome was incorporated into the genome of an ancestral wasp and is
actively expressed, providing the proteins required for the formation
of the virions and DNA packaging. However, the viral genes (at least,
most of them) are no longer incorporated into the virions, which
instead package random fragments of the host genome. This striking
convergence of the evolution of pseudoviruses in prokaryotes and
eukaryotes implies a powerful selective pressure for the "invention"
of dedicated devices for HGT and genome rearrangement.

These observations on unusual virus genomes point to the
remarkable plasticity but also equally notable resilience of the virus
state: Viruses have the potential to dramatically change their genomic
content, to the point of shedding all genes directly involved in the

viral genome replication, yet remain distinctly viral in terms of their life cycles and intracellular parasitic lifestyle.[6] The GTAs and the polydnaviruses take this plasticity an important step further because, in these cases, the content of the virus genome is completely displaced. Moreover, these pseudoviruses dispense with independent replication; the genes required for virion formation and DNA packaging are replicated only within a host genome, so *the relationship between a virus and a cellular host(s) becomes symbiotic rather than parasitic*. Such symbiosis could be an extremely widespread phenomenon and may assume a variety of forms, as suggested by the demonstration that elimination of all prophages from a bacterial genome substantially decreases the resistance of the bacterium to diverse forms of stress (Wang, et al., 2010). Thus, many of the prophages seem to be not parasites, but rather bona fide symbionts.

In this chapter, I primarily emphasize the continuity and relative autonomy of the Virus World, but paths of genetic information transmission that directly connect the viral and cellular empires certainly abound and are crucial for evolution of both empires. The entire history of the interaction between reverse-transcribing elements and eukaryotes that we have mentioned on multiple occasions could serve as the primary case in point. From the ancient acquisition of introns, the spliceosome, and the telomerase to the very recent recruitment of provirus sequences as promoters, the genomes of eukaryotes have been shaped by numerous, continuous contributions from reverse-transcribing elements. In the course of evolution, these agents lose their selfish character: They have effectively left the Virus World and entered the world of cellular life forms. Clearly, analogous processes are widespread in prokaryotes where prophage genes systematically become lodged in chromosomes and lose their association with viral genomes. Conversely, as also mentioned earlier, viruses with large genomes systematically and continuously acquire numerous host genes that often remain within virus genomes for hundreds of millions of years. The NCLDV—in particular, poxviruses—that possess dozens or even hundreds of genes that were acquired from the host at different stages of evolution and are primarily involved in virus-host interactions represent a perfect case in point. Thus, beyond doubt, multiple two-way routes connect the viral and cellular worlds.

The fundamental inevitability of parasites

In this chapter, we discussed at some length the empirical evidence of the existence of an ancient, semi-autonomous, temporally continuous Virus World that seems to have been a key component of the biosphere since its inception in the precellular era to this day. These observations actually are supported by strong theoretical indications that the emergence of parasites is *an inevitable consequence of any evolutionary process in which there is a distinction between the genome (genotype) and phenotype*. In turn, the phenotype-genotype distinction is part and parcel of the EPR principle that we discussed in Chapter 2. Parasites invariably appear in computer simulations of the evolution of simple replicators (Szathmary and Maynard Smith, 1997; Takeuchi and Hogeweg, 2008); without going into detail, this outcome is easy to rationalize intuitively.

Let us consider the issue in the context of an RNA-protein replicator system resembling the replication of modern RNA viruses—that is, in which each replicator encodes its own replicase (we deliberately disregard the requirements for a translation system and for monomers, or rather implicitly include it all as an unspecified resource—more on this in Chapter 12). It appears inevitable that the evolution of this system will result in differentiation into "hosts" and parasites. The hosts will be regular genomes encoding a replicase, and the parasites typically will be reduced genomes that have lost the replicase gene but have retained all sequence elements required for efficient recognition by the host replicase (see Figure 10-4). This will necessarily happen because of the selective pressure for faster replication—an obvious strategy to increase the replication rate is to dispense with the large replicase gene, to keep only those parts of the genome that recognize the replicase, to utilize the replicase that the host genome produces, and, furthermore, to evolve increasingly efficient recognition sites for that replicase. Actually, computational models suggest that, without compartmentalization, parasites outcompete the hosts, so the entire system collapses and eventually goes extinct; however, when compartmentalization is explicitly included in the model, in an imitation of protocells (see more in Chapters 11 and 12), the host-parasite systems adopts a stable evolutionary strategy (Takeuchi and Hogeweg, 2008).

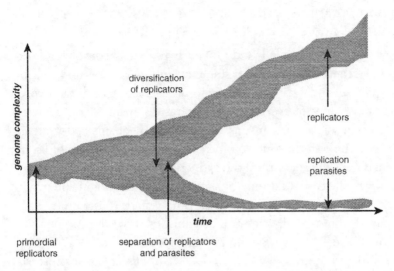

Figure 10-4   Differentiation of a population of evolving replicators into hosts and parasites.

The emergence of parasites as an intrinsic property of replicator systems is not just a matter of theory and simulations. The famed experiments of Spiegelman and colleagues outlined in Chapter 8 demonstrate the same trend in the simplest conceivable situation, in which the only selective force affecting an evolving population of replicators (RNA molecules) is to recognize the replicase and multiply. In these experiments, an RNA bacteriophage genome rapidly loses all coding sequences (including the largest gene, the replicase) and becomes the ultimate selfish replicator, a small RNA that consists almost solely of elements required for replication (Mills, et al., 1973). This outcome is in a full accord with the simple intuition and with the simulation results that reveal the inevitable emergence of parasites. In the present-day Virus World, such selfish replicators continuously emerge *de novo* and are propagated as so-called defective interfering derivatives of viral genomes (producing defective interfering particles) that are particularly common among RNA viruses (Bangham and Kirkwood, 1993). The defective interfering virus derivatives effectively parasitize on the parental virus.

The perennial host-parasite arms races and evolution of defense and counterdefense systems

According to the so-called Red Queen hypothesis,[7] coevolving host-parasite systems can maintain a stable evolutionary trajectory only by perpetually changing in an unceasing arms race. The hosts evolve new defense mechanisms, and the parasites respond by evolving counterdefense mechanisms, as well as new mechanisms for attack that evade the defense, and so on *ad infinitum,* if evolution of life in general is considered, or until the extinction of either host or parasite, in each specific case. Mathematical modeling of the origin and evolution of replicator systems not only inevitably leads to the emergence of parasites, but actually shows that parasites drive the evolution of replication mechanisms (Szathmary and Demeter, 1987). *This arms race is one of the principal driving factors of all evolution* (Forterre and Prangishvili, 2009). The validity of this statement appears obvious when one reviews the known defense and counterdefense mechanisms. Multiple, multilayer defense systems account for substantial parts of the genomes of all cellular organisms, with the sole exception of some intracellular parasites; conversely, counterdefense is among the primary gene functions in viruses with large genomes.

In prokaryotes, the antiviral defense repertoire includes the CRISPR-Cas adaptive immunity systems, which we discussed at some length in Chapter 9; the extremely diverse restriction-modification systems (essential tools of genetic engineering); and an apparently large variety of additional, less thoroughly characterized antivirus systems. Prokaryotic genomes contain multiple "virus defense islands" that are enriched in known defense systems but also contain numerous uncharacterized genes (Merkl, 2006). In many cases, a careful analysis of the protein sequences of the unknowns reveals domains typical of defense functions, such as highly diverged restriction enzymes and other nucleases (Makarova, et al., 2009b). Thus, there is little doubt that quite a few novel defense systems remain to be discovered. There is no good estimate of the characteristic fraction of a bacterial or archaeal genome dedicated to antivirus defense, and these values are likely to be highly variable; however, in most nonparasitic organisms, they are expected to be quite high, on the order of 20% of the total gene repertoire.

The evolutionary dynamics of the defense systems can be extremely complex, especially in prokaryotes, because the defense mechanisms such as CRISPR-Cas or the restriction-modification systems not only abrogate the reproduction of selfish elements, but more generally prevent HGT (Marraffini and Sontheimer, 2008). Given that HGT is the principal route of innovation in archaea and bacteria, and seems to be essential for the survival of asexual microbial populations (see Chapter 5), there is selection pressure against the defense systems. Along with the Red Queen dynamics, this pressure is the probable cause of the evolutionary instability of the defense systems, particularly their extremely frequent loss.

The defense systems of eukaryotes certainly are even more diverse and complex. These include the nearly ubiquitous RNAi machinery and various other mechanisms of innate and adaptive immunity (particularly the dsRNA-stimulated interferon system) that we cannot discuss in detail here. It is worth noting that, in eukaryotes, HGT does not play a role comparable to its role in prokaryotes, hence the lack of selective pressure for the elimination of defense systems that consequently appear to be more stable on the evolutionary scale. Taken together, defense systems occupy a large part of any eukaryotic genome—consider the Major Histocompatibility Complex and immunoglobulin gene clusters in vertebrates, or the huge clusters of stress response genes in plants.

A special and, indeed, radical type of the anti-parasite response is programmed cell death (PCD) that seems to occur in diverse forms in most cellular organisms, with the exception of some bacterial parasites. The best-characterized manifestations of this phenomenon are the elaborate PCD systems of animals and plants (also known as apoptosis, primarily with respect to animals) that involve cascades of suicidal proteolytic and nucleolytic reactions and are routinely triggered by virus infection (as well as by other parasites and other forms of stress). The existence of PCD in unicellular organisms, especially prokaryotes, is a more controversial matter (Bidle and Falkowski, 2004; Koonin and Aravind, 2002), but increasing evidence shows that the toxin-antitoxin systems in bacteria and archaea do trigger PCD in response to virus infection or other forms of stress (Van Melderen, 2010).

Given the Red Queen dynamics, viruses are never far behind (those that were have gone extinct). Eukaryotic viruses with large genomes such as poxviruses or baculoviruses are the best known case in point: Up to half of the genes in these viruses function as counterdefense devices, against all levels of the host defense. The main counter defense strategy these viruses employ is simple and efficient: The virus "steals" a gene encoding a component of a host defense. After mutating in the viral genome, the protein product of this gene morphs from an effector into a dominant-negative inhibitor of the respective defense system. Smaller viruses cannot afford a comparable repertoire of counterdefense genes, but they nevertheless carry genes for "security proteins" that are mostly involved in aggression, such as proteases that cleave protein factors required for the translation of host but not viral RNAs (Agol and Gmyl, 2010). At a different, more fundamental level, a notorious manifestation of the Red Queen effect is the rapid antigenic change in some viruses, such as influenza and HIV, which allows these viruses to evolve ahead of the host immune responses.

It is important to emphasize that parasites and defense systems are linked not only through the "genomic wars," but also in a more direct way: Selfish elements are systematically recruited for defense functions, whereas defense systems may evolve selfish features. The restriction-modification systems are traditionally construed as a means of defense, and they certainly make bacteria resistant to foreign DNA. However, these systems are also selfish elements of a special kind (Kobayashi, 2001, 1998). Although they do not encode any devices for their own replication and, hence, are not bona fide denizens of the Virus World, they often reside on plasmids, thus effectively entering the Virus World as symbionts of plasmids (that do belong to the Virus World). The restriction-modification systems render the cells addicted to themselves and to the carrier plasmids, and hence promote their own propagation. The toxin-antitoxin systems (see Chapter 5) possess similar properties. More generally, it is well known that, both in bacteria and in eukaryotes, viruses tend to protect hosts from superinfection by other viruses. Thus, the interplay between viruses ("genomic parasites") and host defense systems is a highly complex network with numerous feedback loops that is central to the evolution of both parasites and hosts.

Finally, although this chapter is dedicated to the Virus World, at this point, a note on parasitism among cellular life forms is due. This form of parasitism is relatively uncommon among prokaryotes (however, note the remarkable, tiny parasitic archaeon *Nanoarchaeum equitans*), but it has become extremely widespread in eukaryotes whose large, complex cells and especially multicellular organisms are excellent targets for microbial parasites. During their evolution, these parasites, especially intracellular ones, progressively lose their own antivirus defense systems and evolve mechanisms that counter host defenses. In that respect, they start to *resemble* viruses, but they appear to never *become* viruses.

Synopsis and perspective

In this chapter, we discussed the ancient, vast Virus World. The ideas of the place of viruses in the biosphere and its evolution have drastically changed over the last few years, thanks largely to the advances of metagenomics of diverse viromes. We realize now that, rather than being inconspicuous intracellular parasites, viruses are the dominant entities in the biosphere, in both physical and genetic terms. Comparative genomics of viruses and virus-like elements such as plasmids and transposons *reveals a complex network of evolutionary relationships in which the hubs correspond to hallmark genes shared by a variety of diverse groups of viruses*, with only distant homologs in cellular life forms. These findings of comparative genomics do not support the hypothesis of a single ancestor for all viruses; instead, they suggest the "oligophyletic" scenario, under which the major classes of viruses, with all types of replication-expression cycles, evolved directly from a primordial, precellular pool of genetic elements. In retrospect, this conclusion might not be particularly unexpected, given that theory and experiment indicate that emergence of parasites is an intrinsic feature of evolving replicator systems and that the host–parasite competition drives their evolution.

Thus, we recently became aware of the existence of a vast Virus World that evolved continuously and semiautonomously from cellular organisms throughout the history of life on Earth. In a general classification of life forms, *viruses and other selfish elements on one hand and cellular life forms on the other hand represent the two major "empires."* Notwithstanding the importance of the hallmark genes and the relative autonomy of the Virus World, these two empires are connected by multiple two-way routes of genetic exchange so that viruses largely shape the evolution of cellular genomes, and vice versa. Moreover, the interaction between the viral and cellular empires follows the Red Queen dynamics of the perennial arms race, which is one of the primary formative factors in the evolution of all genomes and one of the clearest manifestations of Darwin's *struggle for existence.*

Somewhat paradoxically, recent research advances not only reveal the vast expanse of the Virus World, but also show how little we know of its actual structure and content. The analysis of viromes suggests that the Virus World mostly consists of uncharacterized "dark matter" that could be very different from the known viruses. Understanding this dark matter could well lead to substantial changes in the general picture of the evolution of life.

Recommended further reading

Forterre, P., and D. Prangishvili. (2009) "The Great Billion-Year War Between Ribosome- and Capsid-Encoding Organisms (Cells and Viruses) As the Major Source of Evolutionary Novelties." *Annals of the New York Academy of Sciences* 1,178: 65–77.

Important conceptual article that emphasizes the key role of the arms race between viruses and their hosts in the evolution of complexity.

Holmes, E. C. (2009) *The Evolution and Emergence of RNA Viruses.* Oxford: Oxford University Press.

An overview of the evolution of viruses with an emphasis on the macroevolutionary processes, rapid change, and emergence of new viruses.

Koonin, E. V., T. G. Senkevich, and V. V. Dolja. (2006) "The Ancient Virus World and Evolution of Cells." *Biology Direct* 1: 29.

A conceptual analysis of the relationships among diverse groups of viruses, revealing the wide spread of hallmark viral genes. The notion of the Virus World, a continuous flow of genetic information through diverse selfish genetic elements, is developed. The Virus World evolved concomitantly with cellular life forms throughout the entire history of life on Earth but maintained autonomy from the evolving cellular organisms, despite multiple gene exchanges.

Koonin, E. V., Y. I. Wolf, K. Nagasaki, and V. V. Dolja. (2008) "The Big Bang of Picorna-like Virus Evolution Antedates the Radiation of Eukaryotic Supergroups." *Nature Reviews Microbiology* 6: 925–939.

An evolutionary scenario for the largest group of RNA viruses infecting eukaryotes, the picorna-like superfamily that was vastly expanded through metagenomics. The comparative genomic results point to the assembly of the ancestral picorna-like viral genome from diverse prokaryotic elements, followed by a phase of explosive evolution.

Koonin, E. V., and N. Yutin. (2010) "Origin and Evolution of Eukaryotic Large Nucleo-cytoplasmic DNA Viruses." *Intervirology* 53: 284–292.

An overview of the evolution of the NCLDV, including an analysis of the diverse origins of the core NCLDV genes.

Kristensen, D. M., A. R. Mushegian, V. V. Dolja, and E. V. Koonin. (2010) "New Dimensions of the Virus World Discovered Through Metagenomics." *Trends in Microbiology* 18: 11–19.

An overview and statistical analysis of the virus metagenomic data, arriving at the conclusion that the reported viromes indeed consist primarily of viral sequences instead of random bacterial contamination. The hypothesis is proposed that much of the "dark matter" in the viromes could represent GTAs.

Krupovic, M., and D. H. Bamford. (2008) "Virus Evolution: How Far Does the Double Beta-barrel Viral Lineage Extend?" *Nature Reviews Microbiology* 6: 941–948.

Article that presents the concept of the primacy of the capsid proteins as the markers for evolutionary lineages of viruses and, on the basis of the idea, numerous viruses of bacteria, archaea, and eukaryotes that possess the double beta-barrel (jelly roll) capsid proteins that belong to an ancient, vastly diversified lineage. A counterargument, that the evolution of viruses cannot be reduced to the evolution of capsid proteins, has been published.

Moreira, D., and P. Lopez-Garcia. (2009) "Ten Reasons to Exclude Viruses from the Tree of Life." *Nature Reviews Microbiology* 7: 306–311.

Moreira and Lopez-Garcia argue that, because viruses are obligate intracellular parasites and do not encode all the information that is necessary for their own replication, they effectively lack distinct evolutionary histories, are irrelevant for understanding the evolution of cells, and should not be considered "alive." A series of follow-up letters in *Nature Reviews Microbiology*, including one from my colleagues and myself, counter these propositions (apart from the metaphysical issue of "being alive") by demonstrating the evolutionary coherence of major groups of viruses and the spread of hallmark genes through the Virus World.

Raoult, D., and P. Forterre. (2008) "Redefining Viruses: Lessons from Mimivirus." *Nature Reviews Microbiology* 6: 315–319.

In the wake of the discovery of giant viruses, Raoult and Forterre delineate two fundamental types of organisms: capsid-encoding organisms (viruses) and ribosome-encoding organisms (cellular life forms). However, this classification ignores the numerous evolutionary connections between bona fide viruses and various mobile elements that lack capsid.

Van Etten, J. L., L. C. Lane, and D. D. Dunigan. (2010) "DNA Viruses: The Really Big Ones (Giruses)." *Annual Review of Microbiology* 64: 83–99.

The discovery of giant viruses (sometimes called giruses) in the NCLDV class blurred the boundaries between viruses and cells in terms of genome size and highlighted the role of HGT in the evolution of large viruses.

11

The Last Universal Common Ancestor, the origin of cells, and the primordial gene pool

In the preceding chapters, we addressed the fundamental aspects of the evolution of prokaryotes and eukaryotes, and outlined the ancient Virus World. This lays the groundwork for the discussion of the crucial transition in the evolution of life, the origin of cellular organization and of different types of cells. As Darwin first presciently proposed (Darwin, 1859) and comparative genomics amply vindicated, all extant cells evolved from a common ancestor that has become known as LUCA, after Last Universal Common Ancestor (of cellular life). However, no consensus exists on the nature of the LUCA and the degree to which it resembled modern cells. Arguments for a LUCA that would be indistinguishable from a modern prokaryotic cell have been presented, along with scenarios depicting LUCA as a much more primitive entity (Glansdorff, et al., 2008).

The difficulty of the problem cannot be overestimated. Indeed, all known cells are complex and elaborately organized. The simplest known cellular life forms, the bacterial (and the only known archaeal) parasites and symbionts (see Chapter 5), clearly evolved by degradation of more complex organisms; however, even these possess several hundred genes that encode the components of a fully fledged membrane; the replication, transcription, and translation machineries; a complex cell-division apparatus; and at least some central metabolic pathways. As we have already discussed, the simplest free-living cells are considerably more complex than this, with at least 1,300 genes. The only known autonomously replicating agents that are substantially simpler are viruses, but these are obligate intracellular parasites and do not present anything resembling an intermediate stage between a cell and a virus (whatever the direction of evolution might have been). So considering *Omnis cellula e cellula* and *Omnis virus e virus*, something has to give: The uniformitarian principle cannot apply to the origin of cells that must have evolved through a series of events that were fundamentally different from the familiar evolutionary processes. So here we discuss first the reconstruction of the gene repertoire of LUCA and then the implications of the results for the origin of cells.

Comparative-genomic reconstruction of the gene repertoire of LUCA

Why do we believe that there was a LUCA? More than one argument supports the LUCA conjecture, but the strongest one seems to be the universal evolutionary conservation of the gene expression system. Indeed, all known cellular life forms use essentially the same genetic code (the same mapping of 64 codons to the set of 20 universal amino acids and the stop signal), with only a few minor deviations in highly degraded genomes of bacterial parasites and organelles. All cells use homologous ribosomes that consist of three universally conserved RNA molecules and some 50 proteins, of which about 20 are universally conserved. Additional universally conserved components of the translation system include about 30 tRNAs, several translation factors, 18 aminoacyl-tRNA synthetases, and several tRNA modification enzymes (Anantharaman, et al., 2002). Beyond the translation system, the only universally conserved genes are those for the three core subunits of the RNA polymerase. Thus, altogether, there are about 100 universally conserved genes, all of which are involved in gene expression (in practice, it is rather common for some of these genes, especially those for small ribosomal proteins, to be missing in the annotations of new sequenced genomes [Charlebois and Doolittle, 2004]—however, it seems most likely that all such disappearances are artifacts of sequencing or annotation rather than actual losses). We are already familiar with these (nearly) universal genes from Chapter 6, where we saw that they display consistent (although not identical) phylogenetic tree topologies. The universal conservation of the code and the expression machinery, and the mostly coherent evolutionary history of its components, leave no reasonable doubt that this system is the heritage of some kind of LUCA. The real issue, then, is not whether a LUCA existed, but rather, what it was like, which features of this entity we can infer with reasonable confidence, and which (so far) remain uncertain.[1]

Clearly, no organism, however primitive, could consist of the expression machinery alone, so reconstruction of the rest of the gene repertoire of the LUCA is required. The methods of reconstruction are the same as those outlined in Chapters 5 and 7, and so are the pitfalls. In the context of the evolution of prokaryotes, which is relevant for the reconstruction of LUCA, one has to consider three classes of elementary events: (i) gene "birth"—that is, the emergence of a new gene, typically via gene duplication, followed by radical divergence, (ii) gene acquisition via horizontal gene transfer (HGT), and (iii) gene loss. Let us recall that reliable reconstruction of the course of evolution and of the ancestral gene sets is hampered by the uncertainty associated with the relative probabilities or rates of different events, particularly, gene loss versus HGT. In principle, even a gene that is found in all modern cellular life forms might not be inherited from LUCA: Its ubiquity could instead result from an HGT sweep. Furthermore, the estimates of the gene content of ancestral forms produced by parsimony and even by maximum likelihood are conservative, and the extent of the underestimate is uncertain. All the difficulties and uncertainties of evolutionary reconstructions notwithstanding, parsimony analysis combined with less formal efforts on the reconstruction of the deep past of particular functional systems leaves no serious doubts that LUCA already possessed at least several hundred genes (Mirkin, et al., 2003; Ouzounis, et al., 2006; Snel, et al., 2002). In addition to the aforementioned "golden 100" genes involved in expression, this diverse gene complement consists of numerous metabolic enzymes, including pathways of the central energy metabolism and the biosynthesis of amino acids, nucleotides, and some coenzymes, as well as some crucial membrane proteins, such as the subunits of the signal recognition particle (SRP) and the H+-ATPase.

However, the reconstructed gene repertoire of LUCA also has gaping holes. The two most shocking ones are (i) the absence of the key components of the DNA replication machinery, namely the polymerases that are responsible for the initiation (primases) and elongation of DNA replication and for gap-filling after primer removal, and the principal DNA helicases (Leipe, et al., 1999), and (ii) the absence of most enzymes of lipid biosynthesis (Pereto, et al., 2004). These essential proteins fail to make it into the reconstructed gene repertoire of LUCA because the respective processes in bacteria, on one hand, and archaea, on the other hand, are catalyzed by different, unrelated enzymes and, in the case of membrane phospholipids, yield chemically distinct membranes.[2] Thus, the reconstructed gene set of LUCA seems to be remarkably nonuniform, in that some functional systems appear to have reached complexity that is almost indistinguishable from that in modern organisms, whereas others come across as rudimentary or missing. This strange picture resembles the general concept of asynchronous "crystallization" of different cellular systems at the early stages of evolution that Carl Woese proposed (Woese, 1998) and prompts one to step back and take a more general view of the LUCA problem.

It seems to make sense to think of LUCA in two distinct dimensions (Koonin, 2009c):

- Genetic complexity that can be expressed as the number of distinct genes.
- The degree of organizational and biological similarity to modern cells. For brevity and convenience, this property may be denoted cellularity.

These two characteristics are likely to correlate but are not necessarily tightly coupled, let alone deterministically linked. In principle, it is not inconceivable that LUCA was a cellular entity that was substantially simpler than any modern cell (at least, a free-living one) in terms of its genetic content, or, conversely, that considerable genetic complexity evolved prior to the emergence of the (modern type) cellular organization—for the latter scenario, we will use the term *Last Ancestral Universal Common State (LUCAS)* (see Figure 11-1).

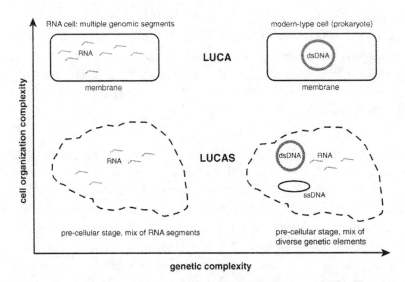

Figure 11-1 The space of logical possibilities for the LUCA(S): genetically complex vs. simple, cellular vs. noncellular. It is assumed that a pool of RNA elements is genetically simpler than a mixed pool of diverse genetic elements (LUCAS), and a putative RNA cell is simpler than a modern-type cell (LUCA).

The "uniformitarian assumption," namely that LUCA was a more or less regular, modern-type prokaryotic cell, is often accepted by default in the discussions of early evolution, even if rarely explicitly stated. However, any reconstruction of LUCA must account for the evolution of the features that are not immediately traceable back to the common ancestor of archaea and bacteria, the two main ones being DNA replication and membrane biogenesis (and chemistry). The uniformitarian hypotheses that explicitly or implicitly are based on the cellular LUCA assumption would explain the lack of conservation of these key systems in one of two ways:

- LUCA somehow combined both versions of these systems, with the subsequent differential loss in the archaeal and bacterial lineages.
- LUCA had a particular version of each of these systems, with subsequent nonorthologous displacement in archaea or bacteria.

More specifically, with regard to membrane biogenesis, it has been proposed that LUCA had a mixed, heterochiral membrane, so that the two versions with opposite chiralities emerged as a result of subsequent specialization in archaea and bacteria, respectively (Pereto, et al., 2004). With regard to DNA replication, a hypothesis has been developed under which one of the modern replication systems is ancestral, whereas the other system evolved in viruses and subsequently displaced the original system in either the archaeal or the bacterial lineage (Forterre, 1999, 2006).

By contrast, more radical proposals on the nature of LUCA adopt a "what you see is what you get" approach and so postulate that LUCA lacked those key features that are not homologous in extant archaea and bacteria, at least in their modern form (Koonin, 2009c; Koonin and Martin, 2005). The possibility that LUCA was dramatically different from any known cells has been brought up, originally in the concept of "progenote," a hypothetical, primitive entity in which the link between the genotype and the phenotype was not yet firmly established (Doolittle and Brown, 1994; Woese and Fox, 1977). In its original form, the progenote idea involves primitive, imprecise translation, a notion that is not viable, given the extensive pre-LUCA diversification of proteins that the analysis of diverse protein superfamilies has demonstrated beyond doubt (see Chapter 12). More realistically, it can be proposed that the emergence of the major features of cells ("crystallization" *sensu* Woese) was substantially asynchronous, so that LUCA closely resembled modern cells in some ways but was distinctly "primitive" in others. The results of comparative genomics provide clues for distinguishing between advanced and primitive features of LUCA. Thus, focusing on the major areas of nonhomology between archaea and bacteria, it has been hypothesized that LUCA(S):

- Did not have a typical, large DNA genome
- Was not a typical membrane-bounded cell (see Figure 11-2; Koonin and Martin, 2005).

With respect to the DNA genome and replication, the conundrum to explain was the combination of non-homologous and conserved components in the DNA replication machineries in archaea and bacteria, along with the universal conservation of the core transcription machinery. To account for this mixed pattern of conservation and diversity, it has been suggested that LUCA had a "retrovirus-like" replication cycle, with the conserved transcription machinery involved in the transcription of provirus-like dsDNA molecules and the conserved components of the DNA replication system playing accessory roles in this process (Leipe, et al., 1999). This speculative scheme combined, in the same hypothetical replication cycle, the conserved proteins that are involved in transcription and replication with proteins, such as the RT, that at least in the modern biosphere clearly belong in the Virus World (see Chapter 10). This putative primordial replication-expression cycle formally accounts for the universal conservation of the involved proteins and the nonuniversality of other key components of DNA replication machineries, but it has no direct analogy in extant genetic systems.

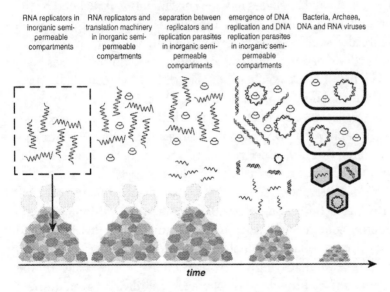

RNA replicators in inorganic semipermeable compartments | RNA replicators and translation machinery in inorganic semipermeable compartments | separation between replicators and replication parasites in inorganic semipermeable compartments | emergence of DNA replication and DNA replication parasites in inorganic semipermeable compartments | Bacteria, Archaea, DNA and RNA viruses

time

Figure 11-2   The Virus World scenario of precellular evolution.

The other major area of nonhomology between archaea and bacteria, lipid biosynthesis (along with lipid chemistry), prompted the even more radical hypothesis of a noncellular although compartmentalized LUCA (Martin and Russell, 2003; Koonin and Martin, 2005). Specifically, it has been proposed that LUCA(S) might have been a diverse population of expressed genetic elements that dwelled in networks of inorganic compartments (see Chapter 12 for further discussion of these potential hatcheries of life). A major hurdle for the models of non-membrane-bounded LUCA is that several membrane proteins and even molecular complexes, such as proton ATPase and signal recognition particle (SRP), are nearly universal among modern cellular life forms and, in all likelihood, were present in LUCA.

A more careful consideration of the major challenges to a cellular LUCA—the lack of homology between the core components of the DNA replication systems and the radical difference between the phospholipids and the enzymes of lipid biosynthesis in archaea and bacteria—suggests that the two issues are tightly linked (Koonin, 2009c). A complex LUCA (as suggested by comparative genomic reconstructions) without a large DNA genome comparable to modern bacterial and archaeal genomes would have to carry a genome consisting of several hundred segments of RNA (or provirus-like DNA), each several kilobases in size. This limitation is dictated by the dramatically lower stability of RNA molecules compared to DNA and is empirically supported by the fact that the largest known RNA genomes (those of coronaviruses) are about 30 Kb in size. It has been proposed that LUCA could be a bona fide RNA cell that subsequently radiated into three major RNA cell lineages (the ancestors of bacteria, archaea, and eukaryotes), in which the genome was independently replaced by DNA as a result of acquiring the DNA replication machinery from distinct viruses (Forterre, 2006). However, this scenario appears far-fetched because the necessity to possess hundreds of "chromosomes" seems to raise an insurmountable obstacle for a RNA cell. Indeed, a reasonable accuracy of partitioning such a fragmented genome into daughter cells during cell division would require extremely elaborate mechanisms of genome segregation of a kind not found in modern prokaryotes. Otherwise, the change in the gene complement brought about by each cell division would effectively prevent reproduction. Those segregation mechanisms that operate in modern bacteria (and probably archaea) involve pumping dsDNA into daughter cells with the help of a specific ATPase and, in all likelihood, coevolved with dsDNA genomes. Thus, if LUCA indeed lacked a large dsDNA genome and instead had a "collective" genome comprised of numerous RNA segments, it must have been a life form distinct from modern cells—perhaps a noncellular life form.

Another broadly discussed aspect of early life forms, including LUCA, is the rampant HGT that is often considered a prerequisite for the evolution of complex life (yet another notable proposition of Carl Woese [Woese, 2002]). Indeed, HGT is the route of rapid innovation (see Chapters 5 and 7), and innovation must have been extremely rapid at the early stages of life's evolution. Woese and colleagues have proposed and illustrated through mathematical modeling that the very universality of the genetic code might be linked to the critical role of HGT during the early phase of evolution (Vetsigian, et al., 2006). Given extensive HGT, a single version of the code would necessarily sweep the population of ancestral life forms, whereas any organisms with a different code would be unable to benefit from HGT and, being isolated from other organisms, would be inevitably eliminated by selection.[3] Constant, extensive HGT is an intrinsic feature of the models of noncellular, compartmentalized LUCA(S) but certainly cannot be taken for granted within the framework of the cellular LUCA(S) models. In the next section, I discuss the noncellular LUCA model in some detail.

A noncellular but compartmentalized LUCA(S): A community of diverse replicators and the playground of early evolution

Michael Russell and colleagues proposed that networks of microcompartments that exist at both extant and ancient hydrothermal vents, and consist primarily of iron sulfide, could be ideal habitats for early life (Martin and Russell, 2003; Russell, 2007). These inorganic compartment networks provide gradients of temperature and pH that could fuel primordial energetics, and versatile catalytic surfaces for primitive biochemistry. The details of this environment remain a subject of investigation and debate (see Chapter 12 for some specifics), but there is little doubt that such networks of inorganic compartments are plausible hatcheries for the prebiological and precellular biological evolution, from mixtures of organic molecules to the putative, primordial RNA world, to the origin of cells that would escape the compartments (see Figure 11-2). These compartments would have been inhabited by diverse populations of genetic elements, initially segments of RNA and subsequently, larger and more complex RNA molecules encompassing one or a few protein-coding genes— and, later yet, also DNA segments of gradually increasing size (see Figure 11-2).

Thus, early life forms, possibly including LUCA(S), are perceived as complex ensembles of genetic elements that inhabited networks of inorganic compartments. A key feature of this model is that all replication and expression strategies that extant viruses employ (see Figure 10-1) and, accordingly, genetic elements encoding all the distinct replication and expression machineries would evolve and eventually coexist within a network and in some cases within the same compartment. Thus, the earlier, somewhat artificial scheme, in which the universally conserved components of the DNA replication machinery were implicated in a primordial, retrovirus-like replication cycle (Leipe, et al., 1999), might be superfluous.

This model explains the lack of homology between the membranes, membrane biogenesis systems, and the DNA replication machineries of archaea and bacteria by inferring a LUCA(S) that did not have a single, large DNA genome and was not a membrane-bounded cell. Under this model, the primordial, precellular, communal life forms are envisaged as "laboratories" in which various strategies of genome replication-expression, as well as rudimentary forms of biogenic compartmentalization, were "invented" and tried out (see Figure 11-2 and the discussion later in this chapter).

The central point of this scenario of the early evolution of life is the virus-like nature of the perceived precellular life forms: According to this model, *life started as the primordial Virus World*. As mentioned in Chapter 10, the idea that viruses could be related to the first life forms is almost as old as virology itself. Following the original speculation of d'Herelle, J. B. S. Haldane propounded this view in a more definitive form in his classic 1928 essay on the origin of life (Haldane, 1928). With his trademark prescience and panache, Haldane posited that the first self-reproducing agents were viruses or virus-like agents and that a virus stage in life's evolution preceded the emergence of cells: "[L]ife may have remained in the virus stage for many millions of years before a suitable assemblage of elementary units was brought together in the first cell." Subsequently, however, the concept of the primordial origin of viruses and the more dramatic idea of a primordial viral stage in the evolution of life were effectively abandoned as it became obvious that viruses were obligate intracellular parasites that depend on the host cells for most of their functions. Instead, as discussed in Chapter 10, the scenarios of cell degeneration or escaped cellular genes became dominant in the thinking on the origin and evolution of viruses.

The renaissance of virology in the first decade of the third millennium led to a proliferation of hypotheses and models that revolve around the concept that viruses were important contributors to the origin and evolution of cells. Under Patrick Forterre's scenario of "three RNA cells and three DNA viruses," modern-type DNA-based cells evolved when three distinct DNA viruses displaced the original RNA genomes in three cellular lineages (ancestors of bacteria, archaea, and eukaryotes, respectively; Forterre, 2006). The DNA viruses themselves are postulated to have evolved as parasites of these primordial RNA cells. However, as pointed out earlier in this chapter, RNA cells do not appear to be a viable proposition. The more plausible alternative scenario that seems to reconcile the results of comparative genomics and the general logic of precellular evolution revives Haldane's idea at a new level and involves evolution of diverse virus-like elements and even virus-like particles prior to the advent of modern-type cells.

The emergence of cells is the epitome of the problems encountered by all explanations of the evolution of complex biological structures (see Chapter 8). Indeed, among modern biological entities, we do not see any intermediates between macromolecules and cells, and to imagine how such intermediates might operate is a huge challenge. As repeatedly pointed out in this book, the minimal cell that is not a parasite or symbiont reproducing inside other cells has to carry at least 400 genes, whereas an autotrophic cell can hardly exist with fewer than 1,000 genes. These genes reside on one large chromosome (as in most prokaryotes) or on several smaller chromosomes and/or large plasmids (as in a minority of bacteria and archaea), but never on operon/gene-sized segments of DNA. The selective factor that drives the evolution of the large, contiguous genomes is straightforward, given *Omnis cellula e cellula*: To evolve, dividing cells must accurately segregate their genomes, which is virtually inconceivable with hundreds of segments. This evolutionary logic strongly suggests that the first cells actually would have a single chromosome—not only because this is the case in most of the modern archaea and bacteria, but, more importantly, considering the probable simplicity and relative inaccuracy of the ancestral division machinery. The evolutionary build-up of complex genomes encoding the minimal complement of genes defining a functional cell demands some form(s) of primordial, abiogenic compartmentalization that obviously should be able to function without requiring the complexity of the membrane apparatus of modern cells. This complexity is not to be underestimated: Recall that all cellular membranes are not only elaborate-transport devices, but also energy-transformation machines that convert electrochemical potential (proton or sodium gradient across the membrane) into the chemical energy of ATP.

At present, two forms of primordial, abiogenic compartmentalization can be seriously considered: lipid vesicles and networks of inorganic compartments. The lipid vesicles scenario is attractive because, in this case, the abiogenic membranes would be direct ancestors of the modern biological membranes. This possibility is being extensively studied experimentally, primarily in the laboratory of Jack Szostak, and interesting results on transport of polar compounds, including nucleotides, across lipid membranes have been reported (Mansy, et al., 2008). However, the difficulties this model faces remain formidable. These problems are obvious enough and include not only the transport of monomers at rates sufficient to support the replication of genetic elements and translation inside the vesicles prior to the emergence of protein transporters, but also generation and maintenance of membrane potential for energy production. Furthermore, the vesicle model does not seem to be conducive to extensive HGT, which is an essential aspect of all microbial evolution but would have been especially important at the precellular stage.

Without ruling out the potential relevance of the lipid vesicle model, let us consider the model of the origin of cells from a Virus World evolving within networks of inorganic compartments as an ancestral state. This model might encounter fewer problems than the lipid vesicle model and seems to offer several attractive features, including possible clues to the origin of biological membranes and bioenergetics. As in all biological evolution, precellular evolution was undoubtedly driven by a combination of random drift and natural selection. Opportunities for drift abound under this model—perhaps the most conspicuous one is seeding a virgin compartment with a random genetic element. Selection immediately enters the scene with the appearance of replicating entities (see Chapter 2)—initially, it is currently presumed, RNA molecules replicated by ribozymes and, subsequently, after the emergence of translation, RNA molecules replicating with the aid of proteins (see Chapter 12). One of the central aspects of the model of a virus-like, compartmentalized, pre-cellular stage of evolution is a *gradual transition from selection at the level of individual genetic elements to selection of ensembles of such elements* encoding enzymes directly involved in replication, as well as proteins responsible for accessory functions, such as translation and nucleic acid precursor synthesis. Selection at the level of ensembles of genetic elements obviously is a form of *group selection*, which is the subject of long-standing controversy among evolutionary biologists and sometimes is denied as fictitious. Without delving deep into the theoretical tangles, I posit that primordial evolution leading from small genetic elements to large genomes comparable to the genomes of modern cellular life forms appears to be all but impossible without some form of group selection (Koonin and Martin, 2005). Through mathematical modeling studies, Eors Szathmary and colleagues have demonstrated the feasibility of group selection in ensembles of replicators reproducing within compartments (Fontanari, et al., 2006; Szathmary and Demeter, 1987). Some of the solutions that might have been available to ensembles of "selfish cooperators" are known from group selection theory. The most obvious and important one seems to be reciprocal altruism, in which members of the group provide complementary functionalities enhancing each other's reproduction. Thus, in a primordial ensemble of genetic elements, an element encoding the replicase would catalyze the replication of elements encoding accessory functions for its own replication, such as translation system components and precursor synthesis.

Ensembles of selfish cooperators could potentially evolve via two (not at all mutually exclusive) routes: (i) physical joining of genetic elements and (ii) compartmentalization. The first route would represent the onset of the evolution of operons, including the ribosomal-RNA polymerase superoperon, the only array of genes that is substantially conserved between archaea and bacteria (see Chapter 5). The compartmentalization route would depend on the evolution of virus-like particles that could harbor relatively stable sets of genomic segments resembling the extant RNA viruses with multipartite genomes. Unlike cells, virus-like particles with small genomes, particularly, the nearly ubiquitous icosahedral (spherical) capsids, are simple, symmetrical structures that, in many cases, are formed by self-assembly of a single capsid protein. Thus, it is attractive to speculate that simple virus-like particles were the first form of genuine, biological compartmentalization that were important at the precellular stage of evolution. In addition to the benefit of compartmentalization, virus-like particles would protect genetic elements (especially RNA) from degradation and could serve as vehicles for gene movement between compartments and networks.

Most of the spherical viruses with relatively complex genomes possess molecular motors for DNA or RNA packaging within the capsid; at least in some cases, these molecular machines also work in the opposite direction, mediating extrusion of viral transcripts from the capsid (Rao and Feiss, 2008). The viral packaging and extrusion machines contain motor ATPases of at least three distinct families that seem to share a common architecture, forming hexameric channels through which DNA or RNA is actively translocated. Notably, one of the groups of viral packaging ATPases is a branch of the FtsK-HerA superfamily that also includes prokaryotic ATPases responsible for DNA pumping into daughter cells during cell division, whereas another family is homologous to bacterial twitching mobility ATPases (Iyer, et al., 2004b). In membrane-containing virions of many viruses, the packaging motors translocate the DNA or RNA across both the capsid and the lipid membrane of the virion. It is tempting to hypothesize that viral packaging machines were evolutionary precursors of the cellular pumping and motility ATPases. The H+(Na+)-ATPase/ATP synthase, the key, universal membrane enzyme that is the centerpiece of modern cellular energetics, also forms a similar hexameric channel and might have started out as part of the packaging/extrusion machinery in a still uncharacterized (possibly extinct) class of virus-like agents.

The membrane ion gradient-dependent ATP synthase is a remarkable molecular "dynamo" machine, a rotary motor that transforms an ion gradient into the mechanical energy of rotation, and then into the chemical energy of the β-γ phosphate bond in ATP. Comparative-genomic analysis has suggested that the common ancestor of the two major branches of membrane ATPases/synthases, the so-called F-ATPases typically found in bacteria (and the

endosymbiotic organelles of eukaryotes) and the V-ATPases characteristic of archaea and endomembrane systems of eukaryotes,
evolved from a common ancestor that functioned as a protein or RNA
translocase (Mulkidjanian, et al., 2007). The translocase stage antedates the radiation of bacteria and archaea, but the bona fide membrane ATP synthases are substantially different in archaea and
bacteria, and might have evolved twice, independently in the two
trunks of cellular life. These comparisons allow one to peer even further back into the precellular evolutionary past and to infer the origin
of this ancient translocase from an RNA helicase and a membrane
pore or channel (see Figure 11-3). The reconstruction of the evolution of membrane ATP synthases has a fundamental implication for
the staging of precellular evolution: Extensive diversification of the P-
loop enzymes (see Chapter 12) that yielded, among a variety of
ATPases, a particular family of RNA helicases (those including the
bacterial transcription termination factor Rho) occurred *prior to the
emergence of membrane energetics,* at least in the form that is universal in modern cells.

It is an attractive possibility that primordial viral membranes
could have been intermediate steps in the evolution of biological
compartmentalization that antedated the emergence of full-fledged
cellular membranes. Indeed, the evolution of modern-type, complex
membranes involves a paradox. The membranes of all modern cells
are extremely elaborate devices in which the lipid bilayer is impermeable even to small molecules, and all traffic between the inside of the
cell and the outside environment is mediated by membrane protein
complexes such as channels, pores, translocases, and the aforementioned gradient-dependent ATPases that are responsible for the cell
energetics. This membrane is yet another high-complexity system
whose origin faces the classic Darwinian problem: Viable intermediates
are hard to imagine. A leaky membrane would not help maintain the
integrity of the cell content, whereas a tight lipid-only membrane
would be of no use to a cell because it would prevent the import of
building blocks for replication. Virus-like particles can resolve this paradox because they would benefit from tight membranes as long as the
virion is equipped with a nucleic acid translocase. Just as genome replication of virus-like agents can be viewed as the original test ground for
replication strategies, two of which have been subsequently recruited

for the two major lineages of cellular life, evolving virus particles might have been the "laboratory" for testing molecular devices that were later incorporated into the membranes of emerging cells.

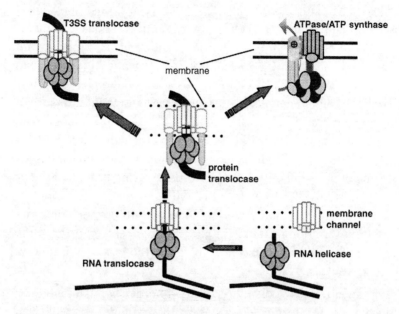

Figure 11-3   The model for the evolution of molecular motors and membrane bioenergetics: from an RNA helicase and a membrane channel to an RNA/protein translocase, to an ion gradient-dependent ATP synthase. Solid lines show modern-type, ion-tight membranes, and dotted lines show hypothetical leaky primordial membranes. The curved arrow shows cation efflux from the cell. T3SS is Type 3 Secretion System, a protein translocase common in modern bacteria. In protein translocases, the position of the central stalk is transiently occupied by the translocated protein, whereas, in the membrane ATPases, dedicated protein subunits take this position. The evolutionary scenario comes from Mulkidjanian, et al., 2007.

From the selection for gene ensembles, there is a direct path to selection for compartment content such that compartments sustaining rapid replication of genetic elements would "infect" adjacent compartments and effectively propagate their collective "genomes"; primordial virus-like particles could facilitate this process (Koonin and Martin, 2005). The precellular equivalent of HGT—that is, the transfer of the genetic content between compartments—is part of this model, in agreement with the idea that rampant HGT was an essential feature of the early stages of life's evolution. After a substantial degree of complexity has been attained through the evolution of self-ish cooperators within the networks of inorganic compartments, repeated escapes of cell-like entities that combined relatively large DNA genomes and membranes containing transport and transloca-tion devices (originally evolved in virus-like agents, under this model) became possible. There is no telling how many such attempts have failed quickly and how many might have lasted longer; only two, archaea and bacteria (assuming a symbiotic scenario for the later ori-gin of eukaryotes, as discussed in Chapter 7), survived to this day. The first successful escapes of cellular life forms from the hypothetical precellular pool would correspond to the *Darwinian threshold* for cellular life that Woese postulated—that is, the threshold beyond which HGT would be substantially curtailed and evolution of distinct lineages (species) of cellular organisms could take off (Woese, 2002).

As in other models of the early stages of evolution of biological complexity, and perhaps even more explicitly, the "primordial Virus World" scenario outlined here faces the problem of takeover by self-ish elements. As discussed in Chapter 10, the emergence of parasites is an intrinsic feature of any evolving replicator system. If the primor-dial parasites became too aggressive, they would kill off their hosts within a compartment and could survive only by infecting a new com-partment (where they could be dangerous again). Devastating "pan-demics" sweeping through entire networks and eventually wiping out their entire content are imaginable; indeed, this would be the likely fate of many, if not most, primordial "organisms." Notably, mathemat-ical modeling of replicator systems suggests that an important driving force behind the emergence of DNA, which led to the separation of the roles of template and catalyst at the precellular stage of evolution, in addition to the high stability of DNA, could have been the higher resistance to parasites in systems with dedicated templates (Takeuchi, et al., 2011). The conditions for the survival of precellular life forms were, first, the emergence of temperate parasites that did not kill the host, and second, the evolution of defense mechanisms, likely based on RNA interference (RNAi). The ubiquity of both temperate selfish elements and RNAi-based defense systems in all major branches of cellular life suggests that these phenomena evolved at a very early, quite possibly precellular stage of evolution.

Under this scenario, in the primordial gene pool, no clear-cut distinction existed between selfish genetic elements that later became viruses and larger gene ensembles that later gave rise to genomes of cellular life forms, although the beginning of such a distinction emerged when parasites started to feed off ensembles of "selfish cooperators." The emergence of cells was also the true beginning of the Virus World as we understand it now.

The primordial Virus World model of precellular evolution sketched here seems to offer plausible, even if largely speculative, solutions to many puzzles associated with the origin of cells. Comparative genomics of viruses and other selfish elements seems to provide substantial empirical support for this model. Considering that, under this scenario, the first cells emerged from a noncellular ancestral state in multiple, independent escapes, it seems sensible to replace the acronym LUCA with LUCAS, for *Last Universal Common Ancestral State—the state that corresponds to the primordial pool of virus-like genetic elements* (Koonin, 2009c).

Synopsis and Perspective

All extant life forms reproduce as cells or within cells. Although in Chapter 10 we considered strong arguments from comparative genomics in favor of the view that the Virus World evolved continuously and quasiautonomously from cellular life forms for the entire duration of life evolution on Earth, the fact remains that viruses cannot reproduce outside cells. We are unaware of any intermediate stages in cell evolution: Even the simplest cells possess the complex energy-coupling membrane, complete with diverse transport systems, as well as large DNA genomes and complex systems for genome replication and cell division. No uniformitarian explanation exists for the evolution of cells—the precellular "biota" necessarily must have been dramatically different from all life known to us. Here we discussed primarily the Virus World scenario for the evolution of both cells and viruses. Under this hypothesis, the precellular stage of the evolution of life took place within networks of inorganic compartments that hosted a diverse mix of virus-like genetic elements that gradually differentiated into ensembles of "selfish cooperators" and true parasites. These ensembles of genetic elements are thought to have been the ancestral state from which cells emerged, probably in multiple, independent escapes; only two of those (the ancestors of bacteria and archaea, respectively) yielded stable cellular lineages that enjoyed long-term evolutionary success.

Considering this hypothetical consortial state of primordial life forms that eventually gave rise to cells, it has been proposed to replace the acronym LUCA with LUCAS, for Last Universal Common Ancestral State. That LUCA(S) might have been quite different from modern cells, as suggested by the lack of homology between the key components of DNA replication and membrane biogenesis (and the different chemical structures of the lipids themselves) in archaea and bacteria. These fundamental differences between the two primary domains of cellular life seem to imply a noncellular LUCAS. However, notes of caution are due: For all its plausibility, the noncellular LUCAS scenario also faces substantial difficulties. For example, the universal conservation of the signal recognition particle, the ribonucleoprotein molecular machine that cotranslationally inserts nascent proteins into membranes, is not easy to explain under this scenario.

However intriguing the possibility of a noncellular LUCAS is and however important it is to reconstruct the details of this key ancestral state, all this is secondary with respect to the Virus World scenario. Even if the noncellular LUCAS model can be convincingly falsified and a compelling argument is made for a cellular LUCA, this will not render obsolete the model of precellular evolution discussed here, rather just push it back and imply a single successful cellular escape. Ditto for the model of inorganic compartment networks (more on this in Chapter 12). Even if this model is shown to be implausible, whereas, say, the model of cell evolution from lipid vesicles gains strong experimental support—this can hardly affect the requirement for a primordial pool of genetic elements. In short, *the virus-like character of the genetic pool at the precellular stage of the evolution of life seems to be a logical necessity.*

348

the logic of chance

Recommended further reading

Doolittle, W. F., and J. R. Brown. (1994) "Tempo, Mode, the Progenote, and the Universal Root." *Proceedings of the National Academy of Sciences USA* 91: 6,721–6,728.

A discussion of the nature of LUCA, particularly whether it is likely to have been a progenote, on the brink of the genomic era.

Glansdorff, N., Y. Xu, and B. Labedan. (2008) "The Last Universal Common Ancestor: Emergence, Constitution, and Genetic Legacy of an Elusive Forerunner." *Biology Direct* 3: 29.

A comprehensive review of the hypotheses and ideas on LUCA. Under the preferred model of Glansdorff, et al., the LUCA was a community of diverse RNA-based cells.

Koonin, E. V. (2009) "On the Origin of Cells and Viruses: Primordial Virus World Scenario." *Annals of the New York Academy of Sciences* 1,178: 47–64.

A conceptual analysis combining the model of precellular evolution in networks of inorganic compartments with the Virus World model, to suggest that the LUCAS was a community of virus-like entities.

Koonin, E. V. (2003) "Comparative Genomics, Minimal Gene-Sets, and the Last Universal Common Ancestor." *Nature Reviews Microbiology* 1: 127–136.

An overview and critical analysis of the reconstructions of minimal and ancestral gene sets.

Koonin, E. V., and W. Martin. (2005) "On the Origin of Genomes and Cells Within Inorganic Compartments." *Trends in Genetics* 21: 647–654.

The model of the early evolution of life, from the formation of first polymers to the emergence of cells. The LUCA(S) is envisaged as a noncellular community of diverse replicators. Multiple cellular escapes are postulated, with the only ones surviving in the long term being the ancestors of archaea and bacteria.

Morange, M. (2010) "Some Considerations on the Nature of LUCA, and the Nature of Life." *Research in Microbiology* 162: 5–9.

A discussion of the epistemological aspects of the research into the early stages of the evolution of life, including LUCA.

Mulkidjanian, A. Y, K. S. Makarova, M. Y. Galperin, and E. V. Koonin. (2007) "Inventing the Dynamo Machine: The Evolution of the F-type and V-type ATPases." *Nature Reviews Microbiology* 5: 892–899.

A scenario for the origin of the membrane ATPases (ATP synthases) that utilize transmembrane ion gradient to synthesize ATP, from helicase and a membrane pore protein complex. The scenario implies that substantial protein diversification, particularly the origin of helicases within the P-loop fold, antedates membrane bioenergetics of the modern type, so a different kind of energy transformation must have been operative at the early stages of evolution.

Mushegian, A. (2008) "Gene Content of LUCA, the Last Universal Common Ancestor." *Frontiers in Bioscience* 13: 4,657–4,666.

An update on the methods and results of the reconstructions of the gene repertoire of LUCA.

Woese, C. R. (2000) "Interpreting the Universal Phylogenetic Tree." *Proceedings of the National Academy of Sciences USA* 97: 8,392–8,396.

In this influential paper, Woese interprets LUCA as the root of the universal Tree of Life, the first stage in evolution when the integration of genetic elements became sufficiently tight to sustain the evolution of a cellular lineage.

Woese, C. R. (2002) "On the Evolution of Cells." *Proceedings of the National Academy of Sciences USA* 99: 8,742–8,747.

A development of the concept put forward in the preceding article. Woese posits that the first cellular life forms were "communal," a stage in evolution when rampant HGT was essential for innovation and no cellular lineages existed. The concept of the "Darwinian threshold" is introduced as a stage in evolution when sufficient genome cohesion was achieved to allow vertical inheritance.

12

Origin of life: The emergence of translation, replication, metabolism, and membranes—the biological, geochemical, and cosmological perspectives

In the preceding chapter, we discussed possible scenarios for the origin of cells and (hopefully) reached some degree of plausibility with the primordial Virus World scenario of cellular evolution. However, this was all about relatively late stages of evolution, at which replication of the genetic material and translation yielding diverse proteins were already well established. All these models seem to be of dubious value unless we develop some kind of explanation for the origin of the fundamental processes of information transmission.

The origin of life is the most difficult problem that faces evolutionary biology and, arguably, biology in general. Indeed, the problem is so hard and the current state of the art seems so frustrating that some researchers prefer to dismiss the entire issue as being outside the scientific domain altogether, on the grounds that unique events are not conducive to scientific study. However, this position appears deeply unsatisfactory, especially because, although life certainly evolved only once on this planet (see Chapter 11), we have no idea just how unique (or otherwise) it is in our universe as a whole. If one does accept the origin of life as a scientific issue, then there seems to be no denying that it is a problem of overwhelming importance before which all other questions in biology are relatively mundane.

It might seem natural to demand that, before one starts to analyze the origin of a particular phenomenon, the phenomenon in question be explicitly defined. A number of definitions of life have been given in the course of history of science and philosophy,[1] and the unmistakable stale smell of essentialism emanates from this whole issue (see Appendix A). However, in the context of the discussion in the preceding chapters, it is surprisingly easy to make a decision on what kind of entity should be considered living. *Any temporally stable replicator system is a life form.* An inalienable feature of any such replicator system is that it can and will evolve via some combination of drift and natural selection (the EPR principle—see Chapter 2). An implicit but important aspect of this definition is the existence of the genotype-phenotype feedback: Some of the mutations (replication errors) should affect the efficiency of replication (see Chapter 2). This feedback is quite conceivable within the hypothetical RNA World. However, in all known life forms, the separation between genotype and phenotype is sharper and more definitive: While the genotype invariably resides in nucleic acid molecules, the phenotype is largely embodied in proteins, molecules that possess exclusively operational (executive) as opposed to informational (template) functions.[2]

Therefore, although the origin of translation is not, in principle, an integral part of the origin of life problem (given that denizens of a full-fledged RNA World would have to be considered bona fide life forms), in practice, the two are tightly and perhaps inseparably linked. In this chapter, we discuss the entire conundrum of the origin of replication and translation. Given the exclusive, universal conservation of the translation machinery, this probably should be considered the core of the origin of life issue.

The origin of life is a problem that, by its very nature, cannot belong entirely in the domain of biology: Before there was life (even in its simplest imaginable embodiment), there must have been "prebiotic" chemistry that has to be analyzed from the chemical, geochemical, and geophysical perspectives. The relevant data is extensive and complex, and largely beyond my professional competence. In this chapter, we attempt only a brief overview, highlighting some of the most relevant findings.

Finally, and not without trepidation, we touch upon extremely general aspects of the probability of "unique events" in the context of the modern theories in cosmology. This discussion should allow us to at least develop some intuitions with regard to the frequency of life in the cosmos.

The origin of replication and translation and the RNA World

The Darwin-Eigen cycle

The primary incentive behind the theory of self-replicating systems that Manfred Eigen outlined (see Chapter 2) was to develop a simple model explaining the origin of biological information and, hence, of life itself. Eigen's theory revealed the existence of the fundamental limit on the fidelity of replication (the *Eigen threshold*): If the product of the error (mutation) rate and the information capacity (genome size) is below the Eigen threshold, there will be stable inheritance and hence evolution; however, if it is above the threshold, the mutational meltdown and extinction become inevitable (Eigen, 1971). The Eigen threshold lies somewhere between 1 and 10 mutations per round of replication (Tejero, et al., 2011); regardless of the exact value, staying above the threshold fidelity is required for sustainable replication and so is a prerequisite for the start of biological evolution (see Figure 12-1A). Indeed, the very origin of the first organisms presents at least an appearance of a paradox because a certain minimum level of complexity is required to make self-replication possible at all; high-fidelity replication requires additional functionalities that need even more information to be encoded (Penny, 2005). However, the replication fidelity at a given point in time limits the amount of information that can be encoded in the genome. What turns this seemingly vicious circle into the (seemingly) unending spiral of increasing complexity—the *Darwin-Eigen cycle*, following the terminology introduced by David Penny (Penny, 2005)—is a combination of natural selection with genetic drift. Even small gains in replication fidelity are advantageous to the system, if only because of the decrease of the reproduction cost as a result of the increasing yield of viable copies of the genome. In itself, a larger genome is more of a liability than an advantage because of higher replication costs. However, moderate genome increase, such as by duplication of parts of the genome or by recombination, can be fixed via genetic drift in small populations. Replicators with a sufficiently high fidelity can take advantage of such randomly fixed and initially useless genetic material by evolving new functions, without falling off the "Eigen cliff" (see Figure 12-1B). Among such newly evolved, fitness-increasing functions will be those that increase replication fidelity, which, in turn, allows a further increase in the amount of encoded information. And so the Darwin-Eigen cycle recapitulates itself in a spiral progression, leading to a steady increase in genome complexity (see Figure 12-1A).

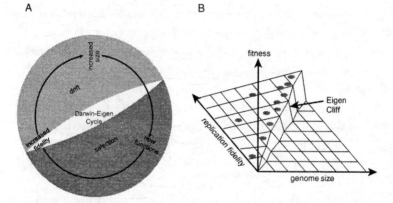

Figure 12-1   Replication fidelity and evolution: (A) The Darwin-Eigen cycle; (B) Evolution at the edge of the Eigen cliff.

The crucial question in the study of the origin of life is how the Darwin-Eigen cycle started—how was the minimum complexity that is required to achieve the minimally acceptable replication fidelity attained? In even the simplest modern systems, such as RNA viruses with the replication fidelity of only about 10^{-3} and viroids that replicate with the lowest fidelity among the known replicons (about 10^{-2}; Gago, et al., 2009), replication is catalyzed by complex protein polymerases. The replicase itself is produced by translation of the respective mRNA(s), which is mediated by the immensely complex ribosomal apparatus. Hence, the dramatic paradox of the origin of life is that, *to attain the minimum complexity required for a biological system to start on the Darwin-Eigen spiral, a system of a far greater complexity appears to be required.* How such a system could evolve is a puzzle that defeats conventional evolutionary thinking, all of which is about biological systems moving *along the spiral*; the solution is bound to be unusual. In the next sections, we first examine the potential of a top-down approach based on the analysis of extant genes, to obtain clues on possible origins of replicator systems. We then discuss the bottom-up approach.

The case for a complex RNA World from protein domain evolution: The top-down view

As pointed out earlier, the translation system is the only complex ensemble of genes that is conserved in all extant cellular life forms. With about 60 protein-coding genes and some 40 structural RNA genes universally conserved, the modern translation system is the best-preserved relic of the LUCA(S) and the strongest available piece of evidence that some form of LUCA(S) actually existed (see Chapter 11). Given this extraordinary conservation of the translation system, comparison of orthologous sequences reveals very little, if anything, about its origins: The emergence of the translation system is beyond the horizon of the comparison of extant life forms. Indeed, comparative-genomic reconstructions of the gene repertoire of LUCA(S) point to a complex translation system that includes at least 18 of the 20 aminoacyl-tRNA synthetases (aaRS), several translation factors, at least 40 ribosomal proteins, and several enzymes involved in rRNA and tRNA modification. It appears that the core of the translation system was already fully shaped in LUCA(S) (Anantharaman, et al., 2002).

Fortunately, sequence and structure comparisons of protein and RNA components within the translation system itself are informative, thanks to the extensive paralogy among the respective genes. Whenever a pair of paralogous genes is assigned to LUCA(S), the respective duplication must have been a more ancient event, so reconstruction of the series of ancient duplications opens a window into very early stages of evolution. The story of the paralogous aaRS is particularly revealing. The aaRS form two distinct classes of ten specificities each (that is, each class is responsible for the recognition and activation of ten amino acids), with unrelated catalytic domains and distinct sets of accessory domains. The catalytic domains of the Class I and Class II aaRS belong to the Rossmann fold and the biotin synthase fold, respectively. The analysis of the evolutionary histories of these protein folds has far-reaching implications for the early evo-lution of the translation system and beyond (Aravind, et al., 2002). The catalytic domains of Class I aaRS form but a small twig in the evolutionary tree of the Rossmann fold domains (see Figure 12-2A). Thus, the appearance of the common ancestor of the aaRS is pre-ceded by a number of nodes along the evolutionary path from the primitive, ancestral domain to the highly diversified state that corre-sponds to LUCA(S). So a substantial diversity of Rossmann fold domains evolved prior to the series of duplications that led to the emergence of the aaRS of different specificities, which itself ante-dates LUCA(S) (see Figure 12-2A). A similar evolutionary pattern is implied by the analysis of the biotin synthase domain that gave rise to Class II aaRS. Thus, even within these two folds alone, *remarkable structural and functional complexity of protein domains had evolved before the full-fledged RNA-protein machinery of translation resem-bling the modern system was in place.*

The evolutionary analysis of the vast class of P-loop GTPases, within which a variety of translation factors comprise distinct, tight families, leads to essentially the same conclusions: In the succession of evolutionary bifurcations (tree branchings) that constitute the his-tory of the GTPase domain, the translation factors are relatively late arrivals (see Figure 12-2B; Leipe, et al., 2002). The GTPases taken together are but one of the several major branches of the P-loop fold, which includes a huge variety of protein domains that bind NTP (nucleoside triphosphates—most often, the substrate is ATP, in a sub-stantial minority of cases GTP, and rarely others) and cleave the β-γ phosphodiester bond (see Figure 12-2B). The P-loop fold is the most abundant domain in all prokaryotes (Wolf, et al., 1999b), and in any reconstruction of the gene repertoire of the LUCA(S), several dozen P-loop proteins come up (see Figure 12-3; Mirkin, et al., 2003). Thus, extensive evolution of the P-loop domain obviously antedates not only LUCA(S), but also—much more surprisingly—the modern-type translation system. The P-loop itself (see Figure 12-3), the glycine-rich loop that wraps around the phosphate tail of the NTP substrate (also known as the Walker A motif[3]), is the most conserved element among all protein sequences, one that undoubtedly was fixed at the earliest stages of protein evolution (Gorbalenya and Koonin, 1989; Trifonov, et al., 2006).

So an inevitable (even if perhaps counterintuitive) conclusion from the comparative analysis of ancient paralogous relationship between protein components of the translation system is that, with the interesting exception of the core ribosomal proteins, *all proteins that play essential roles in modern translation are products of a long and complex evolution of diverse protein domains.* Here comes the Catch-22: For all this protein evolution to occur, an accurate and efficient translation system is required. This primordial translation system might not need to be quite as good as the modern version, but it seems a safe bet that is must have been within an order of magnitude from the modern one in terms of fidelity and translation rates to make protein evolution possible. However, from all we know about the modern translation system, this level of precision is unimaginable without a complex, dedicated protein apparatus.

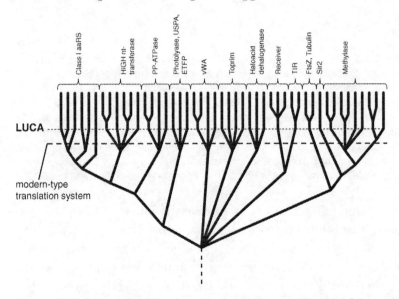

Figure 12-2A   Diversification of protein domains, crystallization of the translation system, and the LUCA(S): Evolution of the Rossmann fold–like nucleotide-binding domains. Based on data from Aravind, et al., 2002. Only the better-known proteins are indicated. USPA = Universal stress protein A; ETFP = electron transfer flavoprotein; vWA = Von Willebrand A factor; Toprim = catalytic domain of topoisomerases, primases, and some nucleases; Receiver = a component of prokaryotic two-component signaling systems; TIR = a widespread protein-protein interaction domain in prokaryotic and eukaryotic signaling systems; Sir2 = protein (in particular, histone) deacetylase; Methylase = diverse methyltransferases. For details, see (Aravind, et al., 2002) and references therein.

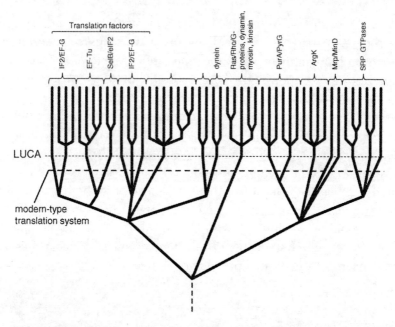

Figure 12-2B   Evolution of the P-loop GTPase domains. Based on data from Leipe, et al, 2002. Only the better-known proteins are indicated. Dynein, dynamin, kinesin, and myosin are cytoskeleton-associated motor GTPases and ATPases; Ras/Rho are singaling GTPases associated in particular with the endomembrane system in eukaryotes; G-proteins are membrane-associated GTPases that function jointly with G-protein-coupled receptors; PurA and PyrG are enzymes of nucleotide metabolism; ArgK, arginine kinase, is an enzyme of amino acid metabolism; Mrp and MinD are ATPases involved in cell division in prokaryotes; and SRP is signal recognition particle. For details, see Leipe, et al., 2002.

Thus, the translation system presents us with the "Darwin-Eigen paradox" that is inherent to all thinking on the emergence of complex biological entities: For a modern-type, efficient, and accurate translation system to function, many diverse proteins are required, but for those proteins to evolve, a translation system almost as good as the modern one would be necessary. There seems to be only one conceivable solution to this paradox—namely, a (partial) refutation of the first part of the opposition: We are forced to conclude that *a translation system comparable to the modern one in terms of accuracy and speed functioned without many proteins, possibly without any proteins at all.* Hence, the existence of a complex, elaborate RNA World (see the next section), in which a primitive version of the Darwin-Eigen cycle was already operating, can be conjectured from the comparative analysis of the translation system components.

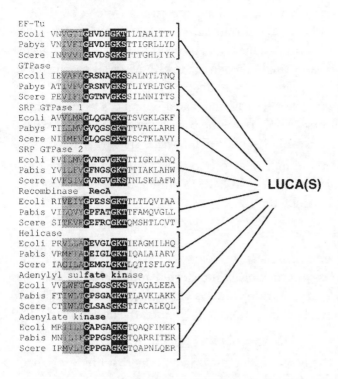

EF-Tu
Ecoli   VNVGT CHVDHGKT TLTAAITTV
Pabys   VNIVFI CHVDHGKS TTIGRLLYD
Scere   INVVV CHVDSGKS TTTGHLIYK
GTPase
Ecoli   IEVAF GRSNAGKS SALNTLTNQ
Pabys   ATIVF CRSNVGKS TLIYRLTGK
Scere   PEVIF CGTNVGKS SILNNITTS
SRP GTPase 1
Ecoli   AVVLMA GLQGAGKT TSVGKLGKF
Pabys   TILLMV GVQGSGKT TTVAKLARH
Scere   NIIMF GLQGSGKT TSCTKLAVY
SRP GTPase 2
Ecoli   FVILMY GVNGVGKT TTIGKLARQ
Pabis   YVLLFV GFNGSGKT TTIAKLAHW
Scere   YVFSIV GVNGVGKS TNLSKLAFW
Recombinase RecA
Ecoli   RIVEIY GPESSGKT TLTLQVIAA
Pabis   VIQVV GPFATGKT TFAMQVGLL
Scere   SITFVF GEFRCGKT QMSHTLCVT
Helicase
Ecoli   PRVLLA DEVGLGKT IEAGMILHQ
Pabys   VRMFIA DEIGLGKT IQALAIARY
Scere   IACILA DEMGLGKT LQTISFLGY
Adenylyl sulfate kinase
Ecoli   VVLWFI GLSGSGKS TVAGALEEA
Pabis   FTIWLI GPSGAGKT TLAVKLAKK
Scere   CTIWLI GLSASGKS TIACALEQL
Adenylate kinase
Ecoli   MRILL GAPGAGKG TQAQFIMEK
Pabis   MNILI GPPGSGKG TQARRITER
Scere   IRMVLI GPPGAGKG TQAPNLQER

LUCA(S)

Figure 12-3   The P-loop, the most common primordial motif in protein sequences. The figure shows the alignment of the P-loops from eight ancient lineages of NTPases, each of which was inferred to have been represented in the LUCA(S) (Mirkin, et al., 2003). For each lineage, three sequences from a bacterium (Escherichia coli, Ecoli), an archaeon (Pyrococcus abyssi, Pabys), and a eukaryote (the yeast Saccharomyces cerevisiae, Scere) are shown for each lineage. The white lettering against the black background shows the amino acid residues that form the flexible loop directly interacting with the phosphate tail of NTP, and shading shows the characteristic hydrophobic β-strand upstream of the P-loop. SRP is a signal recognition particle.

This is not all that the comparative analysis can do: Comparison of RNAs themselves also yields important information and presents startling puzzles. Thus, an analysis of the sequence and structure of the large ribosomal subunit 23S RNA has suggested a hierarchical scenario of a series of duplications that might have led from a simple primordial RNA hairpin to the modern, elaborate, universally conserved RNA structure (Bokov and Steinberg, 2009).

The conservation of the structure, some sequence elements (such as the pseudouridine loop), and even modification sites of the tRNAs of all specificities (and, needless to say, all species) leaves no doubt that they all evolved from a single common ancestor (Eigen, et al., 1989). Hence, the second paradox of translation evolution ensues from the comparison of modern sequences and structures: If, at some point in evolution, there was a single progenitor to tRNAs of all specificities, how could such a translation system function—that is, how could it possibly ensure specific encoding of amino acid sequences by nucleotide sequences? Conversely, if there was no translation system at that stage, what could drive the evolution of the amino acid–specific tRNAs?

We address these and related questions shortly, but before we proceed with this discussion, we have to outline the central concept in the origin of life field: the RNA World.

Ribozymes and the RNA World

The Central Dogma of molecular biology (Crick, 1970) states that, in biological systems, information is transferred from DNA to protein through an RNA intermediate (Francis Crick added the possibility of reverse information flow from RNA to DNA after the discovery of RT):

$$DNA \leftrightarrow RNA \Rightarrow protein$$

Obviously, when considering the origin of the first life forms, one faces the proverbial chicken-and-egg problem: What came first, DNA or protein, the gene or the product? In that form, the problem might be outright unsolvable due to the Darwin-Eigen paradox: To replicate and transcribe DNA, functionally active proteins are required, but production of these proteins requires accurate replication, transcription, and translation of nucleic acids. If one sticks to the triad of the Central Dogma, it is impossible to envisage what could be the starting material for the Darwin-Eigen cycle. Even removing DNA from the triad and postulating that the original genetic material consisted of RNA (thus reducing the triad to a dyad), although an important idea (see the discussion later in this chapter), does not help much because the paradox remains. For the evolution toward greater complexity to take off, the system needs to somehow get started on the Darwin-Eigen cycle before establishing the feedback between the (RNA) templates (the information component of the replicator system) and proteins (the executive component).

The brilliantly ingenious and perhaps only possible solution was independently proposed by Carl Woese, Francis Crick, and Leslie

Orgel in 1967–68 (Crick, 1968; Orgel, 1968; Woese, 1967): *neither the chicken nor the egg, but what is in the middle—RNA alone.* The unique property of RNA that makes it a credible—indeed, apparently, the best—candidate for the central role in the primordial replicating system is its ability to combine informational and catalytic functions. Thus, it was extremely tempting to propose that the first replicator systems— the first life forms—consisted solely of RNA molecules that functioned both as information carriers (genomes and genes) and as catalysts of diverse reactions, including, in particular, their own replication and precursor synthesis. This bold speculation has been spectacularly boosted by the discovery and subsequent study of ribozymes (RNA enzymes), which was pioneered by the discovery by Thomas Cech and colleagues in 1982 of the autocatalytic cleavage of the *Tetrahymena* rRNA intron, and by the demonstration in 1983 by Sydney Altman and colleagues that RNAse P is a ribozyme. Following these seminal discoveries, the study of ribozymes has evolved into a vast, expanding research area (Cech, 2002; Doudna and Cech, 2002; Fedor and Williamson, 2005).

The discovery of ribozymes made the idea that the first replicating systems consisted solely of RNA molecules, which catalyzed their own replication, enormously attractive. In 1986, Walter Gilbert coined the term "RNA World" to designate this hypothetical stage in the evolution of life, and the RNA World hypothesis caught on in a big way; it became the leading and most popular hypothesis on the early stages of evolution. (The diverse aspects of the RNA World hypothesis and the supporting data are thoroughly covered in the eponymous book that in 2010 appeared in its fourth edition: Atkins, et al., 2010.)

The popularity of the RNA World hypothesis has further stimulated ribozyme research aimed at testing the feasibility of various RNA-based catalytic activities—above all, perhaps, an RNA replicase. It is noteworthy that the main experimental approach employed to develop ribozymes with desired activities is *in vitro* selection that, at least conceptually, mimics the Darwinian evolution of ribozymes thought to have occurred in the primeval RNA World (Ellington, et al., 2009). The directed selection experiments are designed in such a way that, from a random population of RNA sequences, only those are amplified that catalyze the target reaction. In multiple-round selection experiments, ribozymes have been evolved to catalyze an extremely broad variety of reactions.

Box 12-1 lists some of the most biologically relevant ribozyme-catalyzed reactions. Notably, all three elementary reactions that are required for translation—namely (i) amino acid activation through the formation of aminoacyl-AMP, (ii) (t)RNA aminoacylation, and (iii) transpeptidation (the peptidyltransferase reaction)—have been successfully modeled with ribozymes. The self-aminoacylation reaction that is key to the origin of the primordial RNA-only adaptors (the RNA analog of aaRS) has been selected *in vitro* with relative ease. Strikingly, the best of the resulting ribozymes catalyze this reaction with a rate and specificity greater than those of the respective aaRS, and very short oligonucleotides possessing this activity have been selected (Turk, et al., 2010).

Understandably, major effort has focused on the demonstration of nucleotide polymerization and, ultimately, RNA replication catalyzed by ribozymes, the central processes in the hypothetical primordial RNA World. The outcome of the experiments aimed at the creation of ribozyme replicases so far has been mixed (Cheng and Unrau, 2010). Ribozymes have been obtained that are capable of extending a primer annealed to a template (Johnston, et al., 2001); initially, the ribozymes with this activity could function only by specific base-pairing to the template, but subsequently general ribozyme polymerases of this class have been evolved through additional selection (Lincoln and Joyce, 2009). The latest breakthrough in the field of polymerase ribozymes has been published at the time of the final editing of this chapter: an active endonuclease ribozyme was produced using a ribozyme polymerase that itself was constructed by recombining two pre-existing ribozymes, potentially, a plausible route for pre-biological evolution (Wochner, et al., 2011). All this progress notwithstanding, the ribozyme polymerases that are currently available are a far cry from processive, sufficiently accurate (in terms of the Eigen threshold) replicases, capable of catalyzing the replication of exogenous templates and themselves. Enzymes with such properties appear to be a *conditio sine qua non* for the evolution of the hypothetical RNA World. Besides, even the available ribozymes with the limited RNA polymerase capacity are rather complex molecules that consist of some 200 nucleotides and could be nontrivial to evolve in the prebiotic setting.

Box 12-1: Some ribozyme activities with a clear relevance for biological evolution

Reaction	Properties of the Ribozyme
Aminoacyl adenylate synthesis	Low-efficiency formation of leucyl and phenylalanyl adenylates observed with a 114-nucleotide ribozyme.
Self-aminoacylation	Self-aminoacylation of a 43-nucleotide ribozyme with phenylalanine, using phe-AMP as the substrate. A 77-nucleotide RNA catalyzed the same reaction with a specificity and aminoacylation rate greater that those of PheRS.
RNA 3'-aminoacylation in-trans	The smallest ribozyme capable of nonspecific tRNA aminoacylation consists of 29 nucleotides. A 45-nucleotide ribozyme has been obtained with a broad spectrum of activity toward diverse tRNAs and amino acids. Larger ribozymes with highly specific and efficient aminoacylation activity have been reported.
In vitro selected peptidyltransferase ribozymes	Several ribozymes selected to form dipeptides from an amino acid esterified to AMP or a oligonucleotide and a free amino acid. Structural similarity was observed between peptidyltransferase ribozymes and the relevant portion of 23S rRNA. Formation of Phe-Phe-tRNA was reported for the 29-nucleotide aminoacylating ribozyme.
Ribosomal peptidyltransferase	In the ribosomal large subunits, the peptidyltransferase center maps to an area containing only RNA, leading to the conclusion that the reaction is catalyzed by a ribozyme; however, identification of the active residues remains elusive.
RNA ligase	Multiple ribozymes are capable of ligating RNA molecules that are juxtaposed through base-pairing to the ribozyme molecule.
RNA polymerase	Ribozymes capable of extending a preannealed RNA primer by 10 to 14 nucleotides selected from a pool of RNA ligase ribozymes.

Adapted and extended from Wolf and Koonin, 2007.

The RNA World is not just a concept supported by the catalytic prowess of ribozymes: Although overshadowed by the multitude of proteins with catalytic and structural functions, the RNA World still lurks within modern life forms (Doudna and Cech, 2002). Reactions catalyzed by ribozymes, although far less numerous and diverse than those catalyzed by protein enzymes, are of crucial importance in modern cells. The foremost case of today's natural ribozyme is the ribosome itself, where the key peptidyltransferase reaction is catalyzed by the large-subunit rRNA without direct participation of proteins (Beringer and Rodnina, 2007). In the nearly ubiquitous tRNA-processing enzyme RNAse P, the catalytic moiety is an RNA molecule, whereas the protein subunits play the role of cofactors stabilizing the RNA catalyst and facilitating the reaction (McClain, et al., 2010). Furthermore, Group I and Group II self-splicing introns, which are widespread in bacteria and in plant, fungal, and protozoan organelles, are ribozymes that catalyze their own excision from RNA transcripts, often facilitated by specific proteins, the so-called maturases (see also Chapter 7). It is essentially certain that the myriad eukaryotic spliceosomal introns, as well as the small nuclear (sn)RNAs that comprise the active moieties of the eukaryotic spliceosomes, have evolved from Group II introns (see Chapter 7). Thus, splicing, the ubiquitous signature process in eukaryotic cells, is based on a ribozyme-catalyzed reaction. Similarly, in the case of the smallest known infectious agents, viroids and virusoids, the ribozyme-catalyzed reactions are directly involved in replication: Although the polymerization of nucleotides is catalyzed by a protein polymerase, processing replication intermediates into genomic units depends on a built-in ribozyme (Flores, et al., 2004). The existence and central importance of these (and probably other, still-undiscovered) RNA-catalyzed reactions in modern cells imply a major role of RNA catalysts in the early evolution of life. All this evidence certainly falls far short of proving the reality of the primordial RNA World, as defined earlier: *a community of diverse RNAs possessing diverse catalytic activities and replicated by ribozyme polymerases*. Nevertheless, these features of modern RNAs, especially the ribozyme activities, are fully compatible with such an evolutionary stage and greatly add to its plausibility. In particular, the fundamental fact that the peptidyltransferase reaction in the ribosome is catalyzed by a ribozyme strongly suggests that the primordial translation system started as a ribozyme machine.

Thus, three independent lines of evidence converge in support of a major role of RNA—and, more specifically, RNA catalysis at the earliest stages of the history of life—and are compatible with the reality of a complex, ancient RNA World that Woese, Crick, and Orgel first postulated on purely logical grounds.

1. Comparative analysis of the protein components of the translation machinery and their homologs involved in other functions strongly suggests that extensive diversification of the protein world took place at the time when the translation system was comprised primarily of RNA.

2. Several classes of ribozymes operate within modern cells, and their properties are compatible with the idea that they are relics of the primordial RNA World.

3. Although limited in versatility and typically vastly inferior to protein enzymes in catalytic activity, ribozymes have been shown—or, more to the point, evolved—to catalyze a remarkable variety of reactions, including those that are central to the evolution of translation (see Box 12-1).

All these arguments in favor notwithstanding, the RNA World hypothesis faces grave difficulties. First, despite all invested effort, the *in vitro* evolved ribozymes remain (relatively) poor catalysts for most reactions; the lack of efficient, processive ribozyme polymerases seems particularly troubling, but there is also a serious shortage of other activities, such as those required for the synthesis of nucleotides. Admittedly, it might be unrealistic to expect that experiments on *in vitro* evolution of ribozymes could easily mimic the actual complexity of the primordial RNA World. Although these experiments harness the power of selection, they are obviously performed on a totally different time scale and under conditions that cannot accurately reproduce the (unknown) conditions at the origin of life (we discuss the potential environmental niches for the origin of life later in this chapter).

A study by Eors Szathmary and coworkers puts some important numbers on the complexity that might be attainable in the RNA World and the replication fidelity that is required to reach this level of complexity (Kun, et al., 2005). An estimate based on the functional tolerance of well-characterized ribozymes to mutations suggests that, at a fidelity of 10^{-3} errors per nucleotide per replicase cycle (roughly, the fidelity of the RNA-dependent RNA polymerases of modern viruses), an RNA "organism" with about 100 "genes" the size of a tRNA (80 nucleotides) would be sustainable. Such a level of fidelity would require only an order of magnitude improvement over the most accurate ribozyme polymerases obtained by *in vitro* selection. This might be an approximate upper bound of complexity on ensembles of co-evolving "selfish cooperators" that would have been the "organisms" of the RNA World.

Even under the best-case scenario, the RNA World hardly has the potential to evolve beyond very simple "organisms." To attain greater complexity, invention of translation and the *Protein Breakthrough* (the relegation of most catalytic activities to proteins) were required. However, the selective forces underlying the emergence of the translation system in the RNA World remain obscure, and tracing the path to translation is extremely difficult. This lack of clarity with respect to the continuity of evolution from the RNA World to an RNA-protein world is the second major problem of the RNA World hypothesis, perhaps even more formidable than the limited catalytic repertoire and the (typically) low efficiency of ribozymes. We next discuss possible ways out of this conundrum.

The nature and origin of the genetic code

To understand how translation might have emerged, the nature and origin of the codon assignments in the universal genetic code are crucial. The evolution of the code fascinated researchers even before the code was fully deciphered, and the earliest treatises on the subject already clearly recognized three not necessarily mutually exclusive evolutionary models: (i) steric complementarity resulting in specific interactions between amino acids and the cognate codon or anticodon triplets, (ii) "frozen accident," fixation of a random code that would have been virtually impossible to significantly change afterward, and (iii) adaptive evolution of the code starting from an initially random codon assignment (Crick, 1968). The structure of the code is clearly nonrandom: Codons for related amino acids are mostly adjacent in the code table, resulting in a high (although not maximum) robustness of the code to mutations and translation errors, as Woese first noticed (Woese, 1967) and Stephen Freeland and Laurence Hurst subsequently demonstrated quantitatively (Freeland and Hurst, 1998). The robustness of the code falsifies the frozen accident scenario in its extreme form (random assignment of codons without any subsequent evolution); however, the stereochemical model, the selection model, a combination thereof, or frozen accident followed by adaptation all could, in principle, explain the observed properties of the code (Koonin and Novozhilov, 2009).

The main dilemma is whether a stereochemical correspondence between amino acids and cognate triplets exists. The answer to this seemingly simple question proved to be surprisingly elusive. The early attempts to establish specificity in the interactions of (poly)amino acids and polynucleotides have been inconclusive, indicating that, if a specific affinity exists, it must be much less than precise, and the interactions involved would be weak and dependent on extraneous factors. Although some tantalizing nonrandomness in amino acid–oligonucleotide interactions has been claimed, in general, the attempts to demonstrate such interactions directly have failed (Saxinger and Ponnamperuma, 1974).

A resurgence of the stereochemical hypothesis was brought about by the selection amplification (SELEX) methodology for isolation of oligonucleotides (aptamers) that specifically bind amino acids (Yarus, et al., 2005, 2009). For eight amino acids with large side chains, aptamers significantly enriched for codon and/or anticodon triplets have been isolated. The results of aptamer experiments are somewhat inconclusive, in that, for some amino acids, the aptamers contain primarily codons, and for other amino acids, they mostly contain anticodons. Taken together, the aptamer binding data is considered to present a serious argument in support of the stereochemical hypothesis of the code origin. Nevertheless, major questions remain as to the ultimate validity and relevance of these results. The presence of both codons and anticodons in aptamers for several amino acids is difficult to interpret in terms of stereochemical complementarity. Furthermore, the amino acids for which detailed aptamer data is available have complex side chains (which, presumably, are required for the specific interaction with the aptamers) and are likely to be late recruitments to the genetic code (Trifonov,

2004). At least until similar results are obtained for simpler, supposedly ancient amino acids, it is hard to view the aptamer selection experiments as a definitive case for the stereochemical hypothesis of code origin.

Thus, the jury is still out on the key question of whether direct interactions between amino acids and cognate triplets played any role in the origin of the code. In our discussion of the origin of translation, we attempt to be objective and consider the origin of the code starting either from a specific interaction between amino acids and the cognate triplets or from an initial random codon-amino acid mapping (frozen accident).

The origin of translation: The key ideas and models

During the 40 years since the discovery of the translation mechanism and deciphering of the genetic code, numerous theoretical (inevitably, speculative, sometimes far-fetched, often highly ingenious) models of the origin and evolution of various components of the translation apparatus and different aspects of the translation process have been proposed. It is unrealistic to provide here a thorough critical review of these models. Instead, I consider a few central ideas that are germane to the thinking about the origin of translation and then discuss in somewhat greater detail the only two coherent scenarios I am aware of.

The main general point about the evolution of translation is that selection for protein synthesis could not have been the underlying cause behind the origin of the translation system. To evolve this complex system via the Darwinian route, numerous steps are required, but proteins appear only at the last steps; until that point, an evolving organism "does not know" how good proteins could be. As discussed in Chapter 9, many situations exist in which evolution seems to exhibit some foresight capability; however, these cases are effectively based on extrapolation, whereas, in the case of translation, there is nothing to extrapolate from. The emergence of the complex translation machinery by random drift is not practical either—at least, not within the regular framework of evolutionary biology (see the discussion at the end of this chapter). Thus, the only conceivable route for the emergence of translation seems to be exaptation: *Intermediate stages in the evolution of the translation system must have been selected for functions other than protein synthesis*. Different scenarios for the origin of translation started with different speculations on the nature of the exapted function.

A simple and potentially fruitful idea is that, in the RNA World, amino acids and peptides would function as cofactors for ribozymes. Szathmary developed the first hypothesis based on this proposition and speculated that "coding coenzyme handles (CCH)" (oligonucleotides with various ribozyme activities using amino acids as cofactors) could be evolutionary progenitors of tRNAs (Szathmary, 1993, 1999). The CCH are thought to have assembled via their proto-anticodons on emerging mRNAs although the details of this stage remain obscure. The CCH hypothesis ties in with the idea that tRNAs evolved by two successive duplications of amino acid–binding hairpins. A modification of the CCH hypothesis proposed by Rob Knight and Laura Landweber involves evolution of aminoacylating ribozymes (a possibility that is well supported by experimental data— see Box 12-1) and the emergence of nontemplated, ribozyme-mediated peptide synthesis as an intermediate stage in the evolution of translation (Knight and Landweber, 2000).

An alternative to the CCH scheme is the direct-RNA-templating hypothesis of translation origin proposed by Michael Yarus (Yarus, 1998). Under this model, the original form of the amino-acid-proto-tRNA interaction was direct binding, presumably via anticodon triplets. Subsequently, direct binding has been supplanted by the adaptor mechanism, probably with the participation of aminoacylating ribozymes, as under the modified CCH hypothesis.

Taking the lead from the CCH hypothesis, Yuri Wolf and I developed a generalized but detailed model for the emergence of the translation system in the RNA World (Wolf and Koonin, 2007). This model includes both Darwinian selection and aspects of constructive neutral evolution (see Chapter 8), along with exaptation and subfunctionalization.

The starting point of all scenarios for the origin of translation is a replicating ensemble of selfish cooperators consisting of RNA molecules with various ribozyme activities and existing within a network of inorganic compartments (see further discussion in the next section).

One of the functions performed by these ribozymes is that of a replicase; other activities, such as RNA precursor synthesis, are likely to be present as well. Our evolutionary scenario includes the following steps (see Figure 12-4).

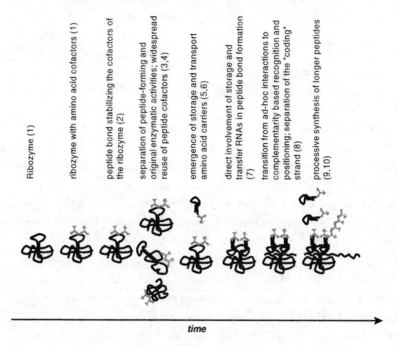

Ribozyme (1)

ribozyme with amino acid cofactors (1)

peptide bond stabilizing the cofactors of the ribozyme (2)

separation of peptide-forming and original enzymatic activities; widespread reuse of peptide cofactors (3,4)

emergence of storage and transport amino acid carriers (5,6)

direct involvement of storage and transfer RNAs in peptide bond formation (7)

transition from ad-hoc interactions to complementarity based recognition and positioning; separation of the "coding" strand (8)

processive synthesis of longer peptides (9,10)

time

Figure 12-4 A conceptual scenario for the origin of the translation system by means of exaptation and subfunctionalization. The stages of the model described in the text are indicated in parentheses.

1. Ribozyme R is part of an ensemble of selfish cooperators within a compartment. This ribozyme is sufficiently complex to catalyze the reaction $(X{\rightarrow}Y)$ the rate of which affects the fitness of the ensemble and includes a certain number of evolvable positions, allowing the evolution of new activities. Two or more abiogenic amino acids present in the compartment bind to R. Specific binding of the amino acids is mediated by an *ad hoc* binding site present in R. Involvement of a stereochemical proto-code (codon or anticodon) at this stage is possible but would not substantially affect the scenario. The bound amino acids stimulate the $X{\rightarrow}Y$ reaction catalyzed by R. Ribozymes strongly stimulated by peptides have been produced by *in vitro*

selection, so there is experimental underpinning behind this crucial step (Robertson, et al., 2004). In the context of a selfish cooperative evolution (see Chapter 11), natural selection would pick up amino acid stimulation of R, resulting in gradual perfection of the spatial alignment of amino acids on R and selection of the optimal sequence and structure for amino acid binding.

2. R evolves an additional peptide ligase activity, yielding oligopeptide P from adjacent amino acids bound to R. Highly active ribozymes with peptide ligase activity, albeit with low specificity, have been evolved through *in vitro* selection. Probably only short peptides consisting of, at most, four or five amino acids could be synthesized by this class of ribozymes. The selective advantage of this innovation would be the increased stability of the reactive complex, resulting in a further boost to the $X \rightarrow Y$ reaction. An inevitable question regarding this step is where the energy required for the peptide bond formation comes from. In experimentally characterized ribozyme peptide ligases, one of the substrates is an activated derivative (aminoacyl adenylate), so the energy of the ester bond is utilized. This mimics modern translation, in which the aaRS use the aminoacyl adenylates to charge the cognate tRNAs, and the high-energy ester bond of the latter is utilized for transpeptidation. The putative primordial peptide ligase might have functioned in the same mode using aminoacyl adenylates or other activated derivatives of amino acids produced by other ribozymes. Indeed, ribozymes that catalyze each of these these reactions, from amino acid adenylation to peptide synthesis, have been reported (see Box 12-1). Certainly, these ribozymes still depend on the energy of a phosphodiester bond in ATP or some other form of energy.

3. Spontaneous disassembly or decay of R would release the peptide P into the compartment. If P has a generic ribozyme-stimulating and/or ribozyme-stabilizing capacity, it might be captured by another ribozyme E, which catalyzes a different reaction $(U \rightarrow V)$. An interesting case in point would be a peptide containing a pair of acidic amino acids and coordinating a divalent cation analogously to a variety of unrelated modern enzymes of

nucleic acid metabolism (polymerases, nucleases, ligases, topoi-somerases, and others). If P boosts the catalytic activity of E, it again increases the fitness of the entire ensemble.

4. With the activity of E dependent on the presence of P, a copy of R (R_L) might lose the original $X{\rightarrow}Y$ activity, with a concomitant enhancement of the amino acid ligase activity, whereas the other copy (R_0) would retain the original activity, still enhanced by the peptide P. Note that this is typical subfunctionalization, the major route of evolution of duplicated genes in modern genomes (see Chapter 8). Subfunctionalization might have been impor-tant already in the RNA World, with the benefit of improved catalysis by R_0 and E outweighing the increased replication cost.

5. Widespread peptide-assisted catalysis in the compartmental-ized prebiological system makes amino acids a useful commod-ity for the evolving selfish cooperatives. Given that amino acids are small polar molecules that would diffuse through compart-ment walls, accumulation of amino acids within a compartment would be beneficial. Thus, small amino acid–binding RNAs (T) evolve under the pressure of selection for amino acid accumu-lation; these molecules may be considered analogs of amino acid–binding aptamers (see the preceding section). Originally, the T RNAs bind amino acids nonspecifically. Autocatalytic aminoacylation of the 3′ end of RNA T evolves, resulting in an increase in affinity and specificity of amino acid binding. As with the peptide ligase in step 2, there should be a source of energy for this reaction; activated amino acid derivatives, such as aminoacyl adenylates, would serve in this capacity.

6. Different species of T RNAs specifically binding different amino acids evolve by duplication and diversification, with the reten-tion of variants driven by selection for efficient accumulation of a broad repertoire of amino acids. The details of the T RNA-amino acid binding would depend on whether there is specific recognition between amino acids and the cognate (anti)codons. If there is no such recognition, the frozen accident scenario would have to be invoked, in which the amino acid recognition site in T RNA is unrelated to either the codon or the anticodon, whereas the sequence of the exposed loop (the ancestor of the anticodon loop) is chosen by chance. Regardless of the specific

model (even if it is just the frozen accident), this is the critical step that establishes the correspondence between amino acids and cognate triplets, creating the basis of the genetic code.

7. Ribozyme R_L evolves the capacity to bind aminoacyl-T RNA complexes instead of individual amino acids, resulting in greater stability and spatial precision of binding. The primary biochemical activity of R_L changes from amino acid ligation to transpeptidation (transfer of a growing peptide from one T RNA species to another), resulting in an increased yield of peptides, thanks to the high energy of the aminoacyl-RNA bond. Notably, the 50S subunit of the bacterial ribosome, of which the ribozyme R_L is deemed to be the ancestor, can catalyze the transpeptidation reaction at a rate comparable to that of the complete ribosome (Wohlgemuth, et al., 2006).

8. An accessory RNA subunit R_S evolves, driven by selection for increasingly efficient binding and positioning of aminoacyl-T complex on R_L. The T RNA recognition switches from a relatively nonspecific interaction between RNA T and R_L to specific base pairing between the proto-anticodon loop of T and an extended RNA strand of R_S. This is the crucial step in the evolution of *bona fide* translation, a mechanism based on the adaptors (proto-tRNAs, the T RNAs in this model) combining amino acids with the cognate codons.

9. The evolutionary path from the set of primitive T RNAs to the modern tRNAs demands a specific explanation, given the obvious common ancestry of tRNAs of all specificities. At the early stages of the translation system evolution outlined earlier, different species of T RNAs might have been evolving along roughly parallel (convergent) paths. However, the common origin of tRNAs implies a subsequent bottleneck through which only a single winner has passed, an L-shaped molecule with the acceptor CCA 3[s] end. Selection for spatial complementarity and efficient interaction between the aminoacylated T RNAs and the peptidyl-transferase R_L could be the driving force behind this selective sweep. This selection originally would affect only one T RNA, perhaps the one chargeable with the most abundant primordial amino acid. Subsequently, the remaining tRNAs would evolve by duplication and specialization.

10. The next step in the evolution of the translation system would be the physical separation of the template strand M from R_S, resulting in further disentanglement of coding and catalysis. At this point, the strand M is freed from evolutionary constraints associated with the binding and catalytic activities involved in the primitive translation because all these functions are provided by physically distinct RNA molecules, R_L, R_S, and the proto-tRNAs. The only requirement for M is to adopt a semi-extended conformation to accommodate the codon-anticodon base pairing involved in binding an aminoacyl-T RNA. The selective benefits of such separation are obvious: The transient association of $R_S R_L$ (which, at this point, can be reasonably denoted *proto-ribosome*) with different oligo/polynucleotides present in the compartment would lead to the production of an increasing variety of peptides, thus enhancing the catalytic potential of the ensemble. Furthermore, this step would enable the selection for improved replication potential (such as high-affinity replicase recognition sites) of those species of M that encode useful peptides, leading to enrichment of these RNA species in the compartment. Thus, a distinct version of the Darwin-Eigen cycle would be effectively established within the selfish cooperative.

11. The release of a discharged (proto)tRNA from $R_S R_L$ upon trans-peptidation would trigger the trinucleotide shift, the signature movement of modern ribosomes, allowing for the synthesis of longer peptides—these would effectively be the first proteins. This is the *Protein Breakthrough*.

Under this type of evolutionary scenario, the path from the break-through stage to the modern-type translation system was largely a story of takeover of the primordial ribozyme functions by evolving proteins. Proteins have an incomparably greater potential for evolution of diverse binding and catalytic capacities than RNA or peptides, so they gradually but irreversibly supplanted the primordial ribozymes.

We now discuss a substantive alternative, an evolutionary model originally sketched by Anatoly Altstein (Altstein, 1987) and, later, independently and more completely developed by Anthony Poole and colleagues (Poole, et al., 1998). In this model, the ribosome and the translation mechanism are derived from an ancient ribozyme replicase. The model postulates that the protoribosome originally functioned as a "triplicase," a complex ribozyme combining the activities of RNA polymerase and RNA ligase that synthesized RNA molecules complementary to a template in three-nucleotide steps. This "triplicase"-proto-ribosome would facilitate the assembly of tRNA-like molecules (analogous to the CCH of the *T* RNAs of the previous model) on the template RNA through base pairing of (proto)anticodons with complementary triplets (codons) on the template, cleaving off the rest of the pre-tRNA, and joining (ligating) adjacent triplets. A replication mechanism based on a complementary interaction of trinucleotides (instead of mononucleotides) with the template was deemed plausible by Poole, et al., given the low catalytic efficiency of ribozymes. A complex of a template RNA with a complementary trinucleotide would persist much longer than a complex with a mononucleotide, giving the triplicase a chance to ligate the adjacent triplets. The triplicase mechanism might seem particularly plausible in view of the experimental results of Fredrick and Noller, which demonstrated that mRNA is threaded through the ribosome in three-nucleotide steps, with concordant movements of tRNAs and without the involvement of translation factors (Frederick and Noller, 2002).

The transition from a triplicase to a modern-type translation-replication system would require several complex steps, namely the emergence of the genetic code (in this case, at the level of amino acid recognition by the proto-tRNAs) and the feedback between translation and RNA replication (the origin of protein RNA polymerases or protein cofactors of a ribozyme polymerase). Furthermore, a sub-functionalization stage would be required in which the triplicase gave rise to separate proto-ribosome and replicase, the latter having to switch from triplet joining to the conventional replication mechanism of one nucleotide at a time.

A skeptical summary of the existing models for the evolution of replication and translation

In the preceding sections, we outlined the status of the RNA World and discussed the origins of replication and translation in some detail. Let us now ask a simple, straightforward question: Is the evidence in support of any of these models and scenarios compelling? Of course, the question already implies a negative answer. We do have some strong hints, even if these are a far cry from a coherent scenario of the earliest stages of evolution of biological information transmission. First, consider the apparent *logical inevitability of an RNA World*: What other starting point for the evolution of the translation system could there be? Second, comparative analysis of the translation system components does point to a much greater role of RNA in ancestral translation, compared to the modern system—notably, the decisive function of RNA as the determinant of amino acid–codon specificity. Third, ribozymes are impressive (if in general far inferior to proteins) in their catalytic versatility and efficiency. Thirty years ago, no catalytic activity was reported for any RNA molecule to catalyze any reaction at all; now we are aware of dozens of ribozyme activities, including some, such as highly efficient aminoacylation, that get the translation system going.

However, this is about all the good news; the rest is more like a sobering cold shower. For all the advances of "ribozymology," no ribozyme polymerase comes close to what is required if we are to accept an RNA-only replicator system as a key intermediate stage in the evolution of life. Nor are any ribozymes capable of catalyzing the synthesis of nucleotides or even their sugar moieties. Even sweeping all these problems under the proverbial rug, the path from a putative RNA World to the translation system is incredibly steep. The general idea of a function(s) for abiogenic amino acids and possibly peptides in the RNA World, such as the role of ribozyme cofactors (see the discussion in the preceding sections), appears fruitful and is compatible with experimental data. Nevertheless, breaking the evolution of the translation system into incremental steps, each associated with a biologically plausible selective advantage, is extremely difficult even within a speculative scheme let alone experimentally. The triplicase/protoribosome hypothesis is attractive as an attempt to explain the origin of translation and replication in one sweep, but is this scenario realistic? The triplicase itself would have to be an extremely complex, elaborate molecular machine, leaving one with the suspicion that, all its attraction notwithstanding, the triplicase might not be the most likely solution to the origin of translation problem.

All things considered, my assessment of the current state of the art in the study of the origins of replication and translation is rather somber. Notwithstanding relevant theoretical models and suggestive experimental results, we currently do not have a credible solution to these problems and do not even see with any clarity a path to such a solution. Granted, the ribozyme field is young, and much progress can be reasonably expected to be achieved soon enough. Nevertheless, toward the end of this chapter, we discuss a radical alternative. First, however, we need to look into the origin of life field proper, primarily its chemical, geological, and geochemical aspects; obviously, in a short section, we have to be very perfunctory and can outline only some key ideas and developments.

The origin of life from the chemical and geochemical standpoints

The origin of life emerged as a scientific problem with Louis Pasteur's demonstration of the apparent implausibility of spontaneous generation of life forms. By an uncanny coincidence, the experiment was reported in 1859, the same year Darwin published *The Origin of Species,* which among other seminal ideas, included the proposition on LUCA. Apparently, the first coherent origin-of-life narrative was published in 1924 by the Russian biochemist Alexander Oparin in the form of a semipopular brochure (Oparin, 1924) that was subsequently repeatedly reissued in steadily expanding versions (Oparin and Fesenkov, 1956). Oparin's scenario was naïve and arbitrary (one might even less kindly brand it chemically unsound), yet it included some of the key ideas that persist in the origin of life field to this day.[4] Oparin's key assumption apparently was that, in some way, the environment where life emerged (conceivably, the primitive ocean as a whole, but possibly some version of Darwin's little warm pond) was a complex solution of abiogenic organic molecules, including amino acids and sugars—in other words, all the monomers required for the synthesis of biopolymers. Oparin denoted this hypothetical medium for the origin of life the *primordial broth (or soup).* A similar scenario for the origin of life was proposed later in the article of J. B. S. Haldane that was already quoted in Chapter 11 on the account of the postulated "viral" stage preceding the emergence of the first cells (Haldane, 1928).

Oparin and Haldane maintained that the atmosphere of the primitive
Earth was reductive and that the first organisms were anaerobic het-
erotrophs that lived off the mix of monomers that they postulated to
have been available in abundance in the hypothetical primordial soup.

Oparin and other early origin-of-life researchers realized the
importance of the emergence of cellular (or cell-like) organization
early on in the history of life or even before the appearance of the
first bona fide life forms. A popular idea was that precellular evolution unfolded within so-called coacervate droplets that form as a
result of interaction between certain oppositely charged polymers.
Oparin and his associates and followers were biochemists by trade
and thus gave priority to metabolism in their entire train of thinking.
The early scenarios of the origin of life held that a simple network of
(pre)metabolic reactions emerged in coacervates, providing feedback
to the growth and possibly fission (division) of those vesicles. From
there on, an increasingly complex metabolism would evolve, eventually resulting in the emergence of autotrophy. In these qualitative
models of precellular evolution, no serious attention was paid to the
origin of genetic information and information-transmission processes.
All these processes (not thoroughly understood at the time) were
assumed to *somehow* have evolved as a consequence or by-product of
the evolution of metabolism. Some limited experimentation on coacervate droplets and other similar vesicles, such as so-called
microspheres consisting of irregular amino acid polymers known as
proteinoids, has demonstrated the ability of these vesicles to maintain
simple reaction networks, grow, and divide, but it did not go much
further than that (Fox, 1976).

We should not be too dismissive of the early hypotheses on the
origin of life. The creators of these qualitative models were completely rational in their thinking and realized the importance of
metabolism, energy sources, and cell-like compartmentalization.
However, they woefully underestimated or plainly disregarded the
completely unrealistic character of the "primary broth" as the homog-
enous medium of the primitive ocean. Any even remotely realistic
origin of life scenario must incorporate well-defined *pre-cellular,
abiogenic compartmentalization;* inorganic catalysts to catalyze "pre-
biochemical" reactions prior to the emergence of bona fide enzymes;
thermal and/or electrochemical potential gradients required for the

generation of energy in accessible forms; a solution to the extremely difficult problem of the origin of genetic information (see the discussion earlier in this chapter). In general, the early concepts underestimated the dimensions of the origin of life problem and failed to investigate special abiogenic conditions that must have been a prerequisite for the jump-start of biological evolution. Subsequently, several groups of researchers attempted to get away from the concept of the homogeneous primary soup, replacing it with some form of inorganic compartments, and sought to address all the origin of life problems in conjunction by combination of modeling, experiment, and observation in nature. The common idea of these hypotheses is the existence of a single framework that could simultaneously provide compartmentalization, energy gradients, and catalysts. We cannot discuss in any detail all these studies here, let alone the entire field of prebiotic chemistry; instead, we concentrate on only a few models that appear particularly productive.

Günter Wachtershauser proposed that life emerged at volcanic sites, on surfaces rich in iron, nickel, cobalt, and other transitional metal centers that were catalytically active and promoted CO_2 fixation, leading to the growth of organic superstructures. Some promising results have been reported on peptide synthesis under primordial conditions postulated in this model (Huber, et al., 2003; Wachtershauser, 1997). So even before actual organisms evolved, a form of inorganic chemoautotrophy might have existed. The synthesized organic molecules might have promoted the reactions catalyzed by inorganic catalysts, thereby resulting in a form of chemical selection that presaged biological selection. The emergence of cellular organization and genetic mechanisms is, under this scenario, the result of the primordial chemical evolution.

As already mentioned in Chapter 11, Michael Russell and colleagues came up with what could be the most realistic, coherent geochemical framework for the origin of life and precellular evolution (Martin, et al., 2008; Russell, 2007; Russell and Hall, 1997), building upon an earlier study by John Baross (Baross and Hoffman, 1985). The basic idea is that the optimal conditions for the emergence of life were provided by warm (not hot) alkaline springs at the bottom of the primordial (known as Hadean) ocean. According to Russell and Hall, hydrothermal springs give rise to continuous flow reactors that generate mounds of precipitated carbonate, silica, clays, and iron-nickel sulfides (see Figure 12-5). Originally, Russell and Hall predicted the existence and dimensions of such mounds from geochemical considerations (Russell and Hall, 1997). The prediction was strikingly confirmed by the discovery of the Lost City site in the mid-Atlantic ridge, called so for the giant carbonate spires that adorn it (Kelley, et al., 2005). The structure of the mounds is truly remarkable: Effectively, they are networks of inorganic compartments formed primarily by iron and nickel sulfides, which can catalyze a variety of organic reactions and maintain a constant flow of protons (proton-motive force) that can provide the energy for reactions taking place in the compartments. As Chapter 11 pointed out, the inorganic compartment networks could have been the hatcheries of life, from the simplest organic syntheses to the emergence and escape of the first cells.

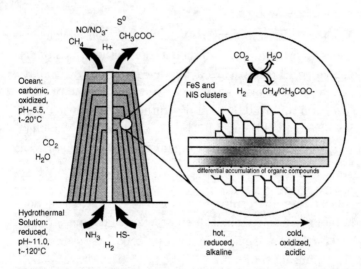

Figure 12-5   Networks of inorganic compartments: flow reactors for primordial chemistry and biochemistry. The data primarily comes from Martin and Russell, 2007.

A considerable amount of experimentation and mathematical modeling has been done to analyze the properties of these networks and the compartments of which they consist. It has been shown that the network membranes are semipermeable—that is, complex organic molecules including nucleobases, amino acids, sugars, and fatty acids are trapped within compartments, whereas small organic molecules such as acetate and methane diffuse freely (Mielke, et al., 2010). Thus, the compartments appear to be a perfect environment for diverse organic syntheses, from the simplest and most fundamental, such as the reduction of CO_2 by molecular hydrogen that yields acetate, to complex reactions such as synthesis of polynucleotides and peptides—in other words, the "origin of biochemistry" (Martin and Russell, 2007). The experiments on the chemistry of the inorganic networks so far have been limited, but computer simulations show that, in the presence of a thermal gradient that inevitably exists at a hydrothermal vent, extremely high concentrations of small molecules and polymers could be reached (Baaske, et al., 2007). This concentration of organic molecules would substantially facilitate a variety of otherwise unlikely reactions, including nucleotide polymerization and RNA ligation (Koonin, 2007c). Although the membranes of the inorganic compartments (Russell goes as far as to call them protocells) appear to be impermeable to biologically relevant monomers, the compartment membranes are relatively unstable ("flimsy," quoting Russell once again), so compartments apparently rupture and merge, thus spreading their content. As we discussed in Chapter 11, compartment-level selection then likely ensues.

In the study of events that apparently happened only once (and about 4 billion years ago), it is important not to stick too firmly to any one particular model, but rather to hedge one's bets. Armen Mulkidjanian proposed a substantially different inorganic compartment scenario that attributes the role of the hatcheries for the earliest life forms to a different kind of mounds, those that exist at more shallow depths and cooler temperatures, and consist mostly of zinc sulfide (Mulkidjanian, 2009). In this model, the source of energy for organic reactions is ultraviolet light that can reach the environments containing zinc sulfide mounds. Furthermore, it is argued that zinc sulfide is vastly superior to iron sulfide as a catalyst, while being a much less harsh chemical that would not have destructive effect on labile molecules such as RNA.

Although the models that Wachtershauser, Baross, Russell, and Mulkidjanian propose differ in substantial aspects, they are unified by common features that appear more important than any of the differences. Under each of these models, life evolved in special habitats and under special conditions that *involved inorganic compartments conducive to the accumulation of organic molecules to high concentrations, possessing catalytic surfaces and maintaining utilizable energy gradients that could support primordial organic chemistry.* These generic features probably have to be part of any origin-of-life scenario that strives to be realistic.

A radical alternative: Eternal inflation cosmology, the transition from chance to biological evolution in the history of life, and a reappraisal of the role of extremely rare events in evolution

As pointed out earlier, the overall situation in the origin of life field appears rather grim. Even under the (highly nontrivial) assumption that monomers such as NTP are readily available, the problem of the synthesis of sufficiently stable, structurally regular polymers (RNA) is formidable, and the origin of replication and translation from such primordial RNA molecules could be an even harder problem. As emphasized repeatedly in this book, evolution by natural selection and drift can begin only after replication with sufficient fidelity is established. Even at that stage, the evolution of translation remains highly problematic.

The emergence of the first replicator system, which represented the "Darwinian breakthrough," was inevitably preceded by a succession of complex, difficult steps for which *biological evolutionary mechanisms were not accessible* (see Figure 12-6). Even considering environments that could facilitate these processes, such as networks of inorganic compartments at hydrothermal vents, multiplication of the probabilities for these steps could make the emergence of the first replicators staggeringly improbable (see Appendix B).

This profound difficulty of the origin of life problem might appear effectively insurmountable, compelling one to ask extremely general questions that go beyond the realm of biology. Did certain factors that were critical at the time of the origin of life but that are hidden from our view now significantly change these numbers and make the origin of life much more likely? Or is it possible that the processes that form the foundation for the origin of life are as difficult as we imagine, but the number of trials is so huge that the appearance of life forms in one or more of them is likely or even inevitable? In other words, is it conceivable that our very concepts of probability are inadequate?

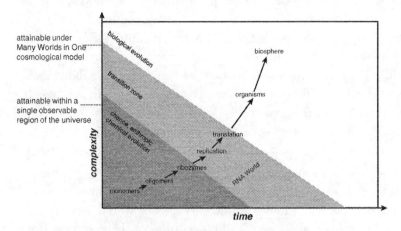

Figure 12-6   The prebiological and biological stages of the origin of life: the transition from anthropic causality to biological evolution.

The first possibility has to do with finding conditions that existed on primitive Earth and somehow made the origin of life "easy." Russell's compartments go some way in that direction, but apparently not far enough: Even in these flow reactors rich in energy and catalysts, the combination of all the necessary processes would be an extreme rarity.

The second possibility may be addressed in the context of the entire universe by asking, how many planets are there with conditions conducive to the origin of life? That is, how many trials for the origin of life were there altogether? In this section, we pursue this second line of inquiry from the perspective of modern physical cosmology.

During the twentieth century, cosmology has undergone a complete transformation, from a quaint (and not particularly reputable) philosophical endeavor to a vibrant physical field deeply steeped in observation. The leading direction in cosmology these days centers on the so-called inflation, a period of exponentially fast initial expansion of a universe (Carroll, 2010; Guth, 1998a; Guth and Kaiser, 2005; Vilenkin, 2007). In the most plausible, self-consistent models, inflation is eternal, with an infinite number of island (pocket) universes (or simply universes) emerging through the decay of small regions of the primordial "sea" of false (high-energy) vacuum and comprising the infinite multiverse (see Appendix B). The many worlds in one (MWO) model makes the startling prediction that all macroscopic, "coarse-grain" histories of events that are not forbidden by conservation laws of physics have been realized (or will be realized) somewhere in the infinite multiverse—and not just once, but an infinite number of times (Garriga and Vilenkin, 2001; Vilenkin, 2007). For example, there are an infinite number of (macroscopically) exact copies of the Earth, with everything that exists on it, although the probability that a given observable region of the universe contains one of these copies is vanishingly small. This picture appears extremely counterintuitive ("crazy"), but it is a direct consequence of eternal inflation, the domi-nant model for the evolution of the multiverse in modern cosmology.

The MWO model is tightly linked to the anthropic principle (sometimes called anthropic selection), a controversial but powerful and popular concept among cosmologists (Barrow and Tipler, 1988; Carter, 1974; Livio and Rees, 2005). According to the anthropic prin-ciple, the only "reason" our universe has its specific parameters is that, otherwise, there would be no observers to peer into that uni-verse.[5] The (weak) anthropic principle can be realistically defined only in the context of a vast (or, better yet, infinite) multiverse. In the MWO model, anthropic selection has a straightforward interpreta-tion: Among the vast number of parameter sets that exist in the mul-tiverse (in an infinite number of copies each), our universe may have only those parameters that are conducive to the emergence and sus-tenance of complex life forms. Sometimes it is said that our universe belongs to the "biophilic domain" of the multiverse (Livio and Rees, 2005). The term "anthropic principle" might be unfortunate as it could be construed to imply some special importance of humans or more generally conscious observers, and worse, might invoke teleo-logical interpretations. Nothing could be further from the correct view of the anthropic principle. At the end of the day, it is nothing more than "observation selection" (Bostrom, 2002): The fact that life exists in this universe severely constraints its characteristics—in the least, our part of the universe must contain galaxies and planetary sys-tems as opposed to only massive black holes or dilute gases of parti-cles that are otherwise much more likely, higher entropy states.

Compared to older cosmological concepts that considered a finite universe, the MWO model changes the very definitions of *possible, likely, and random,* with respect to any historical scenario. Simply put, the probability of the realization of any scenario permitted by the conservation laws in an infinite multiverse is exactly 1. Conversely, the probability that a given scenario is realized in the given universe is equal to the frequency of that scenario in the multiverse and could be vanishingly small. From a slightly different perspective, the well-known idea of the second law of thermodynamics being true only in the statistical sense takes a literal meaning in an infinite multiverse: Any violation of the second law that is permitted by other conservation laws will indeed happen—and on an infinite number of occasions. Thus, *spontaneous emergence of complex systems that would have to be considered virtually impossible in a finite universe becomes not only possible, but inevitable under MWO,* even though the prior probabilities of the vast majority of histories to occur in a given universe are vanishingly small. This new power of chance, buttressed by anthropic reasoning, has profound consequences for our understanding of any phenomenon in the universe, and life on Earth cannot be an exception (Koonin, 2007b).

The history of life is bound to include a crucial transition from *chance to biological evolution* (see Figure 12-6). The synthesis of nucleotides and (at least) moderate-sized polynucleotides could not have evolved biologically and must have emerged abiogenically—that is, effectively by chance abetted by chemical selection, such as the preferential survival of stable RNA species. At the other end of the spectrum, there can be no reasonable doubt that the first cells were brought about by biological evolution at a precellular stage of evolution (see Chapter 11). Somewhere in between is the transition, *the threshold of biological evolution.* Most often, since the advent of the RNA World concept, this threshold is (implicitly) linked to the emergence of replicating RNA molecules. Translation is thought to have evolved later via an *ad hoc* selective process. As discussed in the preceding section, both the ribozyme-catalyzed replication and especially evolution of translation in the RNA World face formidable difficulties. The MWO model dramatically expands the interval on the axis of organizational complexity where the threshold can belong by making the emergence of complexity attainable by chance (see Figure 12-6). In this framework, the possibility that the breakthrough stage for the onset of biological evolution was a high-complexity state cannot be dismissed, however unlikely (that is, extremely rare in the multiverse) and counterintuitive. For example, under this model, the breakthrough could have been brought about by the chance emergence of the core of a coupled system of translation-replication resembling, at least in principle, the present-day RNA viruses (Koonin, 2007b).

The MWO model not only permits but guarantees that, somewhere in the infinite multiverse (moreover, in every single infinite universe), such a complex system would emerge; moreover, there is an infinite number of these systems. Thus, the pertinent question is not whether systems of any complexity have emerged spontaneously by chance alone (the MWO guarantees this), but *what is the most likely breakthrough stage whose appearance on Earth should be attributed to chance under anthropic reasoning?* I submit that, given the severe problems that haunt the evolutionary scenarios developed to explain the origin of replication and translation through biological evolutionary routes, *the possibility that the threshold of biological evolution corresponds to a highly complex stage (possibly a coupled replication-translation system with protein polymerases responsible for RNA replication) should be taken seriously.* This hypothesis (which I refer to as Anthropic Chemical Evolution, or simply ACE) certainly does not rule out the special importance of ribozymes in early biology, particularly in the primordial translation system, as suggested by comparative sequence analysis of protein components of the translation apparatus (see the discussion earlier in this chapter). However, a corollary of the ACE scenario is that the RNA World, as it is currently pictured (as a vast community of replicating RNA molecules endowed by a variety of catalytic activities but containing no translation system and no genetically encoded proteins), might have never existed.

Under the ACE hypothesis, the core elements of the translation system—namely, an RNA-only ribosome and the specific adaptors for at least a subset of the 20 modern protein amino acids—emerged by chance, in accord with the anthropic reasoning. Under this model, the breakthrough system that jump-started biological evolution was a primitive but relatively efficient RNA-based translation machine that was capable of translating exogenous RNAs such that functional proteins, including a replicase, could be generated. The presence of a diversity of randomly synthesized RNAs, including one that encoded a protein with a replicase activity (however low, initially), would be another anthropically determined feature of the sites on the early Earth where life evolved. As discussed in the preceding section, networks of inorganic compartments at hydrothermal vents could play the role of prebiotic chemical reactors. The existence of such networks is itself part of the anthropic scenario.

Under these conditions, the emergence of RNA-based translation machinery would lead to the production of the replicase, and, with the ensuing RNA replication, the fundamental transition from anthropic causality to biological selection would occur (see Figure 12-6). In principle, the start of biological evolution is imaginable with the replicase initially being the only active protein. However, given the plausibility of a RNA-producing "reactor" discussed earlier in this chapter, it seems likely that, upon the advent of translation, other random RNA sequences gave rise to ancestral forms of the other major protein folds, yielding several protein activities (such as RNA-binding proteins and primitive enzymes facilitating nucleotide synthesis), thus conferring the minimal required robustness to the emerging biological system. The emergence of these folds would comprise the Big Bang of protein evolution.

As pointed out earlier, the modern, universal genetic code is far more robust than expected by chance with respect to mutational and probably also translational errors. This robustness is apparent in the well-known nonrandomness of the code structure, such that series of codons that differ only in the third position encode either the same or two similar amino acids, and in other features of the amino acid assignment to codons (Koonin and Novozhilov, 2009). Notably, a putative ancestral "doublet" code in which the third position carried no information could have been even more robust than the modern code (Novozhilov and Koonin, 2009). The robustness of the code is usually assumed to have evolved in the course of the code optimization. However, the ACE model suggests an alternative view under which the basic structure of the code emerged by sheer chance, inasmuch as only codes with a certain minimal level of robustness would allow the appearance of a functional replicase in the breakthrough system. Of course, this scenario does not preclude subsequent adjustments of the code via biological evolution, which, in all likelihood, have indeed happened.

Thus, the ACE hypothesis eliminates the paradoxes of the origin of replication and translation by postulating that both these processes, in their primitive forms, did not evolve biologically, but rather were brought about as a coupled system, by chance abetted with anthropic selection.

The ACE hypothesis certainly should appear outrageous and repugnant to most evolutionary biologists because it shirks the quest for "mechanisms" of precellular evolution. However, mitigating factors exist. First, the postulated chance origin of the replication-translation system does not require any unknown processes. On the contrary, only well-characterized, regular reactions are involved, such as polymerization of nucleotides and amino acids, and nucleotide phosphorylation/dephosphorylation; the only interactions required are those that are common in chemistry and biochemistry. As pointed out earlier in this chapter, the elementary reactions required for translation (amino acid activation, RNA aminoacylation, and transpeptidation) are readily modeled with ribozymes, in a marked contrast with RNA replication that is notoriously hard to achieve without proteins. Second, barring a major inadequacy of the current understanding of the conditions on the primordial Earth, any conceivable scenario for the evolution of life necessarily requires combinations of highly unlikely conditions and events prior to the onset of biological evolution. Such events include the abiogenic synthesis of fairly complex and not particularly stable organic molecules, such as nucleotides, the accumulation of these molecules within appropriate compartments to high concentrations, and their polymerization yielding polynucleotides of sufficient size and diversity. Thus, regardless of the cosmological considerations, some form of anthropic causality appears to be an inevitable aspect of the evolution of life (see Figure 12-6).

I invoked the ACE scenario to suggest that the range of complexity that is open to anthropic causality could be much wider than previously envisaged, so much so that a primitive coupled replication-translation system might have emerged without biological selection. The origin of an elaborate system capable of performing a complex biological function by chance might appear nonsensical. I submit, however, that this is merely a semantic trap. Prior to the onset of biological evolution, there could be no "function"—only complexity, and the MWO model *guarantees the emergence of any level of complexity*. (It is guaranteed to happen "somewhere" in the infinite universe but anthropic reasoning squarely places it on Earth.)

All these considerations trigger a rather nightmarish question: In the infinitely redundant world of MWO, why is biological evolution—and, in particular, Darwinian selection—relevant at all? Will not systems of any, even the highest complexity emerge simply by chance? The answer is yes, but the question misses the point. Under the MWO model, emergence of an infinite number of complex biotas by chance is inevitable, but these would be vastly less common than those that evolved via the ACE scenario, which includes the switch from chance to biological evolution once the breakthrough system is in place (see Figure 12-6). The onset of biological evolution canalizes the historical process by pruning the numerous trajectories that are possible in principle to the relatively few robust ones that are compatible with the Darwinian mode of evolution of complex systems (see Figure 12-7). This transition leads to a much greater rate of evolutionary change than would be achievable by chance such that, as soon as there is an opportunity for biological evolution to take off, anthropic causality is relegated to a secondary role in the history of life. Certainly, "secondary" does not mean unimportant: Contingency and randomness are crucial, especially at transitional stages of evolution (see the discussion earlier in this book, especially in Chapter 7). Thus, in any reconstruction of the origin of life and early evolution, the threshold should be mapped to the lowest possible point, that is, to the minimally complex system capable of biological evolution.

The strong form of the ACE hypothesis, under which the breakthrough stage in the history of life was a primitive coupled replication-translation system (see Figure 12-6), is, in principle, readily falsifiable. Such a system should be construed as the upper bound of complexity for the breakthrough stage. As soon as the possibility of biological evolution at a lower level of complexity, such as in the RNA World, is convincingly demonstrated and the route from the RNA World to the translation system is mapped, either experimentally or in a compelling model, the strong form of the ACE hypothesis will be falsified. A demonstration that life independently emerged on several planets in our universe will have the same effect. In Appendix B, I provide a rough but hopefully instructive calculation of the upper bound of the probability of the emergence of a coupled replication-translation system in the observable part of our universe; this probability is, indeed, vanishingly small. The converse prediction is that any life forms that might be discovered on Mars or perhaps Europa (a satellite of Jupiter where liquid water has been discovered) or even on any extrasolar planets during future planetary explorations will have a common origin with the life on Earth. Any of these falsifications will refute the strong ACE hypothesis but will not make the MWO model irrelevant for our understanding of the origin of life. Indeed, any such discovery (as important as it will be in itself) will simply lower the threshold of biological evolution on the scale of Figure 12-6.

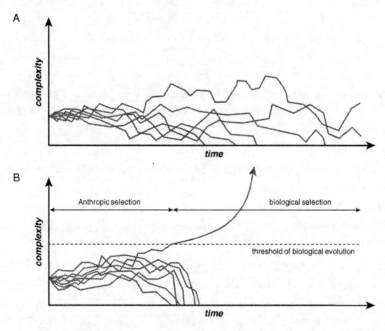

Figure 12-7   Pruning of evolutionary trajectories at the threshold of biological evolution. (A) chemical evolution alone; (B) emergence of biological evolution.

The most straightforward and powerful falsification of the ACE hypothesis would be disproval of the MWO itself. However, an important disclaimer is due. It is not crucial for the validity of the conceptual framework presented here that MWO be correct in all detail.

Only two general assumptions are essential: (i) a spatially infinite universe such as any (island) universe in MWO; the multiverse, while integral to eternal inflation, is not actually required for the argument, and (ii) the finiteness of the number of distinct macroscopic histories. Even the strong form of the ACE hypothesis presented here will not be falsified if some specific details of the MWO turn out to be wrong, but only if one of these general assumptions fails.

Synopsis and perspective

The origin of life is one of the hardest problems in all of science, but it is also one of the most important. Origin-of-life research has evolved into a lively, interdisciplinary field, but other scientists often view it with skepticism and even derision. This attitude is understandable and, in a sense, perhaps justified, given the "dirty," rarely mentioned secret: Despite many interesting results to its credit, when judged by the straightforward criterion of reaching (or even approaching) the ultimate goal, the origin of life field is a failure—we still do not have even a plausible coherent model, let alone a validated scenario, for the emergence of life on Earth. Certainly, this is due not to a lack of experimental and theoretical effort, but to the extraordinary intrinsic difficulty and complexity of the problem. A succession of exceedingly unlikely steps is essential for the origin of life, from the synthesis and accumulation of nucleotides to the origin of translation; through the multiplication of probabilities, these make the final outcome seem almost like a miracle.

Not everything is bleak: Major props for the origin of life have been discovered. Certain environments that exist even now, such as networks of inorganic compartments at hydrothermal vents, were likely present 4 billion years ago as well and could be suitable hatcheries for all the earliest steps of the evolution of life, from the synthesis and concentration of monomers to the origin of translation. The RNA World hypothesis that the impressive body of data on the catalytic activities of ribozymes strongly, if not necessarily directly, supports is an attractive—and apparently the only conceivable—way out of the paradoxes associated with the origin of translation.

Still, the difficulties remain formidable. For all the effort, we do not currently have coherent and plausible models for the path from simple organic molecules to the first life forms. Most damningly, the powerful mechanisms of biological evolution were not available for all the stages preceding the emergence of replicator systems. Given all these major difficulties, it appears prudent to seriously consider radical alternatives for the origin of life. The Many Worlds in One version of the cosmological model of eternal inflation might suggest a way out of the origin of life conundrum because, in an infinite multiverse with a finite number of distinct macroscopic histories (each repeated an infinite number of times), the emergence of even highly complex systems by chance is not just possible, but inevitable. Thus, the interval on the scale of organizational complexity to which the transition from anthropic selection to biological evolution might belong expands dramatically. Specifically, it becomes conceivable that the breakthrough stage for the onset of biological evolution could have been a primitive coupled replication-translation system that emerged by chance. That this extremely rare event occurred on Earth and gave rise to life as we know it might be attributed to anthropic causality alone. Under this model, a bona fide RNA World, with a diverse population of replicating RNA molecules but without translation, was never a stage in the early evolution of life on Earth. However, this scenario by no means defies the central role of RNA in the emergence of biological evolution and early evolution of life. Indeed, the Anthropic Chemical Evolution model includes a complex ensemble of nonreplicating RNA molecules emerging by chance that enabled the onset of biological evolution.

Given the enormous complexity and difficulty of the origin of life problem, and the unavailability of biological evolution mechanisms (selection and drift) for any stage that antedates fairly elaborate replicator systems, I suggest that the possibility that life emerged through a combination of exceedingly unlikely events that the MWO theory renders inevitable, however rare, should not be dismissed. This possibility is counterintuitive in the extreme, but we know only too well that intuition is a poor guide when temporal and spatial scales far outside human experience are involved. Furthermore, the ACE model is no idle speculation. On the contrary, it is a readily falsifiable hypothesis, and the falsification, whether it comes in the form of a demonstration of the feasibility of an RNA World in which translation evolves or as the discovery of independent life in our universe, will be a truly momentous achievement.

Recommended further reading

Aravind, L., R. Mazumder, S. Vasudevan, and E. V. Koonin. (2002) "Trends in Protein Evolution Inferred from Sequence and Structure Analysis." *Current Opinion in Structural Biology* 12: 392–399.

This article summarizes the evidence that major protein folds diversified before the advent of the modern-type translation system. The reconstruction of the evolution of the Rossmann fold present in the Class I aaRS is used to show that a largely RNA-based translation system must have been efficient enough to allow protein evolution.

Crick, F. H. (1968) "The Origin of the Genetic Code." *Journal of Molecular Biology* 38: 367–379.

This amazingly prescient article remains relevant more than 40 years after its publication. In it, Crick delineates the idea of the RNA World (without using this term) and the possible scenarios for the evolution of the genetic code that still define this field of study.

Koonin, E. V. (2007) "The Cosmological Model of Eternal Inflation and the Transition from Chance to Biological Evolution in the History of Life." *Biology Direct* 2: 15.

This article puts the problem of the origin of life into the context of inflational cosmology, according to which every macroscopic state that is physically possible exists in an infinite number of copies in the infinite multiverse. Under this concept, extremely unlikely events, such as the origin of a complete coupled system of replication and translation by chance, cannot be ruled out as crucial steps in the prebiological evolution.

Koonin, E. V., and W. Martin. (2005). "On the Origin of Genomes and Cells Within Inorganic Compartments." *Trends in Genetics* 21: 647–654.

This article develops the ideas of the preceding one into a coherent scenario for the precellular evolution of genetic elements, ensembles of "selfish cooperators," and increasingly large genomes within inorganic compartments.

Martin, W., J. Baross, D. Kelley, and M. J. Russell. (2008) "Hydrothermal Vents and the Origin of Life." *Nature Reviews Microbiology* 6: 805–814.

An overview of the conditions at hydrothermal vents and the features of the inorganic compartment networks that make the suitable hatcheries for life.

Martin, W., and M. J. Russell. (2003) "On the Origins of Cells: A Hypothesis for the Evolutionary Transitions from Abiotic Geochemistry to Chemoautotrophic Prokaryotes, and from Prokaryotes to Nucleated Cells." *Philosophical Transactions of the Royal Society London B Biological Sciences* 358: 59–83.

A seminal article that, for the first time, put into the biological context the idea that the early stages in the evolution of life could have been confined to networks of inorganic compartments at hydrothermal vents.

Robertson, M. P., and G. F. Joyce. (2010) "The Origins of the RNA World." *Cold Spring Harbor Perspectives in Biology* [Epub ahead of print].

This review article describes the RNA World concept, the activities of ribozymes relevant for the origin of replicator systems. The authors concede the difficulties encountered by the RNA World hypothesis and suggest the possibility that an unrelated class of replicators preceded RNA.

Russell, M. J. (2007) "The Alkaline Solution to the Emergence of Life: Energy, Entropy, and Early Evolution." *Acta Biotheoretica* 55: 133–179.

This detailed article shows that the "flow reactors" at hydrothermal vents are low-entropy states conducive to the origin of complex formations such as replicator systems and, eventually, cells.

Vetsigian, K., C. Woese, and N. Goldenfeld. (2006) "Collective Evolution and the Genetic Code." *Proceedings of the National Academy of Sciences USA.* 103: 10,696–10,701.

A compelling argument, supported by mathematical modeling, to the effect that the universality of the genetic code results from the essential role of extensive HGT at the early stages of the evolution of life.

Wolf, Y. I., and E. V. Koonin. (2007) "On the Origin of the Translation System and the Genetic Code in the RNA World by Means of Natural Selection, Exaptation, and Subfunctionalization." *Biology Direct* 2: 14.

A detailed hypothetical scenario for the origin of the translation system by exaptation of ancestral ribozymes stimulated by amino acids and peptides.

13

The postmodern state of evolutionary biology

In the preceding 12 chapters, we discussed a variety of aspects of the evolution of life. By no account could this discussion be comprehensive, and that has never been my goal. However, what is presented in these chapters is sufficient (and necessary) to convey the main point of the book: In the 50 years that have elapsed since the crystallization of Modern Synthesis, evolutionary biology has dramatically changed and entered a new, "postmodern" era.

According to Modern Synthesis, the evolution of life is a process of active adaptation of populations to changing environments. We now realize that although such adaptation is undoubtedly an essential component of the evolutionary process, it is not quantitatively dominant. Although fully aware of the oversimplification inherent in any attempts at grand definitions, I submit this:

The evolution of life is largely a stochastic process based on historical contingency, substantially constrained by various requirements for the maintenance of basic biological organization, and modulated by adaptation.

The constraints that shape evolution should be understood most broadly to include all forms of damage control and local optimization, such as the decrease of the error rates of all information processes, as well as energy expenditure, and the perennial arms race between parasites and hosts that fuels the evolution of diverse adaptations through the Red Queen effect. In this final chapter, I briefly summarize various aspects of the postmodern state of evolutionary biology and discuss the feasibility and possible contours of a "Postmodern

Synthesis." Box 13-1 summarizes the postmodern reappraisal of some
fundamental tenets of Darwin's concept and Modern Synthesis.

Box 13-1: Postmodern reassessment of some central propositions of Darwin and Modern Synthesis

Proposition	Postmodern Status
The material for evolution is provided primarily by random, heritable variation.	**Only partly true.** The repertoire of relevant random changes greatly expanded to include duplication of genes, genome regions, and entire genomes; loss of genes and, generally, genetic material; HGT, including massive gene flux in cases of endosymbiosis; invasion of mobile selfish elements and recruitment of sequences from them; and more. More importantly, (quasi) directed (Lamarckian) variation is recognized as a major factor of evolution.
Fixation of (rare) beneficial changes by natural selection is the main driving force of evolution.	**Only partly true.** Natural (positive) selection is important but is only one of several fundamental factors of evolution and is not quantitatively dominant. Neutral processes combined with purifying selection dominate evolution, and direct effects of environmental cues on the genome ([quasi] Lamarckian phenomena) are important as well.
The variations fixed by natural selection are "infinitesimally small." Evolution adheres to gradualism.	**False.** Even single gene duplications and HGT of single genes are by no means "infinitesimally small," nor are deletion or acquisition of larger regions, genome rearrangements, whole-genome duplication, and, most dramatically, endosymbiosis. Gradualism is not the principal regime of evolution.
Uniformitarianism: Evolutionary processes have remained largely the same throughout the evolution of life.	**Only partly true.** Present-day evolutionary processes were important since the origin of replication. However, major transitions in the evolution, such as the origin of eukaryotes, could be brought about by (effectively) unique events such as endosymbiosis, and the earliest stages of evolution (pre-LUCA) partially relied on distinct processes not involved in subsequent "normal" evolution.

Proposition	Postmodern Status
Evolution by natural selection tends to produce increasingly complex adaptive features of organisms, hence progress is a general trend in evolution.	**False.** Genomic complexity probably evolved as a "genomic syndrome" caused by weak purifying selection in small population, not as an adaptation. There is no consistent trend toward increasing complexity in evolution, and the notion of evolutionary progress is unwarranted.
The entire evolution of life can be depicted as a single "big tree."	**False.** The discovery of the fundamental contributions of HGT and mobile genetic elements to genome evolution invalidates the TOL concept in its original sense. However, trees remain essential templates to represent evolution of individual genes and many phases of evolution in groups of relatively close organisms. The possibility of salvaging the TOL as a central trend of evolution remains.
All extant cellular life forms descend from very few ancestral forms (and probably one, LUCA).	**True.** Comparative genomics leaves no doubt of the common ancestry of cellular life. However, it also yields indications that LUCA(S) might have been very different from modern cells.

Pattern and process pluralism in evolution: The changing concepts of selection, variation, and the Tree of Life

The role and status of selection

The double meaning of *postmodern* in the Preface to this book could not have escaped the reader's attention. Whatever one thinks of the postmodern philosophy (see Appendix A), its worldview certainly emphasizes the richness and extreme diversity of the processes and patterns that constitute reality. Such is the complexity of these multiple trends that, to some philosophers of the post-modern ilk, any major generalization is anathema. In today's evolutionary biology, the plurality of processes and patterns is arguably the main theme; if we want to speak in paradoxes, it could be said that *"the main theme is the absence of an overarching main theme."*

The exclusive focus of Modern Synthesis on natural selection acting on random genetic variation has been replaced with a plurality of complementary, fundamental evolutionary processes and patterns (see Figure 13-1). In the new evolutionary biology, natural selection is but one of the processes that shape evolving genomes—and, apparently, not the quantitatively dominant one. To a large extent, neutral processes such as genetic drift and draft define evolution.

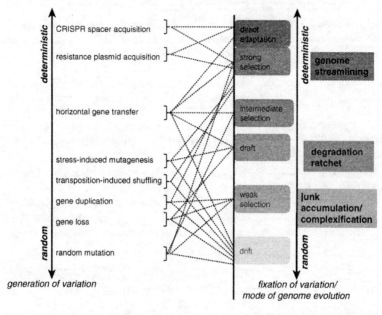

Figure 13-1 The multiple processes that generate genomic variation, affect its fixation in the evolution of life, and shape the evolution of genomes.

To expand on the subject of plurality, the relative contributions of adaptive and neutral processes are far from constant throughout the spectrum of life forms. As most aptly articulated by Michael Lynch (in an obvious paraphrase of Dobzhansky), "Nothing in evolution makes sense except in light of population genetics" (Lynch, 2007b). Indeed, population dynamics—or, simply put, the effective population size over the short and long terms—is the key determinant of the selection pressure. Effective population size may differ by orders of magnitude even in rather closely related organisms, hence dramatic differences in the intensity of selection (for example, among insects and among mammals). These differences dictate distinct evolutionary

regimes: At high *Ne,* evolution is governed primarily (if not exclusively) by selection; at low *Ne,* drift becomes a prominent factor. In
the actual course of evolution, (almost) all lines of descent pass
through multiple population bottlenecks, which are the phases when
evolution is dominated by random drift, hence the *inescapable major
contribution of chance to the evolution of all organisms.*

The role of selection in the evolution of phenotypes might be
greater than in the evolution of genomes. However, the advances of
systems biology have substantially expanded the concept of
phenotype. Along with traditional organismal traits, we now can study
the evolution of molecular phenotypic features such as gene expression and protein abundance, and molecular phenotypic evolution
turns out to include a major neutral component. Moreover, *the ratchet
of constructive neutral evolution* seems to instigate non-adaptive evolution of complex phenotypic traits that, from the traditional (neo)
Darwinian standpoint, appear to be typical adaptations.

Even when selection and adaptation clearly are involved, the
manifestation of these factors of evolution often (possibly, most of the
time) is quite different from the (neo) Darwinian idea of "improvement." Often *adaptations have to do with maintaining the integrity of
cellular organization, preventing malfunction, and performing damage control.* In a sense, this is a trivial statement of fact, considering
the expanse and complexity of the molecular machinery that is dedicated to quality control of each of the major information transfer
processes: Systems of DNA repair and protein degradation, and
molecular chaperones are all cases in point. Moreover, much, if not
most, of the evolution of protein-coding genes appears to be driven
by selection for robustness to misfolding. In multicellular organisms,
the importance of the selection for prevention of malfunction is
apparent at the level of cell and tissue interactions, as illustrated by
the highly complex systems of programmed cell death.

In retrospect, all these findings may appear quite intuitive, considering how advanced, complex, and, in a variety of ways, optimized cells and even individual protein or RNA molecules are.
Once these complex systems are in place—and evolutionary reconstructions clearly show that they have been in place for most of the
history of life, that is, more than 3.5 billion years—quality control

and damage prevention indeed become the bulk of the "work" of evolution, the importance of occasional new adaptations notwithstanding. This realization places an enormous burden on the early, precellular stages of evolution when change must have been rapid and the roles of positive selection along with constructive neutral evolution must have been much greater than they were during the subsequent 3.5 billion years of evolution. In a sense, almost everything "really interesting" in the evolution of life occurred during its relatively brief, earliest stages antedating the "crystallization" of the basic cellular organization (see Chapters 11 and 12, and more discussion later in this chapter). Certainly, major exceptions exist, such as the emergence of eukaryotic cells or multicellular eukaryotic organisms, but there is no doubt that most of the fundamental evolutionary innovations are crammed into the earliest 5% of the history of life.

The changing concepts of variation and the death of gradualism

Complementary to the transformed concept of selection and its role in evolution, the ideas of what constitutes evolutionarily important genomic and phenotypic variation have substantially evolved (see Figure 13-1). Random variation leading to infinitesimally small, beneficial phenotypic changes that Darwin viewed as the key to all evolution remain important, but this is only one class of relevant changes—and, at least quantitatively, not the dominant one.

To begin with, nearly neutral mutations that are not "seen" by selection and that are fixed through drift or that persist in a population without being fixed appear to be more common than slightly beneficial "Darwinian" mutations. The important point that was not clearly realized until recently is that nearly neutral mutations are far from being indifferent in evolutionary terms. Indeed, these mutations give rise to nearly neutral networks that comprise the essential reservoir of evolutionary plasticity.

Moreover, forms of genetic variation that by no account can be viewed as leading to infinitesimally small effects are crucial for evolution. This realization puts to rest the gradualism that Darwin and the architects of Modern Synthesis considered central to all evolution. The nongradualist types of evolutionarily important genetic change include gene and whole genome duplication, gene loss, and HGT, particularly ratchets of extensive, directional gene transfer instigated by endosymbiosis. Horizontal gene transfer is the most prevalent evolutionary process in prokaryotes, and specific adaptations seem to exist that support an optimal level of HGT. The rate of HGT is much reduced in eukaryotes, but extensive duplication followed by subfunctionalization compensates for the curtailment of HGT. Furthermore, endosymbiosis and the ensuing gene transfer from the symbionts to the host were decisive in the evolution of eukaryotes.

Not only the infinitesimal effect of variation, but also its exclusive randomness is gone. Mechanisms such as stress-induced mutagenesis, for which highly evolved, elaborate systems exist in all cellular life forms, are adaptive and nonrandom, and the adaptive immunity system in prokaryotes appears to function through bona fide Lamarckian inheritance. Generally, evolutionary processes span the continuum of the Lamarckian, Darwinian, and Wrightian modes of evolution, and the relative contribution of each in any particular episode of life's history depends on the population dynamics and environmental pressure. Finally, it is becoming increasingly clear that phenotypic mutations are not necessarily irrelevant for evolution and could contribute to adaptation in conjunction with genetic mutations, particularly through the look-ahead effect. Taken together, the importance of nonrandom mutations and (quasi) Lamarckian mechanisms of evolution, along with the contribution of phenotypic mutations, shows that *evolvability is evolvable and seems to invalidate, at least in part, one of the most cherished beliefs of evolutionary biologists, that evolution has no foresight.*

From the Tree of Life to the Web of Life

The commitment of evolutionary biology to the Tree of Life as a single definitive representation of the history of life forms on Earth has given way to a pluralist picture in which diverse web-like processes complement tree-like processes of gene evolution. These processes include HGT that is particularly widespread in prokaryotes but also made pivotal contributions to the evolution of eukaryotes, especially through endosymbiosis, as well as various forms of genome fusion and exchange of genetic material between hosts and parasites (see Chapters 5 and 7). For a first approximation description of the change that occurred in our concepts of the history of life, it may be said that *the Tree of Life has been replaced with a Web of Life (or per-haps a Rhizome of Life* [Raoult, 2010].[1])

When we look deeper into the evolutionary processes, it becomes clear that evolution of life can be meaningfully depicted only as a complex, dynamic representation of interacting processes among which tree-like evolution actually could be the most fundamental one. This is because tree-like evolution is a direct consequence of the binary replication of the genetic material, the universal core process of all life (Chapter 6). Moreover, there is undeniable dynamical coherence between evolutionary histories of large gene sets (the largest of such coevolving gene sets are commonly known as genomes) that might even give a new life to TOL, now construed as a statistical central trend in the "forest" of phylogenetic trees of individual genes. The statistical coherence of gene histories notwithstanding, the core of universally conserved, ubiquitous genes of cellular life forms is tiny due to lineage-specific gene loss and nonorthologous gene displacement, major evolutionary phenomena whose importance could not have been appreciated in the pregenomic era. All things considered, *the web of evolution represents the highly dynamic genomic space-time* in which the genome of each species is only a transient, metastable constellation of genes.

The surprising relevance of simple physical and mathematical models for understanding evolution: Biological evolution as a subject of statistical physics

In the preceding section, I outlined the plurality of patterns and processes, which is the defining aspect of the new vision of evolution. Speaking loosely, this plurality greatly increases the entropy of evolutionary biology. However, analysis of the data produced by genomics and systems biology made the opposite, "antientropic" trend toward structuring of evolutionary theory equally apparent and prominent. A number of universal distributions and dependencies have been discovered, such as the distribution of gene evolution rates, the connection between gene evolution and expression, and the node degree distribution of diverse networks. Moreover, at least some of these universals can be readily derived from simple mathematical models of evolution that are quite similar to models employed in statistical physics. These models are becoming increasingly general as they connect and explain jointly universal dependencies that initially appeared unrelated, such as the distribution of evolutionary rates together with the anticorrelation between evolutionary rate and expression, or scaling laws for gene families together with the scaling for functional classes.

A startlingly simple general explanation for this tractability of genome-wide patterns seems to exist. Evolutionary genomics deals with large ensembles of objects (genes, proteins) that, for many purposes, can be treated as weakly interacting and moving (evolving) along independent trajectories. Accordingly, *the principles of statistical physics apply to ensembles of genes much as they apply to ensembles of molecules.* Certainly, statistical treatment of evolutionary phenomena is subject to the same limitations as the analogous approaches in physics: These patterns and models are hardly sufficient to explain specific biological phenomena that often have to do with a small set of genes rather than a large ensemble. Furthermore, interactions between genes (epistasis) often substantially constrain evolution. These limitations notwithstanding, it is remarkable that the advances of genomics and systems biology, while revealing an extremely complex, multifaceted picture of evolution, at the same time allow us to derive powerful and simplifying generalizations. It is tempting to offer yet another version of the famous phrase: *Nothing in evolution—and in population genetics—makes sense except in light of statistical physics.*

Replaying the tape: Determinism and stochasticity in evolution

The space of genotypes, even if one considers only relatively simple, small genomes, is unimaginably vast (for example, for a prokaryote with a 1Mb genome, there are $4^{1,000,000}$ possible sequences, a number that vastly exceeds anything that actually exists in the observable part of the universe, such as the total number of protons or electrons). What fraction of these genotypes are actually viable and thus could have played a role in evolution? Or, to ask the question in a way that makes more sense in the context of evolution, what is the fraction of all possible trajectories in the genotype space that are open for exploration by the evolutionary process? This is a more technical reformulation of the favorite question of Stephen Jay Gould (Gould, 1997b): What would we observe if we had a chance to replay the tape of evolution? The answer Gould gave, as did Francois Jacob in his famous "tinkering" article (Jacob, 1977), Dan Dennett in "Darwin's Dangerous Idea" (Dennett, 1996), and many others, was that we would not see anything like our present biosphere because evolution is all historical contingency. Dennett fittingly invoked the physical phenomenon of deterministic chaos to account for this general picture of evolution: Each event that occurs during evolution certainly has specific physical causes, but small perturbations may cause large changes in the course of evolution so that distant outcomes become completely unpredictable.

It remains difficult to give a strongly supported general answer to this key question about evolution; however, the limited available results of direct exploration of evolutionary trajectories for both individual proteins and bacterial populations yielded unexpected results (O'Maille, et al., 2008; Ostrowski, et al., 2008; Weinreich, et al., 2006). It appears that, in most cases, only a small proportion of the theoretically possible paths are actually accessible to evolution, so evolution seems to be less stochastic, more deterministic, and more predictable than previously suspected (see Figure 13-2). These findings suggest that the fitness landscapes for at least some evolving genes and genomes are rugged, so that the majority of paths are interrupted by deep ravines of low fitness and thus are forbidden (O'Maille, et al., 2008). The primary underlying reason is likely to be epistasis, the interaction between different parts of the same gene or between different genes: on a rugged landscape, one mutation often leads to a prohibitive drop in fitness, but a second one, through epistatis, might lead to a high fitness area of the landscape. Epistasis seems to be one of the important factors that hold together evolving biological systems, so that many aspects of their evolution reflect an integral whole (Kogenaru, et al., 2009). As pointed out in the previous section, epistasis certainly limits the applicability of the representation of evolving genomes as ensembles of weakly interacting "particles." The epistatic interactions severely constrain the range of available evolutionary trajectories—just how severely remains to be determined by further modeling and experimental evolution studies. It may well be the case that the deterministic chaos view is valid and the discovered constraints in practice have little bearing on the predictability of evolution, or the outcome of the metaphorical tape replaying. The available trajectories, even if they are only a small fraction of those theoretically possible, could still be numerous and diverse enough to render evolution effectively unpredictable. A crucial and so far unresolved problem is the relationship between the accessible trajectories: If these trajectories cluster in a small area of the genomic space-time, evolution might be quasideterministic; in contrast, if the accessible trajectories are randomly scattered, the (un)predictability of evolution would not be much affected by the constraints (see Figure 13-2).

A

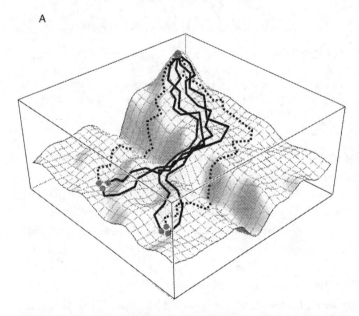

Figure 13-2A The rugged fitness landscape and accessible evolutionary trajectories. Quasideterministic evolution: canalization of the accessible trajectories. Solid lines show monotonic ascending trajectories that are accessible to evolution driven solely by selection. Broken lines show nonmonotonic trajectories that are accessible only with the involvement of genetic drift.

B

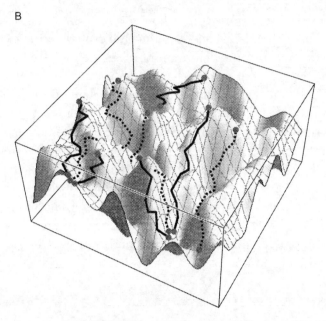

Figure 13-2B The rugged fitness landscape and accessible evolutionary trajectories. Stochastic evolution: random scattering of accessible trajectories.
Solid lines show monotonic ascending trajectories that are accessible to evolution driven solely by selection. Broken lines show nonmonotonic trajectories
that are accessible only with the involvement of genetic drift.

Most likely, the results of the detailed analysis of evolutionary
landscapes and trajectories on them will differ for the evolution at different levels and in different situations, in line with the pattern pluralism discussed earlier. Furthermore, it must be re-emephasized
that the balance between determinism and stochasticity critically
depends on the pressure of selection—that is, on effective population
size. In an effectively infinite population, evolution is essentially
deterministic, whereas in very small populations, evolution is stochastic within fundamental constraints. To avoid any possibility of misun-
derstanding, let us note that even if evolution can be legitimately
described as quasideterministic, this has nothing to do with any teleological notions. However, canalization *sensu* Waddington (see
Chapter 2) does appear an interesting analogy.

The nondeterministic and complex genome-to-phenotype mapping

It is generally accepted that the genome (genotype) determines the phenotype of an organism (with some epigenetic contribution), the phenotype is tightly controlled by selection, and phenotypic variation has no evolutionary consequences. Comparative genomics and systems biology show that none of this is strictly true, and such oversimplifying generalizations miss key aspects of biology. Two complementary facets of the genome-phenotype relationship defy the simple, deterministic link:

1. Phenotypic mutations and other forms of noise, such as the genome-wide spurious transcription in eukaryotes, are intrinsic to biological systems and contribute to their evolution (see Chapter 9). This evolutionarily important phenotypic variation is partially controlled by the genome, but the link between the genome and the noise is intrinsically stochastic.

2. The genome-to-phenotype mapping is intrinsically nonisomorphous and complex (in simplistic terms, not a one-to-one, but a many-to-many correspondence); all genes are pleiotropic, and all phenotypic traits ("functions" or spandrels) are multigenic—they depend on the activities of multiple genes. Altogether, the genome-to-phenotype mapping is an extremely complex network (see Chapter 5, particularly Figure 5-9). The edges in the graph depicting this mapping have different weights, which is a reflection of different contributions of multiple genes to the same trait.

The ubiquity and evolutionary importance of phenotypic variation make the genome-phenotype relationship fundamentally nondeterministic. The many-to-many mapping constrains evolution, perhaps substantially (see the preceding section), but it makes the genome-phenotype relationships harrowingly complex. Together, these two features render phenotype reconstruction from the genome sequence extremely difficult. Some simple phenotypic features are certainly predictable: For example, if a bacterium has no *lac* operon, it will be unable to grow on lactose. However, even for such simple traits, multiple pathways often exist. Any complex phenotype is extremely difficult to predict, as we have seen in Chapter 5 for thermophily and radiation resistance in prokaryotes. The complexity of the genome-phenotype relationship and the consequent difficulty of functional inferences from genome sequences are further exacerbated in eukaryotes, particularly the multicellular forms. The surprising and counterintuitive but by now well-established lack of a strong connection between the apparent biological importance of a gene and the rate of its evolution emphasizes the emerging understanding that the phenotypic consequences of genome evolution are nontrivial and, in general, are difficult to predict (see Chapter 4). A lack of appreciation of this complexity may lead to unrealistic hopes for quick success in projects aimed at dissection of and manipulation with complex phenotypes, such as whole-genome association studies, the "war on cancer," or personalized medicine.

The rise of experimental evolution

This book is primarily about concepts, ideas, and models rather than methods. Nevertheless, before ending this final chapter, I find it necessary to say a few words on the new generation of approaches that have already provided remarkable insights into key evolutionary processes but that should start really changing the face of evolutionary biology in the next decade or so. These research strategies fall under the umbrella of "experimental evolution." In today's evolutionary experiments, the course of evolution of a population of organisms or molecules can be traced directly by applying new generation-sequencing methods to sequence thousands and potentially millions of DNA or RNA molecules. The experiments of Richard Lenski and colleagues on the long-term laboratory evolution of *E. coli* populations, to which we have referred more than once in this book, are the prime case in point (Ostrowski, et al., 2008; Barrick, et al., 2009; Woods, et al., 2011). These experiments have already yielded invaluable information on different regimes of selection and drift, the prevalence of parallel mutations, the evolution of evolvability, and more. However, with the now realistic possibility of sequencing thousands of complete bacterial genomes, the main promise lies in the not-so-remote future, when evolutionary trajectories of populations under different environmental conditions and selective pressures will be studied comprehensively. Conceptually, these experiments continue the line of research started with the prescient experiments of Spiegelman and colleagues with RNA bacteriophages in the 1960s (see

Chapter 8). Spiegelman's experiments were about half a century ahead of their time and had relatively little impact, but in the first decades of the twenty-first century, the status of experimental evolution studies is rapidly changing.

Another line of evolutionary experimentation includes the study of fitness landscapes for evolving proteins or RNA molecules that we briefly discussed earlier in this chapter. The currently available experimental data describes only tiny fractions of the landscapes, but the possibility to explore larger areas is already realistic (Kogenaru, et al., 2009; Loewe, 2009). Ultimately, explicit reconstruction of complete fitness landscapes will change our ideas of what it means to "understand" the evolutionary process.

The brave new worlds of viruses and prokaryotes

Modern Synthesis focused exclusively on the evolution of animals and plants, the multicellular eukaryotes that mostly reproduce sexually. Unicellular eukaryotes and prokaryotes, let alone viruses, were not considered important for evolutionary biology. Perhaps the incorporation of the vast microbial world into the evolutionary framework is the most momentous development that led to the transition from Modern Synthesis to the current "postmodern state." The initial attempts to decipher the evolutionary relationships among bacteria have been frustrating in the extreme, but subsequent sequence analysis of conserved genes such as rRNA and then of complete genomes led to the most dramatic reverse. Comparative genomics of bacteria and archaea have transformed the central concepts of evolutionary biology, including that of the Tree of Life (see Chapter 6), and have revealed the highly dynamic character of genomes and pangenomes (see Chapter 5).

The study of the Virus World has led to an equally important, or perhaps even more dramatic, shift in our views of the evolution of life on Earth. Far from being fundamentally inconsequential (even if medically important) tiny parasites, viruses are the most physically abundant and genetically diverse biological entities on the planet. The Virus World has existed, in all likelihood, since the earliest, precellular stage of evolution and constantly interacts with the world of cellular life forms, substantially contributing to their evolution while maintaining its autonomy.

The empires and domains of life

Carl Woese used the rRNA tree to introduce the three domains of life, a huge conceptual breakthrough for evolutionary biology and biology in general. However, things have changed drastically in the past 30 years, and this classification of life forms does not reflect the complex realities of evolution uncovered by comparative genomics. The first major discovery overturning the three-domain schema is the demonstration of the chimeric nature of the eukaryote genomes. The three-domain tree reflects only the evolution of a subset of genes involved in information processing, which quantitatively make up a small minority of eukaryotic genes, even within the group traced back to the last common ancestor of eukaryotes. Certainly, the domain classification of life forms is only a convention, so the classification of eukaryotes as a distinct domain is not right or wrong in itself. However, this classification is potentially misleading, especially when accompanied by the tripartite schematic Tree of Life, because the fusion of two organisms and their initially distinct genomes that apparently gave rise to eukaryotes is ignored. The scheme with an explicit fusion that mathematically is not a tree is much preferable (see Figure 13-3).

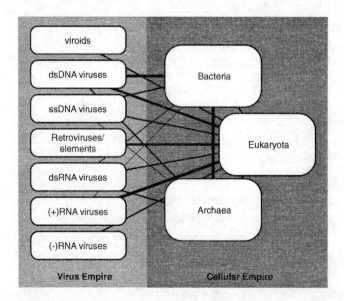

Figure 13-3   The empires and domains of life. The connecting lines show the fluxes of genetic information between domains, including both virus infection and all routes of horizontal gene transfer.

Probably an even more fundamental departure from the three-domain schema is the discovery of the Virus World, with its unanticipated, astonishing expanse and the equally surprising evolutionary connectedness. As discussed in Chapter 10, virus-like parasites inevitably emerge in any replicator systems, so there is no exaggeration in the statement that *there is no life without viruses*. Moreover, it seems almost inevitable that the precellular evolution of life went through a virus-like state. And in quite a meaningful sense, not only viruses taken together, but also major groups of viruses seem to be no less (if not more) fundamentally distinct as the three (or two) domains of cellular life forms, given that viruses employ different replication-expression cycles, unlike cellular life forms which, in this respect, are all the same (see Chapter 10 for details).

All classification is convention, whereas evolutionary scenarios strive to reconstruct, however crudely, the history that actually happened. Obviously, the classification of life forms that most accurately reflects the best available evolutionary scenario is the classification of choice. From this perspective, it makes the most sense to separate all known life into two "empires," namely viruses and cellular life forms (Koonin, 2010d). The two empires are distinct but constantly exchange genes (see Figure 13-3). One may choose to speak of three or only two domains of cellular life, but the archaeo-bacterial fusion is a necessary part of the classification scheme. Although it is not my current intent to make any formal proposals regarding new "viral domains," distinct large groups of viruses that share sets of conserved genes certainly appear comparable in their status to the domains of cellular life (see Figure 13-3).

The paradox of biological complexity, the progress fallacy, and the importance of non-adaptive ratchets

Cells and organisms as devices for gene replication

Many scientists and educated lay readers alike seem to disdain Richard Dawkins's selfish gene concept (Dawkins, 2006)—at least, in its more extreme aspects—probably because it is so counterintuitive and "undignified." Yet once one realizes that replication of the genetic material is the single central property of living systems, there is no logical escape from the selfish gene perspective. In particular, Dawkins provocatively claimed that organisms are but vehicles for replicating and evolving genes, and I believe that this simple concept captures a key aspect of biological evolution. Of course, this is not meant here in any metaphysical or teleological sense at all, such as a claim that cells and organisms exist "for the purpose" of enabling gene replication. The "purpose-oriented" view in general is not constructive (see Appendix A). Thus, suggesting that the phenotype exists for the purpose of replication is as pointless as proposing the opposite. Nevertheless, a tangible, logically inevitable asymmetry exists between the genome and the phenotype: All phenotypic features of organisms—indeed, cells and organisms themselves as complex physical entities—emerge and evolve only inasmuch as they are conducive to genome replication. That is, they enhance the rate of this process, or, at least, do not impede it.

As discussed in some detail in Chapter 10, a large fraction of the core phenotype consists of *antientropic devices* that lower the error rate of information transmission in the replication process itself, as well as the ancillary processes of transcription, translation, and protein and RNA folding, and keep in check the deleterious effects of those errors that do occur. Most of the rest of the phenotypic core are *supplying devices* that obtain and produce building blocks for replication and for maintaining the phenotype. Thus, it is difficult to deny that the evolution of phenotypes centers on gene (genome) replication.

Under this replication-centered perspective, the emergence of complexity is an enigma: Why are there numerous life forms that are far more complex than the minimal, simplest device for replication? We cannot know "for sure" what these minimally complex devices are, but there are excellent candidates—namely, the simplest autotrophic bacteria and archaea, such as *Pelagibacter ubique* or *Prochlorococcus sp.* These organisms get by with about 1,300 genes without using any organic molecules, and generally without any dependence on other life forms. Incidentally, these are also the most "successful" organisms on Earth. They have the largest populations that have evolved under the strongest selection pressure—and consequently have the most "streamlined" genomes. A complete biosphere consisting of such highly effective unicellular organisms is easily imaginable; indeed, the Earth biota prior to the emergence of eukaryotes (that is, probably for

the 2 billion years of the evolution of life or so) must have resembled
this image much closer than today's biosphere (although more complex prokaryotes certainly existed even at that time).

So why complex organisms?

One answer that probably appeared most intuitive to biologists and
to everyone else interested in evolution over the centuries is that
the more complex organisms are also the more fit. This view is
demonstrably false. Indeed, to accentuate the paradox of complexity, the general rule is the opposite: The more complex a life form is,
the smaller effective population size it has, and so the less successful it is, under the only sensible definition of evolutionary success.
This pattern immediately suggests that the answer to the puzzle of
complexity emergence could be startlingly simple: Just turn this
trend around and posit that the smaller the effective population
size, the weaker the selection intensity, hence the greater the
chance of non-adaptive evolution of complexity. This is indeed the
essence of the population-genetic non-adaptive concept that Lynch
propounded.

   We now can formulate a more specific answer to the question in
the title of this section—or, rather, a set of complementary answers:

1. In the most general sense, complexity evolves "simply because
   it can": Under the universal "drunkard's walk" perspective,
   given enough time, the probability that complex biological
   organization evolves steadily increases, by sheer chance.

2. The other, more concrete manifestation of chance as the
   defining factor of evolution comes through the population
   genetic view: Complexity can evolve through random fixation
   of effectively neutral (slightly deleterious) mutations via
   genetic drift in populations with a small effective size. Thus,
   combining (1) and (2), complexity evolves because it can,
   provided a weak purifying selection pressure that cannot
   wipe out slightly deleterious changes such as gene duplications, insertion of mobile elements in many genomic sites,
   and others.

3. With these random factors in the background, evolution of complex organization is made possible by the *ratchet of constructive neutral evolution*: Once two or more genes become dependent on each other as a result of differential accumulation of slightly deleterious mutations in both, they are both fixed in evolution and interlocked, which leads to increased complexity. This is the mechanism of evolution of duplicated genes as well as many horizontally transferred genes via subfunctionalization, apparently a major route of evolution.

4. Complex forms are not generally fitter than simpler forms; however, complexity can facilitate adaptation to new niches, as for example is the case with land plants. The emergence of complexity thus could have a distinct adaptive component, in addition to the major nonadaptive factors mentioned earlier.

The essential role of neutral ratchets in creating apparent directionality in evolution without any involvement of actual "improvement" needs to be emphasized. The ratchet of constructive neutral evolution might be the key to the emergence of a variety of complex biological features; the ratchets of gene transfer from endosymbionts to hosts substantially contributed to eukaryogenesis, and the ratchet of irreversible gene loss is the leading factor in the reductive evolution of parasites and symbionts. Ratchets may be viewed as narrow, steep ridges on fitness landscapes: Once an evolving population finds itself on such a ridge, it starts following a quasideterministic course because falling off the ridge results in a dramatic decrease in fitness and imminent extinction. In the process, complexity may emerge without any contribution from adaptation.

The ultimate enigma of the origin of life

Thanks to the advances of genomics and systems biology, we have learned more about the key aspects of evolution in the first decade of the twenty-first century than in the preceding century and a half. Although major transitions in evolution, such as the origin of eukaryotes or the origin of animals, remain extremely difficult problems, more and more clues appear. Beyond a doubt, substantial progress has been achieved even in these difficult areas of evolutionary biology.[2] We are even starting to develop scenarios for the origin of cells that may go beyond sheer speculation.

However, the origin of life—or, to be more precise, the origin of the first replicator systems and the origin of translation—remains a huge enigma, and progress in solving these problems has been very modest—in the case of translation, nearly negligible. Some potentially fruitful observations and ideas exist, such as the discovery of plausible hatcheries for life, the networks of inorganic compartments at hydrothermal vents, and the chemical versatility of ribozymes that fuels the RNA World hypothesis. However, these advances remain only preliminaries, even if important ones, because they do not even come close to a coherent scenario for prebiological evolution, from the first organic molecules to the first replicator systems, and from these to bona fide biological entities in which information storage and function are partitioned between distinct classes of molecules (nucleic acids and proteins, respectively).

In my view, all advances notwithstanding, evolutionary biology is and will remain woefully incomplete until there is at least a plausible, even if not compelling, origin of life scenario. The search for such a solution to the ultimate enigma may take us in unexpected (and deeply counterintuitive for biologists) directions, particularly toward a complete reassessment of the relevant concepts of randomness, probability, and the possible contribution of extremely rare events, as exemplified by the cosmological perspective given in Chapter 12.

Is a new theory of biological evolution necessary and feasible? Will there be a Postmodern Synthesis?

Bohr's complementarity principle seems to be central to our understanding of evolution—above all, in the sense of *the complementarity between chance and deterministic factors (necessity)*, which is the leitmotif of this book. Complementarity is apparent at all levels and in all aspects of evolution and could be a major guiding principle en route to a new theoretical biology. The clear-cut cases discussed in this book include the complementarity between

- Random and (quasi) directed mutations
- Selection and drift
- Selfish and altruistic behavior of a variety of genetic elements (temperate viruses, retroelements, toxin-antitoxin and restriction-modification systems, and more)
- Robustness and evolvability

Complementarity is also essential for the epistemology of the new evolutionary biology. Given the key role of historical contingency (including Jacob's "tinkering") and the enormous complexity of biological phenomena, it appears inconceivable that any single set of equations will ever qualify as a general theory of biological evolution, even in the limited sense in which Einstein's general relativity is the theory of gravitation or the Standard Model of particle physics is the theory of matter and energy. Furthermore, no combination of simple physical and mathematical models can capture evolution because the historical, contingent component is not directly formalizable. The best we may hope for and work on is a new, genomic incarnation of the population genetic theory that will be the necessary framework for all subsequent studies on genome and phenome evolution. Such a theory will be no small feat, and as I attempted to show here, the prospect of achieving it is becoming increasingly realistic. Moreover, this theory will be buttressed by a comprehensive, explicit description of the fitness landscapes for a variety of evolutionary regimes that will be studied in direct experiments. However, all its importance notwithstanding, such a theory will never explain "all of evolution," no more than, say, statistical physics can "explain" geology. It seems that Postmodern Synthesis can come only in the form of a *complex network of the complementary perspectives that are provided, respectively, by models of the statistical physics/population genetics type and by reconstructions of the actual evolutionary past.*

The phrase *Postmodern Synthesis* repeatedly used in this book is not simply a manifestation of arrogance, but also an obvious oxymoron, because the philosophy of postmodernism is all about the negation of the very possibility of any synthesis (see Appendix A). However, this choice of words is quite deliberate because the complexity of the evolution of life does invoke the specter of the post-modern worldview, however disturbing this might seem. Nevertheless, an increasingly deep understanding can be expected to result from the evolving network of complementary, interacting models, theories, and generalizations. It is interesting to note that some of the leading theoretical physicists of today contemplate the future of physics in a similar light.[3]

Further recommended reading

Goldenfeld, N., and C. Woese. (2007) "Biology's Next Revolution." *Nature* 445: 369.

> A brief summary of the major impact of the "lateral genomics" of archaea and bacteria on the general concepts of evolutionary biology.

Gould, Stephen Jay. (1997) *Full House: The Spread of Excellence from Plato to Darwin*. New York: Three Rivers Press.

> A delightful book whose main subject is debunking the progress fallacy in biology. Gould also discusses the origin of complexity by a purely stochastic route (the drunk walk) and the fundamental unpredictability of evolution (replaying the tape).

Koonin, E. V. (2009) "Darwinian Evolution in the Light of Genomics." *Nucleic Acids Research* 37: 1,011–1,034.

> This article, published on Darwin's 200th anniversary, summarizes the new discoveries in several fields, particularly the reappraisal of the Tree of Life, and the demonstration of the inadequacy of pan-adaptationism and gradualism, that call for a new evolutionary synthesis. A "preview" of the present book.

Koonin, E. V. (2010) "The Two Empires and the Three Domains of Life in the Postgenomic Age." *Nature Education* 3: 27.

> This popular article introduces the division of life forms into the viral and cellular empires.

Lewontin, R. C. (2002) "Directions in Evolutionary Biology." *Annual Review of Genetics* 36: 1–18.

> Article that emphasizes the plurality of evolutionary factors and the insufficiency of natural selection as the explanation of evolution.

Lynch, M. (2007) "The Frailty of Adaptive Hypotheses for the Origins of Organismal Complexity," *Proceedings of the National Academy of Sciences USA*, 104 Supplement 1: 8,597–8,604.

A compelling refutation of pan-adaptationism from the standpoint of the non-adaptive theory of complexity evolution.

Rose, M. R., and T. H. Oakley. (2007) "The New Biology: Beyond the Modern Synthesis." *Biology Direct* 2: 30.

The authors of this conceptual article conclude: "The new biology knits together genomics, bioinformatics, evolutionary genetics, and other such general-purpose tools to supply novel explanations for the paradoxes that undermined Modernist biology."

Woese, C. R., and N. Goldenfeld. (2009) "How the Microbial World Saved Evolution from the Scylla of Molecular Biology and the Charybdis of the Modern Synthesis." *Microbiology and Molecular Biology Review* 73: 14–21.

A discussion of the importance of the developments in microbiology, particularly the ubiquity of HGT, for broadening the conceptual basis of evolutionary biology.

Postmodernist philosophy, metanarratives, and the nature and goals of the scientific endeavor

I am not just a dilettante, but effectively an ignoramus in the issues discussed in this appendix. Yet this brief, unprofessional overview begged to be written. Indeed, the word *postmodern* repeatedly used in this book seems to demand some (pseudo)philosophical discussion. Evolutionary biology in general is the kind of field where some epistemological discourse is unavoidable.

Postmodernist philosophy, (dis)trust of metanarratives, and the (in)feasibility of synthesis

The "postmodern synthesis" in the preface to this book is a deliberate oxymoron. Indeed, much of the pathos of post-modernist philosophy is its distrust of any generalization, any "big picture," any over-arching story, and any "grand scheme of things" that scientists and especially philosophers are tempted to concoct. Quoting Jean-Francois Lyotard, one of the prominent figures in the post-modernist creed of philosophers: *"Simplifying to the extreme, I define postmodern as incredulity toward metanarratives. This incredulity is undoubtedly a product of progress in the sciences: but that progress in turn presupposes it"* (Lyotard, 1979). Postmodern philosophy has been the target of much disdain and more sarcasm—and, indeed, attempts to read through the postmodernist oeuvre provides for a somewhat vertiginous experience. Yet the postmodern emphasis on pattern plurality and diversity meshes perfectly with the latest results of evolutionary biology, which reveal an extremely complex gamut of diverse evolutionary processes and defy all attempts to explain evolution by any straightforward schema, such as natural selection of random variants. The eminent postmodern philosophers Gilles Deleuze and Felix Guattari spoke of the rhizome as a key metaphor of the pattern pluralism that permeates the world (Deleuze and Guattari, 1987). Microbiologist Didier Raoult borrowed this term to describe the newly discovered complexity of evolutionary processes as the *rhizome of life* (Raoult, 2010), and it is hard to deny that the image fits, especially considering the overhaul of the Tree of Life (TOL) concept in the face of the ubiquitous HGT.

The problem with postmodernism is common to many philosophical systems. It is quite efficient at deconstruction but fails to offer any constructive alternative. Actually, perhaps to their credit, postmodern philosophers deny the very need for such alternatives and seem to be happy to just contemplate the rhizome. However, science—and evolutionary biology, in particular—cannot operate that way. To achieve any progress, we have no alternative other than to construct narratives and assemble them into metanarratives, which philosophers of science (especially, physics) often call paradigms, following Thomas Kuhn's classic *Structure of Scientific Revolutions* (Kuhn, 1962). Paradigms and metanarratives are necessary, even if only to pit new observations against them and assess the continuing validity of the existing paradigms and the need of new ones.

All I have said so far on paradigms is quite trivial and generic, but there is something special about the case of evolutionary biology. Given that so much of evolutionary biology is about the unique history of a single instantiation of life known to us and that so much of this history depends on chance and contingency, a concise metanarrative seems to be impossible in principle. The best one may hope for is a tapestry of multiple narratives at different levels of generality and abstraction. Speaking somewhat metaphorically, any description of the course of evolution has an extremely high algorithmic (Kolmogorov) complexity (see Chapter 9) and so is mostly refractory to generalization. Evolutionary biology has another face, though. It has been long known that straightforward population genetic theory can describe microevolutionary processes quite well. Comparative genomics and systems biology have added new classes of quantifiable variables that can be used to test models of evolution. Taken together, these models provide for a different type of metanarrative, one that may qualify as theory in the sense this term is used in physics (see Chapter 4). If not at this time, then in the future, perhaps this direction in evolutionary biology actually will be a legitimate part of physics.

However, a "complete physical theory of evolution" remains an illusory goal in principle. Physical-type theories—again, multiple ones for different aspects of the evolutionary process rather than a single overarching theory—can capture only generic aspects of evolution, the "necessity" component of Monod's opposition (even when this necessity takes on a stochastic form and so itself depends on chance). Our best chance to "understand" evolution is to embrace the complementarity of the physical (theoretical) and historical metanarratives of evolution—as a lasting, intrinsic feature of evolutionary biology rather than a temporary situation caused by the imperfection of theory. A widespread view holds that historical narratives with their inevitable descriptive aspect are, at best, scientifically marginal, a kind of "stamp collection"; and, at worst, nonscientific speculation, given that rare and unique events are critical in evolution, as emphasized in this book. I find this position deeply unsatisfactory and untenable in the context of evolutionary biology. Notwithstanding all the difficulties and the inescapable uncertainty associated with unique events, these are central to the evolution of life, so evolutionary biologists should make a concerted effort to decipher as much as possible about each major evolutionary transition that involves this type of events. I believe time has come to accept that the physical and historical perspectives on evolutionary processes are fundamentally distinct, and the latter is not "inferior," but complementary to the former.

Coming back to the postmodern discourse, should we trust the evolutionary metanarratives even as these are germane to the advancement of research? In a sense, the answer is trivially negative: Any paradigm includes oversimplification, and old paradigms are necessarily replaced with new ones under the pressure of accumulating new findings. This is what we are now witnessing in evolutionary biology as the paradigm of Modern Synthesis is crumbling and is being replaced by a new vision, the subject of this book. There is no reason to believe that any paradigm(s) can be the truth, let alone the final reflection of "reality." This certainly applies to all scientific endeavors, but evolutionary narratives seem to possess additional features that require extra caution in interpretation. We briefly address these problems in the following sections .

"Why?" questions and semantic traps: What are we really saying about evolution?

Many studies in evolutionary biology center on asking and attempting to answer various "Why?" questions. These questions abound at all levels of evolution scholarship, from classical organismal biology (Why are males bigger and stronger than females in some animal species but not in others?) to genomics (Why are there so many introns in the genes of some eukaryotes but not in others?), to systems biology (Why are some proteins so much more abundant than others?), to the origin of life problem (Why are there 20 amino acids in proteins of all organisms?). Evolutionary biologists often (although by no means always) shy away from asking "Why?" questions up front, but the "why's" seem to lurk in the background and affect the very logic of the investigation. Until recently, and sometimes even these days, any "Why?" question almost automatically triggered the concoction of an adaptationist ("just so") story. The San Marco critique of Gould and Lewontin, the neutral theory, and, later, Lynch's non-adaptive theory of the evolution of complexity have changed this, so now we tend to come up with more balanced, complex stories that, in addition to selection, include non-adaptive factors such as drift, draft, and various neutral ratchets.

Are the new stories any better than the previous ones? On one hand, that seems to be the case because they take into account the contributions of multiple processes and, at least in the more careful studies, these contributions are inferred from measurements of specific quantities rather than from qualitative reasoning. Nevertheless, all scenarios composed to answer "Why?" questions should be taken for what they are: narratives that scientists construct. By their very nature, narratives are bound to oversimplify and reduce the complexity of the phenomena under study—in this case, the evolutionary process—to a small number of discrete "factors." These factors, such as natural selection itself, are abstractions derived from observation. What evolutionary biologists actually measure is not selection, but the values of certain specific variables, such as *Ka* or *Ks*. From the relationships between these measured values, conclusions are made on purifying selection, positive selection, and neutrality, and the (meta)narrative is constructed.

The "dialectics" of this situation is that the evolutionary narratives certainly are oversimplified "myths" that have the unfortunate (and, in modern studies, unintended) teleological flavor (as in "selected for" or, worse, "selected for the purpose of"), yet the language of these narratives seems best suited to describing evolution and formulating falsifiable hypotheses that propel further research. At present, we hardly can give up these stories (indeed, much of this book is written in this very manner) precisely because they are necessary means for the advancement of research, even though they tend to leave a scientist (definitely, the author of this book) with feelings of uneasiness and dissatisfaction. It seems important not to forget that evolutionary narratives effectively are semantic devices that are constructed to structure and simplify our thinking about evolution and to facilitate the generation of hypotheses. These narratives should be prudently distrusted and by no account should be construed as "accurate representations of reality" (whatever that might mean—see the next section).

It is an interesting question whether, in a not-too-remote future, evolutionary biology might be able to develop a new language that will have less to do with myth and more to do with measurable quantities. Such a prospect does not appear implausible. After all, the language of today's evolutionary narratives that firmly incorporates, say, the distinction between purifying and positive selection, the formalism of population genetics, the structure of fitness landscapes, or the rate of horizontal gene transfer is much more closely linked to specific measurement than the language of Modern Synthesis, let alone that of Darwin.

Thus, we cannot expect that evolutionary biology (or, for that matter, any branch of science) purges (meta)narratives. However, new narratives seem to be demonstrably "better" than old ones—that is, less simplistic than the preceding ones and more directly linked to actual observation.

The nature and goals of science: Why study evolution at all?

The question in the title of this section might seem preposterous, but although at the end of this book it can be asked only in jest, in principle, it is a legitimate, important question that should be pondered in earnest. Let us first give a trivial but necessary answer: The study of evolution is essential for the progress of biology if only because evolutionary concepts such as variable constraints and purifying selection underlie a huge fraction of experiments in modern biology. Indeed, every experiment on site-directed mutagenesis is based on evolutionary reasoning: Only evolutionary analysis can tell a researcher which positions in a gene are to be mutagenized to affect its activity in a certain way, even if the researcher does not explicitly think in terms of evolution. Even more sophisticated evolutionary analysis is involved, say, in the study of virus evolution or progression of cancer, so knowledge of certain aspects of evolution literally saves thousands of lives and millions of dollars (for example, through the prediction of influenza epidemics and improvement of vaccines).

However, all its biological sensibility notwithstanding, this is a perfunctory answer. In all these studies, the models of evolution can be—and actually often are—used like any other tools or heuristics, without concern for the "evolutionary reality." So are we interested in how evolution "really occurs" and in what actually happened in the deep past of life on Earth? In asking these questions, we hit on the deepest problems of the nature and goals of all science. It may be a common perception that science strives to understand the workings of the world in which we live. However, the very meaning of "understanding reality" is less than clear. All science can actually do is develop models, often (but not necessarily) in the form of equations, and to see whether observations falsify these models—or, in other words, to assess the predictive power of the models. The scientific process does not tell us anything directly about the world; it tells us only about the compatibility of certain observations with the adopted models. All aspects of any worldview ("the picture of reality") can be considered *metaphysical implications* of the models and, as such, inconsequential.

In the case of evolutionary biology, one can use mathematical theory describing the relationship between the data (primarily sequences, but also, for example, comparative expression or proteomics data) to predict the phenotypic effects of mutations or the appearance of new virus isolates with particular properties, without recourse to any "realistic" depiction of the process of evolution. This is not mockery—I am trying to fairly represent the way many, if not most, physicists and philosophers of science, including undisputed leading figures, view the nature of the scientific process and the scientific worldview. Stephen Hawking and Leonard Mlodinow aptly dubbed this position *model-dependent realism* in their latest popular physics and cosmology book (Hawking and Mlodinow, 2010). Under this view, scientists construct models, and competing models are compared with respect to their ability to explain data and predict experimental outcomes. The model that most accurately accounts for the largest body of observations and does so with the maximum possible simplicity (but not simpler) becomes the winner (usually, until it loses to a new, even more accurate and elegant model). Hawking and Mlodinow coined the catchy term *model-dependent realism* in 2010, but this worldview is, of course, much older and common among physicists. Niels Bohr, for example, is quoted saying, *"There is no quantum world. There is only an abstract physical description. It is wrong to think that the task of physics is to find out how nature is. Physics concerns what we can say about nature"* (Pais, 1994).

Model-dependent realism has a lot going for it. To advocate this no-nonsense view of science, physicists (including Richard Feynman as well as Hawking and Mlodinow) turn to a comparison between mythological ideas on the stability of the solar system and the accounts given by Ptolemaic and then Newtonian physics. The Hellenistic astronomers replaced mythology with a rational but *ad hoc* concept of epicycles, multiple spheres that rotated around an immobile Earth and carried fixed celestial bodies. Copernicus and Kepler replaced the epicycle schema with the model of planets orbiting the immobile sun in elliptical orbits. Newton provided the theoretical foundation for that model in his law of gravity, according to which the force of gravity holds bodies in stable trajectories. One can argue, as Feynman did, that the Newtonian worldview is as much of a myth as the Ptolemaic version and that we think otherwise only because we are used to Newton's version. Indeed, what are these "forces" that are supposed to act in empty space at a distance, and how are they any better than, say, gods undertaking certain periodic actions to keep the world going? If one thinks about it soberly, forces are every bit as incomprehensible as gods. Indeed, Newton himself famously pronounced that he did not "invent hypotheses," a statement that is to be understood in the sense that Sir Isaac deliberately declined to say anything about "how the world really is"—and good for him, as Feynman contends. Under model-dependent realism, science develops models that are compared to observations; among competing models, the one that fits the observations most closely, predicts new experimental outcomes most accurately, and is as simple as possible (but not simpler) becomes the winner. The "truth value" of a model (its ability to describe "reality") is not part of this conception of science—only predictive power, elegance, and simplicity matter.

I take exception to the worldview of model-dependent realism. Although all "pictures of reality" are myths, Newton's myth is nevertheless not as bad as the Ptolemaic one because it includes fewer arbitrary assumptions. After all, Newton postulates a small number of entities, such as gravitation and mass, which he openly admits he is unable to interpret, in contrast to the endless succession of *ad hoc* entities such as epicycles postulated by the Ptolemaic cosmogony. Newton's worldview, although it includes entities that are not "understood," appears more parsimonious and less far-fetched than the preceding views. I further submit that Einstein's reinterpretation of gravity in the general relativity theory is another step forward: Einstein introduced a physically plausible description of the formerly mysterious "force" in terms of space-time warping—not just because general relativity is better at explaining certain subtle effects of gravity. I believe this aspect of the evolution of physics is relevant and important for our understanding of the functioning of science in general. No model can claim to accurately represent "reality," which is unknowable in principle; nevertheless, successive models of the world offer not only increasingly precise predictions, but also descriptions of that elusive reality that are progressively less ridiculous and more physically plausible. In other words, to put it simply and bluntly, the phrase "new models are better depictions of reality than preceding ones" makes certain sense.

Interestingly, physicists that fully adhere to model-dependent realism in their view of science and nature seem to express, on different grounds and, perhaps, in an even stronger form, the same position as the (antiscientific) postmodern philosophers: "Big pictures" (metanarratives) are deemed outright irrelevant (or, at least, are not considered part of science itself). I believe that this position takes skepticism too far, to the point of being unrealistic and counterproductive to the advancement of science. A stronger form of realism than that embodied in the model-dependent version is particularly important for the fields of science that are partially historical—those that deal with events that are not reproducible in direct experiments and some of which may have been unique (at least, in the observable part of the universe). To wit, we study evolution not only for the sake of specific predictions, however important these might be, but rather to gain some "understanding" of the history of life and its fundamental trends, those that might be inherent to life in general and would reappear in other instantiations of life, should these ever be discovered.

The philosopher Sir Karl Raymund Popper, the founder of the falsificationist paradigm in epistemology, was initially extremely skeptical of Darwin's theory because of its apparent unfalsifiability, to the point that he declared Darwinism "unscientific." Later, however, Popper changed his stance and suggested that, although Darwinism is not a falsifiable theory, per se, it is a metaphysical program capable of spawning a great number of falsifiable hypotheses. In this context, Popper did not use *metaphysical* as a derogatory term; on the contrary, he considered this program to be scientifically fruitful and productive, and even indispensable. He only meant to say that the Darwinian concept of evolution was not falsifiable (and probably not verifiable) in its entirety. Popper was quite eloquent on this account, even if his understanding of evolutionary biology was rather perfunctory:

> [T]he theory is invaluable. I do not see how, without it, our knowledge could have grown as it has done since Darwin. In trying to explain experiments with bacteria which become adapted to, say, penicillin, it is quite clear that we are greatly helped by the theory of natural selection. Although it is metaphysical, it sheds much light upon very concrete and very practical researches. It allows us to study adaptation to a new environment (such as a penicillin-infested environment) in a

rational way: it suggests the existence of a mechanism of adaptation, and it allows us even to study in detail the mechanism at work. And it is the only theory so far which does all that." (Popper, 1982)

I believe there is considerable merit in Popper's position. It is as true of the "postmodern synthesis" of evolutionary biology as it was of Darwin's theory that falsification of the entire conceptual framework or all of its propositions is hardly feasible. However, its general metaphysical character notwithstanding, the theory yields many specific falsifiable propositions, especially given the rapidly expanding data sets of genomics and systems biology. Moreover, I suggest that considering the evolving, complementary perspectives on the processes of evolution (see Chapter 13) can bring us as close as possible to understanding the way evolution "really occurs" and life "really evolved."

Evolution of the cosmos and life: Eternal inflation, "many worlds in one," anthropic selection, and a rough estimate of the probability of the origin of life[1]

A brief nontechnical introduction to inflationary cosmology

The "many worlds in one" (MWO) model that is essential to the cosmological perspective on the origin of life introduced in Chapter 12 is a consequence of inflational cosmology. At the end of the twentieth century, it had replaced the classical Big Bang model of the evolution of the universe. Inflation is the period of the exponentially fast initial expansion of a universe (Guth, 2001, 1998b). Alan Guth developed the inflation model to account for several key astronomic observations for which the Big Bang cosmology had no explanation:

- The flatness of space in the observable region of the universe (our O-region)
- The overall uniformity of the cosmic microwave background radiation (CMBR)
- The local nonhomogeneities of the CMBR
- The absence of observable magnetic monopoles

In the most plausible, self-consistent inflationary models, inflation is eternal, with an infinite number of island (pocket) universes (hereinafter, simply universes) emerging through the decay of small regions of the primordial "sea" of false (high-energy) vacuum and comprising the infinite multiverse (Guth, 2001; Vilenkin, 2007). The predictions of the eternal inflation model are in excellent quantitative agreement with these observations (Guth and Kaiser, 2005). Furthermore, the "populated landscape" version of string theory independently yields a similar model of the multiverse (Bousso, 2006; Bousso and Polchinski, 2004; Susskind, 2003, 2006a). Thus, although the model of eternal inflation cannot be considered proved, this is the strongly preferred current scenario of cosmic evolution. To observers within each universe, it appears self-contained and infinite, and it contains an infinite number of O-regions. For such observers (like us), their universe is expanding from a singularity (Big Bang) that corresponds to the end of inflation in the given part of the multiverse.

The "many worlds in one" (MWO) model holds that all macroscopic, "coarse-grain" histories of events that are not forbidden by conservation laws of physics have been realized (or will realize) somewhere in the infinite multiverse (and even in an pocket universe)— and not just once, but an infinite number of times (Garriga and Vilenkin, 2001; Vilenkin, 2007). For example, there are an infinite number of (macroscopically) exact copies of the Earth with everything that exists on it, although the probability that a given observable region of the universe (hereinafter, O-region) carries one of such copies is vanishingly tiny. This picture seems counterintuitive in the extreme, but it is a direct consequence of eternal inflation (Guth, 2001; Linde, 1986; Vilenkin, 1983).

Garriga and Vilenkin showed that, in a finite time, the content of each O-region can assume only a finite number of states; accordingly, any O-region has a finite, even if unimaginably vast (on the order of $10^{\wedge}10^{150}$), number of unique macroscopic, coarse-grain histories (Garriga and Vilenkin, 2001). Effectively, the finiteness of the number of coarse-grain histories appears to be a straightforward corollary of quantum uncertainty (the Heisenberg principle; Carroll, 2010; Vilenkin, 2007). The same conclusion is independently reached through a completely different approach: the so-called holographic bound on the amount of entropy that can be contained in any finite region of the universe ('t Hooft, 1993; Bousso, 2006; Carroll, 2010; Garriga and Vilenkin, 2001). Combined, eternal inflation, the finiteness of the number of unique coarse-grain histories, and the inevitable quantum randomness at the Big Bang (the beginning of time for each universe) led to the straightforward and striking conclusion that each history permitted by conservation laws of physics is repeated an infinite number of times in the multiverse and, actually, in each of the infinite number of infinite (island) universes (Bousso, 2006; Vilenkin, 2007).

The MWO model is tightly linked to the anthropic principle (anthropic selection), a controversial but increasingly popular concept among cosmologists. According to the anthropic principle, the only "reason" our O-region has its specific parameters is that, otherwise, there would be no observers to peer into the universe (Barrow and Tipler, 1988; Livio and Rees, 2005; Rees, 2001). It should be emphatically stressed that I discuss here only what is often called "weak" anthropic principle and is the only acceptable scientific rendering of this concept. The so-called "strong" anthropic principle is the teleological notion that our (human) existence is, in some mysterious sense, the "goal" of the evolution of the universe (Barrow and Tipler, 1988); as such, this idea does not belong in the scientific domain. It appears that the anthropic principle can be realistically defined only in the context of a vast (or infinite) multiverse (Susskind, 2006b). In particular, in the MWO model, anthropic selection has a straightforward interpretation: The parameters of our O-region are selected among the vast number of parameter sets existing in the multiverse (in an infinite number of copies each) by virtue of being conducive to the emergence and sustenance of complex life forms.

Compared to older cosmological concepts that considered a finite universe, the MWO model changes the very notions of "possible," "likely," and "random" with respect to any historical scenario (see Box B-1). Simply put, the probability of the realization of any scenario permitted by the conservation laws in an infinite universe (and, of course, in the multiverse) is exactly 1. Conversely, the probability that a given scenario is realized in the given O-region is equal to the frequency of that scenario in the universe. From a slightly different perspective, the usual adage about the second law of thermodynamics being true in the statistical sense takes a literal meaning in an infinite universe: Any violation of this law that is permitted by other conserva-tion laws will happen—and on an infinite number of occasions. Thus, spontaneous emergence of complex systems that would have to be considered virtually impossible in a finite universe becomes not only possible, but inevitable under MWO, even though the prior probabil-ities of the vast majority of histories to occur in a given O-region are vanishingly small. This new power of chance, buttressed by anthropic selection, is bound to have profound consequences for our under-standing of any phenomenon in the universe, and life on Earth can-not be an exception.

Probabilities of the emergence, by chance, of different versions of the breakthrough system in an O-region: A back-of-the-envelope calculation of the upper bounds

General assumptions: An O-region contains 10^{22} stars, and every tenth star has a habitable planet; hence, 10^{21} habitable planets (undoubtedly, a gross overestimation because, in reality, most stars have no planets, let alone habitable ones). Each planet is the size of Earth and has a 10km-thick (10^6 cm) habitable layer; hence, the vol-ume of the habitable layer is $4/3\pi [R^3 - (R - l)^3] \approx 5 \times 10^{24}$ cm^3, where R is the radius of the planet and l is the thickness of the habitable layer. RNA synthesis occurs in 1% of the volume of the habitable layer—that is, a volume $V \approx 5 \times 10^{22}$ cm^3 is available for RNA synthe-sis (a gross overestimation—in reality, there would be very few "RNA-making reactors"). Let the concentration of nucleotides in volume V and the rate of the synthesis of RNA molecules of size n (a free parameter that depends on the specific model of the break-through stage, hereafter n-mer) be 1 molecule/cm^3/second (a gross overestimate for any sizable molecule; furthermore, the inverse dependence on n, which is expected to be strong, is disregarded). The time available after the Big Bang of the given O-region (as an upper bound) of all planets in it is 10^{10} years $\approx 3 \times 10^{17}$ seconds. Then the number of unique n-mers "tried out" during the time after the Big Bang is this:

$$S \approx 5 \times 10^{22} \times 10^{21} \times 3 \times 10^{17} \approx 1.5 \times 10^{61}$$

Let us assume that, for the onset of biological evolution, a unique n-mer is required. The number of such sequences is $N = 4^n \approx 10^{0.6n}$.

Then the expectation of the number of times a unique n-mer emerges in an O-region is this:

$$E = S/N = 1.5 \times 10^{61}/10^{0.6n} \text{ and } n = \log\left(E \times 1.5 \times 10^{61}\right)/0.6.$$

Substituting $E = 1$, we get $n \approx 102$ (nucleotides). Note that, because n is proportional to $\log S$, the estimate is highly robust to the assumptions on the values of the contributing variables; for example, an order of magnitude change in S will result in an increase or decrease of n by less than 2 nucleotides.

A ribozyme replicase consisting of approximately 100 nucleotides is conceivable, so, in principle, spontaneous origin of such an entity in a finite universe consisting of a single O-region cannot be ruled out in this toy model (again, the rate of RNA synthesis considered here is a deliberate, gross overestimate).

The requirements for the emergence of a primitive, coupled replication-translation system, which is considered a candidate for the breakthrough stage in this paper, are much greater. At a minimum, spontaneous formation of the following is required:

- Two rRNAs, with a total size of at least 1,000 nucleotides.
- Approximately 10 primitive adaptors of about 30 nucleotides each, for a total of approximately 300 nucleotides.
- At least one RNA encoding a replicase, about 500 nucleotides (low bound) required. Under the notation used here, $n = 1,800$, resulting in $E < 10^{-1.018}$.

In other words, even in this toy model that assumes a deliberately inflated rate of RNA production, the probability that a coupled translation replication emerges by chance in a single O-region is $P < 10^{-1.018}$. Obviously, this version of the breakthrough stage can be considered only in the context of a universe with an infinite (or, at the very least, extremely vast) number of O-regions.

The model considered here is not supposed to be realistic, by any account. It only illustrates the difference in the demands on chance for the origin of different versions of the breakthrough system and, hence, the connections between this version and different cosmological models of the universe.

Box B-1: Some central new definitions and reinterpretation of familiar definitions in the MWO model

Term(s)	Definition
Inflation	Exponential expansion of the multiverse driven by the repulsive gravity of the false (high-energy) vacuum. Inflation is likely to be eternal—that is, once started, it will never end.
Multiverse (megaverse, master universe)	The entire fabric of reality that consists of an eternally inflating false vacuum with an infinite number of small decaying regions, giving rise to universes.
Universe (island universe, pocket universe, bubble universe)	Part of the multiverse that expands from a Big Bang event resulting from a decay of a region of false vacuum into a low-energy (true) vacuum. A universe is infinite from the point of view of an internal observer, but finite to an imaginary external observer.
Observable (O) region	A finite region within a universe that can be observed from any given point (the interior of the past light cone of the given point). Our O-region contains approximately 10^{20} stars.
Big Bang	In the traditional twentieth-century cosmology, expansion of the universe from a singularity. The nature of the "bang" has never been elucidated. In the eternal inflation cosmology, Big Bang corresponds to the end of inflation in the given region of the multiverse as a result of false vacuum decay and the formation of a universe in the form of an expanding bubble of low-energy (true) vacuum.
Macroscopic (coarse-grained) history	Any combination of physical events permitted by the laws of physics, characterized to the limit of quantum uncertainty and occurring in an O-region within a finite time. The number of all possible macroscopic histories has been shown to be finite, although vast. Hence, even within a single universe, each history is repeated an infinite number of times.

Term(s)	Definition
Probability/chance/ randomness	Textbooks define probability as the limit to which frequency of a specific outcome tends when the number of trials tends to infinity. In an infinite universe (and, obviously, in the multiverse) with a finite number of histories, the infinite number of trials is realized, hence probability equals frequency. The probability of any permissible history, including origin of life, then, is $P = 1$. However, the probability p of observing any particular history in a given O-region lies in the interval between 0 and 1, as in the textbook definition of probability, and can be extremely small for a vast number of histories, including the origin of life. Thus, the notions of chance and randomness apply only to finite regions of a universe, whereas in an infinite universe as a whole, the realization of all permitted histories is a necessity.
Anthropic principle/anthropic selection/anthropic reasoning	The notion that the history of our world (our O-region, our galaxy, our solar system, and so on) prior to the onset of biological evolution does not depend on any special "mechanism" but was simply "selected" from the finite ensemble of all histories that are guaranteed to realize in an infinite universe, by virtue of being conducive to the emergence of complex life. Anthropic selection is an epistemological, not ontological, principle and should not be misconstrued for any kind of active process. This is a formulation of the "weak" anthropic principle adopted for the context of this paper. The "strong" anthropic principle is the notion that the emergence of consciousness somehow is a goal of the cosmic history. This is a teleological, nonscientific concept.

References

(1948) *O Polozhenii V Biologicheskoi Nauke. Stenograficheskii Otchet Sessii Vsesoyuznoi Akademii Selskohozyastvennyh Nauk Imeni V. I. Lenina. (On the Situation in Biological Science. A Transcript of the Session of the V. I. Lenin All-Union Academy of Agricultural Sciences, July 31–August 7, 1948.* Moscow, USSR: The State Agricultural Literature Publishers.

't Hooft, G. (1993) "Dimensional Reduction in Quantum Gravity." gr-qc/9310026.

Adami, C. (2002) "What Is Complexity?" *Bioessays* 24: 1,085–1,094.

Adl, S. M., A. G. Simpson, M. A. Farmer, R. A. Andersen, O. R. Anderson, J. R. Barta, S. S. Bowser, G. Brugerolle, R. A. Fensome, S. Fredericq, T. Y. James, S. Karpov, P. Kugrens, J. Krug, C. E. Lane, L. A. Lewis, J. Lodge, D. H. Lynn, D. G. Mann, R. M. McCourt, L. Mendoza, O. Moestrup, S. E. Mozley-Standridge, T. A. Nerad, C. A. Shearer, A. V. Smirnov, F. W. Spiegel, and M. F. Taylor. (2005) "The New Higher Level Classification of Eukaryotes with Emphasis on the Taxonomy of Protists." *J Eukaryot Microbiol* 52: 399–451.

Agol, V. I. (1974) "Towards the System of Viruses." *Biosystems* 6: 113–132.

Agol, V. I., and A. P. Gmyl. (2010) "Viral Security Proteins: Counteracting Host Defences." *Nat Rev Microbiol* 8: 867–878.

Ahlquist, P. (2006) "Parallels among Positive-Strand RNA Viruses, Reverse-Transcribing Viruses and Double-Stranded RNA Viruses." *Nat Rev Microbiol* 4: 371–382.

Alic, N., N. Ayoub, E. Landrieux, E. Favry, P. Baudouin-Cornu, M. Riva, and C. Carles. (2007) "Selectivity and Proofreading Both Contribute Significantly to the Fidelity of RNA Polymerase III Transcription." *Proc Natl Acad Sci USA* 104: 10,400–10,405.

Allen, E. E., and J. F. Banfield. (2005) "Community Genomics in Microbial Ecology and Evolution." *Nat Rev Microbiol* 3: 489–498.

Alperovitch-Lavy, A., I. Sharon, F. Rohwer, E. M. Aro, F. Glaser, R. Milo, N. Nelson, and O. Beja. (2011) "Reconstructing a Puzzle: Existence of Cyanophages Containing Both Photosystem-I and Photosystem-II Gene Suites Inferred from Oceanic Metagenomic Datasets." *Environ Microbiol* 13: 24–32.

Altschul, S. F., T. L. Madden, A. A. Schaffer, J. Zhang, Z. Zhang, W. Miller, and D. J. Lipman. (1997) "Gapped Blast and Psi-Blast: A New Generation of Protein Database Search Programs." *Nucleic Acids Res* 25: 3,389–3,402.

Altstein, A. D. (1987) "Origin of the Genetic System." *Mol Biol* 21: 309–322.

Amaral, P. P., M. E. Dinger, T. R. Mercer, and J. S. Mattick. (2008) "The Eukaryotic Genome as an RNA Machine." *Science* 319: 1,787–1,789.

Anantharaman, V., E. V. Koonin, and L. Aravind. (2002) "Comparative Genomics and Evolution of Proteins Involved in RNA Metabolism." *Nucleic Acids Res* 30: 1,427–1,464.

Anfinsen, C. B. (1973) "Principles That Govern the Folding of Protein Chains." *Science* 181: 223–230.

Aquadro, C. F. (1997) "Insights into the Evolutionary Process from Patterns of DNA Sequence Variability." *Curr Opin Genet Dev* 7: 835–840.

Aravind, L., V. Anantharaman, S. Balaji, M. M. Babu, and L. M. Iyer. (2005) "The Many Faces of the Helix-Turn-Helix Domain: Transcription Regulation and Beyond." *FEMS Microbiol Rev* 29: 231–262.

Aravind, L., L. M. Iyer and E. V. Koonin. (2006) "Comparative Genomics and Structural Biology of the Molecular Innovations of Eukaryotes." *Curr Opin Struct Biol* 16: 409–419.

Aravind, L., and E. V. Koonin. (1999) "DNA-Binding Proteins and Evolution of Transcription Regulation in the Archaea." *Nucleic Acids Res* 27: 4,658–4,670.

Aravind, L., and E. V. Koonin. (2001) "The DNA-Repair Protein Alkb, Egl-9, and Leprecan Define New Families of 2-Oxoglutarate- and Iron-Dependent Dioxygenases." *Genome Biol* 2: RESEARCH0007.

Aravind, L., R. Mazumder, S. Vasudevan, and E. V. Koonin. (2002) "Trends in Protein Evolution Inferred from Sequence and Structure Analysis." *Curr Opin Struct Biol* 12: 392–399.

Aravind, L., R. L. Tatusov, Y. I. Wolf, D. R. Walker, and E. V. Koonin. (1998) "Evidence for Massive Gene Exchange Between Archaeal and Bacterial Hyperthermophiles." *Trends Genet* 14: 442–444.

Aravind, L., D. R. Walker, and E. V. Koonin. (1999) "Conserved Domains in DNA Repair Proteins and Evolution of Repair Systems." *Nucleic Acids Res* 27: 1,223–1,242.

Assis, R., and A. S. Kondrashov. (2009) "Rapid Repetitive Element-Mediated Expansion of piRNA Clusters in Mammalian Evolution." *Proc Natl Acad Sci USA* 106: 7,079–7,082.

Assis, R., A. S. Kondrashov, E. V. Koonin, and F. A. Kondrashov. (2008) "Nested Genes and Increasing Organizational Complexity of Metazoan Genomes." *Trends Genet* 24: 475–478.

Atkins, J. F., R. F. Gesteland, and T. R. Cech, eds. (2010) *RNA Worlds: From Life's Origins to Diversity in Gene Regulation.* Cold Spring Harbor, NY: Cold Spring Harbor Laboratory Press.

Auster, P. (1991) *The Music of Chance.* New York: Penguin.

Autexier, C., and N. F. Lue. (2006) "The Structure and Function of Telomerase Reverse Transcriptase." *Annu Rev Biochem* 75: 493–517.

Baaske, P., F. M. Weinert, S. Duhr, K. H. Lemke, M. J. Russell, and D. Braun. (2007) "Extreme Accumulation of Nucleotides in Simulated Hydrothermal Pore Systems." *Proc Natl Acad Sci USA* 104: 9,346–9,351.

Babu, M., N. Beloglazova, R. Flick, C. Graham, T. Skarina, B. Nocek, A. Gagarinova, O. Pogoutse, G. Brown, A. Binkowski, S. Phanse, A. Joachimiak, E. V. Koonin, A. Savchenko, A. Emili, J. Greenblatt, A. M. Edwards, and A. F. Yakunin. (2011) "A Dual Function of the CRISPR-Cas System in Bacterial Antivirus Immunity and DNA Repair." *Mol Microbiol* 79: 484–502.

Ball, P. (2011) "A Metaphor Too Far." *Nature* doi:10.1038/news.2011.115.

Bailey, K. A., F. Marc, K. Sandman, and J. N. Reeve. (2002) "Both DNA and Histone Fold Sequences Contribute to Archaeal Nucleosome Stability." *J Biol Chem* 15: 9293–9301.

Baltimore, D. (1971) "Expression of Animal Virus Genomes." *Bacteriol Rev* 35: 235–241.

Bangham, C. R., and T. B. Kirkwood. (1993) "Defective Interfering Particles and Virus Evolution." *Trends Microbiol* 1: 260–264.

Barabasi, A. L. (2002) *Linked: The New Science of Networks*. New York: Perseus Press.

Barabasi, A. L., and Z. N. Oltvai. (2004) "Network Biology: Understanding the Cell's Functional Organization." *Nat Rev Genet* 5: 101–113.

Barbrook, A. C., C. J. Howe, D. P. Kurniawan, and S. J. Tarr. (2010) "Organization and Expression of Organellar Genomes." *Philos Trans R Soc Lond B Biol Sci* 365: 785–797.

Baross, J. A., and S. A. Hoffman. (1985) "Submarine Hydrothermal Vents and Associated Gradient Environments As Sites for the Origin and Evolution of Life." *Origins of Life* 15: 327–345.

Barrangou, R., C. Fremaux, H. Deveau, M. Richards, P. Boyaval, S. Moineau, D. A. Romero, and P. Horvath. (2007) "CRISPR Provides Acquired Resistance Against Viruses in Prokaryotes." *Science* 315: 1,709–1,712.

Barrick, J. E., D. S. Yu, S. H. Yoon, H. Jeong, T. K. Oh, D. Schneider, R. E. Lenski, and J. F. Kim. (2009) "Genome Evolution and Adaptation in a Long-Term Experiment with Escherichia coli." *Nature* 461: 1243–1247.

Barrow, J. D., and F. J. Tipler. (1988) *The Anthropic Cosmological Principle*. Oxford: Oxford University Press.

Barrowman, J., D. Bhandari, K. Reinisch, and S. Ferro-Novick. (2010) "Trapp Complexes in Membrane Traffic: Convergence Through a Common Rab." *Nat Rev Mol Cell Biol* 11: 759–763.

Barton, N. H. (2000) "Genetic Hitchhiking." *Philos Trans R Soc Lond B Biol Sci* 355: 1,553–1,562.

Barton, N. H., and J. B. Coe. (2009) "On the Application of Statistical Physics to Evolutionary Biology." *J Theor Biol* 259: 317–324.

Basu, M. K., E. Poliakov, and I. B. Rogozin. (2009) "Domain Mobility in Proteins: Functional and Evolutionary Implications." *Brief Bioinform* 10: 205–216.

Beardmore, R. E., I. Gudelj, D. A. Lipson, and L. D. Hurst. (2011) "Metabolic Trade-Offs and the Maintenance of the Fittest and the Flattest." *Nature* [Epub ahead of print].

Beeby, M., B. D. O'Connor, C. Ryttersgaard, D. R. Boutz, L. J. Perry, and T. O. Yeates. (2005) "The Genomics of Disulfide Bonding and Protein Stabilization in Thermophiles." *PLoS Biol* 3: e309.

Begley, T. J., and L. D. Samson. (2003) "Molecular Biology: A Fix for RNA." *Nature* 421: 795–796.

Behe, M. J. (2006). *Darwin's Black Box: The Biochemical Challenge to Evolution.* New York: Free Press.

Behm-Ansmant, I., I. Kashima, J. Rehwinkel, J. Sauliere, N. Wittkopp, and E. Izaurralde. (2007) "mRNA Quality Control: An Ancient Machinery Recognizes and Degrades mRNAs with Nonsense Codons." *FEBS Lett* 581: 2,845–2,853.

Beiko, R. G., and N. Hamilton. (2006) "Phylogenetic Identification of Lateral Genetic Transfer Events." *BMC Evol Biol* 6: 15.

Beiko, R. G., T. J. Harlow, and M. A. Ragan. (2005) "Highways of Gene Sharing in Prokaryotes." *Proc Natl Acad Sci USA* 102: 14,332–14,337.

Beringer, M., and M. V. Rodnina. (2007) "The Ribosomal Peptidyl Transferase." *Mol Cell* 26: 311–321.

Bezier, A., M. Annaheim, J. Herbiniere, C. Wetterwald, G. Gyapay, S. Bernard-Samain, P. Wincker, I. Roditi, M. Heller, M. Belghazi, R. Pfister-Wilhem, G. Periquet, C. Dupuy, E. Huguet, A. N. Volkoff, B. Lanzrein, and J. M. Drezen. (2009) "Polydnaviruses of Braconid Wasps Derive from an Ancestral Nudivirus." *Science* 323: 926–930.

Bhattacharya, D., J. M. Archibald, A. P. Weber, and A. Reyes-Prieto. (2007) "How Do Endosymbionts Become Organelles? Understanding Early Events in Plastid Evolution." *Bioessays* 29: 1,239–1,246.

Bidle, K. D., and P. G. Falkowski. (2004) "Cell Death in Planktonic, Photosynthetic Microorganisms." *Nat Rev Microbiol* 2: 643–655.

Biebricher, C. K., and M. Eigen. (2005) "The Error Threshold." *Virus Res* 107: 117–127.

Bigot, Y., S. Samain, C. Auge-Gouillou, and B. A. Federici. (2008) "Molecular Evidence for the Evolution of Ichnoviruses from Ascoviruses by Symbiogenesis." *BMC Evol Biol* 8: 253.

Bjedov, I., O. Tenaillon, B. Gerard, V. Souza, E. Denamur, M. Radman, F. Taddei, and I. Matic. (2003) "Stress-Induced Mutagenesis in Bacteria." *Science* 300: 1,404–1,409.

Blanchard, S. C., R. L. Gonzalez, H. D. Kim, S. Chu, and J. D. Puglisi. (2004) "tRNA Selection and Kinetic Proofreading in Translation." *Nat Struct Mol Biol* 11: 1,008–1,014.

Blasius, M., S. Sommer, and U. Hubscher. (2008) "Deinococcus Radiodurans: What Belongs to the Survival Kit?" *Crit Rev Biochem Mol Biol* 43: 221–238.

Blencowe, B. J. (2006) "Alternative Splicing: New Insights from Global Analyses." *Cell* 126: 37–47.

Bokov, K., and S. V. Steinberg. (2009) "A Hierarchical Model for Evolution of 23s Ribosomal RNA." *Nature* 457: 977–980.

Bolotin, A., B. Quinquis, A. Sorokin, and S. D. Ehrlich. (2005) "Clustered Regularly Interspaced Short Palindrome Repeats (Crisprs) Have Spacers of Extrachromosomal Origin." *Microbiology* 151: 2,551–2,561.

Bonner, J. T. (2004) "Perspective: The Size-Complexity Rule." *Evolution* 58: 1,883–1,890.

Bostrom, Nick. (2002) *Anthropic Bias: Observation Selection Effects in Science and Philosophy*. New York and London: Rutledge.

Bourc'his, D., and O. Voinnet. (2010) "A Small-RNA Perspective on Gametogenesis, Fertilization, and Early Zygotic Development." *Science* 330: 617–622.

Bousso, R. (2006) "Holographic Probabilities in Eternal Inflation." *Phys Rev Lett* 97: 19,1302.

Bousso, R., and J. Polchinski. (2004) "The String Theory Landscape." *Sci Am* 291: 78–87.

Bowman, G. R., V. A. Voelz, and V. S. Pande. (2011) "Taming the Complexity of Protein Folding." *Curr Opin Struct Biol* 21: 4–11.

Branco, M. R., and A. Pombo. (2007) "Chromosome Organization: New Facts, New Models." *Trends Cell Biol* 17: 127–134.

Brinkmann, H., and H. Philippe. (2007) "The Diversity of Eukaryotes and the Root of the Eukaryotic Tree." *Adv Exp Med Biol* 607: 20–37.

Brisson, D. (2003) "The Directed Mutation Controversy in an Evolutionary Context." *Crit Rev Microbiol* 29: 25–35.

Brochier-Armanet, C., B. Boussau, S. Gribaldo, and P. Forterre. (2008) "Mesophilic Crenarchaeota: Proposal for a Third Archaeal Phylum, the Thaumarchaeota." *Nat Rev Microbiol* 6: 245–252.

Bromham, L., and D. Penny. (2003) "The Modern Molecular Clock." *Nat Rev Genet* 4: 216–224.

Brookfield, J. F. (2009) "Evolution and Evolvability: Celebrating Darwin 200." *Biol Lett* 5: 44–46.

Brouns, S. J., M. M. Jore, M. Lundgren, E. R. Westra, R. J. Slijkhuis, A. P. Snijders, M. J. Dickman, K. S. Makarova, E. V. Koonin, and J. van der Oost. (2008) "Small CRISPR RNAs Guide Antiviral Defense in Prokaryotes." *Science* 321: 960–964.

Bruggeman, F. J., and H. V. Westerhoff. (2007) "The Nature of Systems Biology." *Trends Microbiol* 15: 45–50.

Burger, R., M. Willensdorfer, and M. A. Nowak. (2006) "Why Are Phenotypic Mutation Rates Much Higher Than Genotypic Mutation Rates?" *Genetics* 172: 197–206.

Burki, F., K. Shalchian-Tabrizi, and J. Pawlowski. (2008) "Phylogenomics Reveals a New 'Megagroup' Including Most Photosynthetic Eukaryotes." *Biol Lett* 4: 366–369.

Bushman, F. (2001) *Lateral DNA Transfer: Mechanisms and Consequences.* Cold Spring Harbor, NY: Cold Spring Harbor Laboratory Press.

Cairns, J., J. Overbaugh, and S. Miller. (1988) "The Origin of Mutants." *Nature* 335: 142–145.

Cairns, J., G. S. Stent, and J. D. Watson, eds. (1966) *Phage and the Origins of Molecular Biology,* Cold Spring Harbor, NY: CSHL Press.

Carmel, L., Y. I. Wolf, I. B. Rogozin, and E. V. Koonin. (2007) "Three Distinct Modes of Intron Dynamics in the Evolution of Eukaryotes." *Genome Res* 17: 1,034–1,044.

Carroll, L. (1872). *Through the Looking Glass and What Alice Found There.* London: Macmillan.

Carroll, S. (2010) *From Eternity to Here: The Quest for the Ultimate Theory of Time.* New York: Penguin.

Carter, B. (1974) "Large Number Coincidences and the Anthropic Principle in Cosmology" in *Confrontation of Cosmological Theories with Observational Data.* M. S. Longair, editor. Dordrecht: D. Reidel, pp. 291–298.

Casino, P., V. Rubio, and A. Marina. (2010) "The Mechanism of Signal Transduction by Two-Component Systems." *Curr Opin Struct Biol* 20: 763–771.

Cavalier-Smith, T. (1998) "A Revised Six-Kingdom System of Life." *Biol Rev Camb Philos Soc* 73: 203–266.

Cech, T. R. (2002) "Ribozymes, the First 20 Years." *Biochem Soc Trans* 30: 1,162–1,166.

Chan, C. X., A. E. Darling, R. G. Beiko, and M. A. Ragan. (2009) "Are Protein Domains Modules of Lateral Genetic Transfer?" *PLoS One* 4: e4524.

Chargaff, E. (1978) *Heraclitean Fire: Sketches from a Life Before Nature.* New York: Rockefeller University Press.

Charlebois, R. L., and W. F. Doolittle. (2004) "Computing Prokaryotic Gene Ubiquity: Rescuing the Core from Extinction." *Genome Res* 14: 2,469–2,477.

Chen, I., P. J. Christie, and D. Dubnau. (2005) "The Ins and Outs of DNA Transfer in Bacteria." *Science* 310: 1,456–1,460.

Cheng, L. K., and P. J. Unrau. (2010) "Closing the Circle: Replicating RNA with RNA." *Cold Spring Harb Perspect Biol* 2: a002204.

Chernikova, D., S. Motamedi, M. Csuros, E. V. Koonin, and I. B. Rogozin. (2011) "A Late Origin of the Extant Eukaryotic Diversity: Divergence Time Estimates Using Rare Genomic Changes." *Biol Direct,* in press.

Ciccarelli, F. D., T. Doerks, C. von Mering, C. J. Creevey, B. Snel, and P. Bork. (2006) "Toward Automatic Reconstruction of a Highly Resolved Tree of Life." *Science* 311: 1,283–1,287.

Clapier, C. R., and B. R. Cairns. (2009) "The Biology of Chromatin Remodeling Complexes." *Annu Rev Biochem* 78: 273–304.

Claverie, J. M. (2006) "Viruses Take Center Stage in Cellular Evolution." *Genome Biol* 7: 110.

Collins, L., and D. Penny. (2005) "Complex Spliceosomal Organization Ancestral to Extant Eukaryotes." *Mol Biol Evol* 22: 1,053–1,066.

Cortez, D., P. Forterre, and S. Gribaldo. (2009) "A Hidden Reservoir of Integrative Elements Is the Major Source of Recently Acquired Foreign Genes and Orfans in Archaeal and Bacterial Genomes." *Genome Biol* 10: R65.

Costanzo, G., S. Pino, F. Ciciriello, and E. Di Mauro. (2009) "Generation of Long RNA Chains in Water." *J Biol Chem* 284: 33,206–33,216.

Costanzo, G., R. Saladino, C. Crestini, F. Ciciriello, and E. Di Mauro. (2007) "Nucleoside Phosphorylation by Phosphate Minerals." *J Biol Chem* 282: 16,729–16,735.

Cox, C. J., P. G. Foster, R. P. Hirt, S. R. Harris, and T. M. Embley. (2008) "The Archaebacterial Origin of Eukaryotes." *Proc Natl Acad Sci USA* 105: 20,356–20,361.

Cox, M. M., and J. R. Battista. (2005) "Deinococcus Radiodurans—the Consummate Survivor." *Nat Rev Microbiol* 3: 882–892.

Crick, F. (1970) "Central Dogma of Molecular Biology." *Nature* 227: 561–563.

Crick, F. H. (1958) "On Protein Synthesis." *Symp Soc Exp Biol* 12: 138–163.

Crick, F. H. (1968) "The Origin of the Genetic Code." *J Mol Biol* 38: 367–379.

Csuros, M., and I. Miklos. (2009) "Streamlining and Large Ancestral Genomes in Archaea Inferred with a Phylogenetic Birth-and-Death Model." *Mol Biol Evol* 26: 2,087–2,095.

Csuros, M., I. B. Rogozin, and E. V. Koonin. (2011) "Resonstructed Human-Like Intron Density in the Last Common Ancestor of Eukaryotes." *PLoS Comput Biol*, in press.

D'Herelle, F. (1922). *The Bacteriophage; Its Role in Immunity*. Baltimore: Williams and Wilkins.

Dagan, T., and W. Martin. (2006) "The Tree of One Percent." *Genome Biol* 7: 118.

Daly, M. J. (2009) "A New Perspective on Radiation Resistance Based on Deinococcus Radiodurans." *Nat Rev Microbiol* 7: 237–245.

Danchin, A. (2003). *The Delphic Boat: What Genomes Tell Us*. Cambridge, MA: Harvard Univ Press.

Darwin, C. (1859) *On the Origin of Species*. London: Murray.

Darwin, C. (1872) *Origin of Species*. New York: The Modern Library.

Daubin, V., and H. Ochman. (2004) "Bacterial Genomes As New Gene Homes: The Genealogy of Orfans in E. Coli." *Genome Res* 14: 1,036–1,042.

Davidov, Y., and E. Jurkevitch. (2009) "Predation Between Prokaryotes and the Origin of Eukaryotes." *Bioessays* 31: 748–757.

Daviter, T., K. B. Gromadski, and M. V. Rodnina. (2006) "The Ribosome's Response to Codon-Anticodon Mismatches." *Biochimie* 88: 1,001–1,011.

Dawkins, R. (1996) *The Blind Watchmaker: Why the Evidence of Evolution Reveals a Universe Without Design*. London: W.W. Norton & Co.

Dawkins, R. (2006) *The Selfish Gene: 30th Anniversary Edition—with a New Introduction by the Author*. Oxford: Oxford University Press.

Dayhoff, M. O., W. C. Barker, and L. T. Hunt. (1983) "Establishing Homologies in Protein Sequences." *Methods Enzymol* 91: 524–545.

de Nooijer, S., B. R. Holland, and D. Penny. (2009) "The Emergence of Predators in Early Life: There Was No Garden of Eden." *PLoS ONE* 4: e5507.

de Souza, S. J., M. Long, R. J. Klein, S. Roy, S. Lin, and W. Gilbert. (1998) "Toward a Resolution of the Introns Early/Late Debate: Only Phase Zero Introns Are Correlated with the Structure of Ancient Proteins." *Proc Natl Acad Sci USA* 95: 5,094–5,099.

de Visser, J. A., and S. F. Elena. (2007) "The Evolution of Sex: Empirical Insights into the Roles of Epistasis and Drift." *Nat Rev Genet* 8: 139–149.

Deckert, G., P. V. Warren, T. Gaasterland, W. G. Young, A. L. Lenox, D. E. Graham, R. Overbeek, M. A. Snead, M. Keller, M. Aujay, R. Huber, R. A. Feldman, J. M. Short, G. J. Olsen, and R. V. Swanson. (1998) "The Complete Genome of the Hyperthermophilic Bacterium Aquifex Aeolicus." *Nature* 392: 353–358.

Delarue, M., O. Poch, N. Tordo, D. Moras, and P. Argos. (1990) "An Attempt to Unify the Structure of Polymerases." *Protein Eng* 3: 461–467.

Delaye, L., and A. Moya. (2010) "Evolution of Reduced Prokaryotic Genomes and the Minimal Cell Concept: Variations on a Theme." *Bioessays* 32: 281–287.

Deleuze, G., and F. Guattari. (1987) *Thousand Plateaus: Capitalism and Schizophrenia*. Minneapolis, MN: University of Minnesota Press.

Denamur E. and I. Matic. (2006) "Evolution of Mutation Rates in Bacteria." *Mol Microbiol* 60: 820–827.

Dennett, D. C. (1996) *Darwin's Dangerous Idea: Evolution and the Meanings of Life*. New York: Simon & Schuster.

Deppenmeier, U., A. Johann, T. Hartsch, R. Merkl, R. A. Schmitz, R. Martinez-Arias, A. Henne, A. Wiezer, S. Baumer, C. Jacobi, H. Bruggemann, T. Lienard, A. Christmann, M. Bomeke, S. Steckel, A. Bhattacharyya, A. Lykidis, R. Overbeek, H. P. Klenk, R. P. Gunsalus, H. J. Fritz, and G. Gottschalk. (2002) "The Genome of Methanosarcina Mazei: Evidence for Lateral Gene Transfer Between Bacteria and Archaea." *J Mol Microbiol Biotechnol* 4: 453–461.

Deveau, H., J. E. Garneau, and S. Moineau. (2010) "CRISPR/Cas System and Its Role in Phage-Bacteria Interactions." *Annu Rev Microbiol* 64: 475–493.

Diaz, R., C. Vargas-Lagunas, M. A. Villalobos, H. Peralta, Y. Mora, S. Encarnacion, L. Girard, and J. Mora. (2011) "argC Orthologs from Rhizobiales Show Diverse Profiles of Transcriptional Efficiency and Functionality in Sinorhizobium Meliloti." *J Bacteriol* 193: 460–472.

Ding, S. W. (2010) "RNA-Based Antiviral Immunity." *Nat Rev Immunol* 10: 632–644.

Dlakic, M. and A. Mushegian. (2011) "Prp8, the Pivotal Protein of the Spliceosomal Catalytic Center, Evolved from a Retroelement-Encoded Reverse Transcriptase." *RNA* [Epub ahead of print].

Dobzhansky, T. (1951). *Genetics and the Origin of Species*. New York: Columbia University Press.

Dobzhansky, T. (1973) "Nothing in Biology Makes Sense Except in the Light of Evolution." *The American Biology Teacher* 35: 125–129.

Dong, H., and C. G. Kurland. (1995) "Ribosome Mutants with Altered Accuracy Translate with Reduced Processivity." *J Mol Biol* 248: 551–561.

Doolittle, R. F. (1995) "The Multiplicity of Domains in Proteins." *Annu Rev Biochem* 64: 287–314.

Doolittle, W. F. (1999a) "Lateral Genomics." *Trends Cell Biol* 9: M5–8.

Doolittle, W. F. (1999b) "Phylogenetic Classification and the Universal Tree." *Science* 284: 2,124–2,129.

Doolittle, W. F. (2000) "Uprooting the Tree of Life." *Sci Am* 282: 90–95.

Doolittle, W. F., and E. Bapteste. (2007) "Pattern Pluralism and the Tree of Life Hypothesis." *Proc Natl Acad Sci USA* 104: 2,043–2,049.

Doolittle, W. F., and J. R. Brown. (1994) "Tempo, Mode, the Progenote, and the Universal Root." *Proc Natl Acad Sci USA* 91: 6,721–6,728.

Doolittle, W. F., and C. Sapienza. (1980) "Selfish Genes, the Phenotype Paradigm and Genome Evolution." *Nature* 284: 601–603.

Doolittle, W. F., and O. Zhaxybayeva. (2009) "On the Origin of Prokaryotic Species." *Genome Res* 19: 744–756.

Doudna, J. A., and T. R. Cech. (2002) "The Chemical Repertoire of Natural Ribozymes." *Nature* 418: 222–228.

Douzery, E. J., E. A. Snell, E. Bapteste, F. Delsuc, and H. Philippe. (2004) "The Timing of Eukaryotic Evolution: Does a Relaxed Molecular Clock Reconcile Proteins and Fossils?" *Proc Natl Acad Sci USA* 101: 15,386–15,391.

Draghi, J. A., T. L. Parsons, G. P. Wagner, and J. B. Plotkin. (2010) "Mutational Robustness Can Facilitate Adaptation." *Nature* 463: 353–355.

Drake, J. W. (1991) "A Constant Rate of Spontaneous Mutation in DNA-Based Microbes." *Proc Natl Acad Sci USA* 88: 7,160–7,164.

Drake, J. W., and J. J. Holland. (1999) "Mutation Rates Among RNA Viruses." *Proc Natl Acad Sci USA* 96: 13,910–13,913.

Dronamraju, K. R., ed. (1968) *Haldane and Modern Biology*. Baltimore, Johns Hopkins University Press.

Drummond, D. A., J. D. Bloom, C. Adami, C. O. Wilke, and F. H. Arnold. (2005) "Why Highly Expressed Proteins Evolve Slowly." *Proc Natl Acad Sci USA* 102: 14,338–14,343.

Drummond, D. A., A. Raval, and C. O. Wilke. (2006) "A Single Determinant Dominates the Rate of Yeast Protein Evolution." *Mol Biol Evol* 23: 327–337.

Drummond, D. A., and C. O. Wilke. (2008) "Mistranslation-Induced Protein Misfolding As a Dominant Constraint on Coding-Sequence Evolution." *Cell* 134: 341–352.

Drummond, D. A., and C. O. Wilke. (2009) "The Evolutionary Consequences of Erroneous Protein Synthesis." *Nat Rev Genet* 10: 715–724.

Dupuy, C., E. Huguet, and J. M. Drezen. (2006) "Unfolding the Evolutionary Story of Polydnaviruses." *Virus Res* 117: 81–89.

Echols, H. (1981) "SOS Functions, Cancer, and Inducible Evolution." *Cell* 25: 1–2.

Eckerle, L. D., X. Lu, S. M. Sperry, L. Choi, and M. R. Denison. (2007) "High Fidelity of Murine Hepatitis Virus Replication Is Decreased in nsp14 Exoribonuclease Mutants." *J Virol* 81: 12,135–12,144.

Eddy, S. R. (2002) "Computational Genomics of Noncoding RNA Genes." *Cell* 109: 137–140.

Edwards, R. A., and F. Rohwer. (2005) "Viral Metagenomics." *Nat Rev Microbiol* 3: 504–510.

Eigen, M. (1971) "Self-organization of Matter and the Evolution of Biological Macromolecules." *Naturwissenschaften* 58: 465–523.

Eigen, M., B. F. Lindemann, M. Tietze, R. Winkler-Oswatitsch, A. Dress, and A. von Haeseler. (1989) "How Old Is the Genetic Code? Statistical Geometry of tRNA Provides an Answer." *Science* 244: 673–679.

Eisen, J. A., J. F. Heidelberg, O. White, and S. L. Salzberg. (2000) "Evidence for Symmetric Chromosomal Inversions Around the Replication Origin in Bacteria." *Genome Biol* 1: RESEARCH0011.

Eldredge, N., and S. J. Gould. (1997) "On Punctuated Equilibria." *Science* 276: 338–341.

Ellington, A. D., X. Chen, M. Robertson, and A. Syrett. (2009) "Evolutionary Origins and Directed Evolution of RNA." *Int J Biochem Cell Biol* 41: 254–265.

Ellis, R. J. (2003) "Protein Folding: Importance of the Anfinsen Cage." *Curr Biol* 13: R881–883.

Embley, T. M., and W. Martin. (2006) "Eukaryotic Evolution, Changes and Challenges." *Nature* 440: 623–630.

Esser, C., N. Ahmadinejad, C. Wiegand, C. Rotte, F. Sebastiani, G. Gelius-Dietrich, K. Henze, E. Kretschmann, E. Richly, D. Leister, D. Bryant, M. A. Steel, P. J. Lockhart, D. Penny, and W. Martin. (2004) "A Genome Phylogeny for Mitochondria Among Alpha-Proteobacteria and a Predominantly Eubacterial Ancestry of Yeast Nuclear Genes." *Mol Biol Evol* 21: 1,643–1,660.

Esser, C., W. Martin, and T. Dagan (2007) "The Origin of Mitochondria in Light of a Fluid Prokaryotic Chromosome Model." *Biol Lett* 3: 180–184.

Falkowski, P. G., T. Fenchel, and E. F. Delong. (2008) "The Microbial Engines That Drive Earth's Biogeochemical Cycles." *Science* 320: 1,034–1,039.

Falnes, P. O. (2005) "RNA Repair—the Latest Addition to the Toolbox for Macro-molecular Maintenance." *RNA Biol* 2: 14–16.

Fedor, M. J., and J. R. Williamson. (2005) "The Catalytic Diversity of RNAs." *Nat Rev Mol Cell Biol* 6: 399–412.

Felsenstein, J. (1996) "Inferring Phylogenies from Protein Sequences by Parsimony, Distance, and Likelihood Methods." *Methods Enzymol* 266: 418–427.

Felsenstein, J. (2004). *Inferring Phylogenies*. Sunderland, MA: Sinauer Associates.

Feschotte, C. (2010) "Virology: Bornavirus Enters the Genome." *Nature* 463: 39–40.

Field, M. C., and J. B. Dacks. (2009) "First and Last Ancestors: Reconstructing Evolution of the Endomembrane System with Escrts, Vesicle Coat Proteins, and Nuclear Pore Complexes." *Curr Opin Cell Biol* 21: 4–13.

Fields, B. N., P. M. Howley, D. E. Griffin, R. A. Lamb, M. A. Martin, B. Roizman, S. E. Straus, and D. M. Knipe. (2001). *Fields Virology*. New York: Lippincott Williams & Wilkins.

Fisher Box, J. (1978) *R. A. Fisher: The Life of a Scientist*. New York: Wiley.

Fisher, R. A. (1930) *The Genetical Theory of Natural Selection*. New York: Oxford University Press.

Fitch, W. M. (1970) "Distinguishing Homologous from Analogous Proteins." *Syst Zool* 19: 99–113.

Fleischmann, R. D., M. D. Adams, O. White, R. A. Clayton, E. F. Kirkness, A. R. Kerlavage, C. J. Bult, J. F. Tomb, B. A. Dougherty, J. M. Merrick, et al. (1995) "Whole-Genome Random Sequencing and Assembly of Haemophilus Influen-zae Rd [See Comments]." *Science* 269: 496–512.

Flores, R., S. Delgado, M. E. Gas, A. Carbonell, D. Molina, S. Gago, and M. De la Pena. (2004) "Viroids: The Minimal Non-Coding RNAs with Autonomous Replication." *FEBS Lett* 567: 42–48.

Fontanari, J. F., M. Santos, and E. Szathmary. (2006) "Coexistence and Error Prop-agation in Pre-Biotic Vesicle Models: A Group Selection Approach." *J Theor Biol* 239: 247–256.

Forterre, P. (1999) "Displacement of Cellular Proteins by Functional Analogues from Plasmids or Viruses Could Explain Puzzling Phylogenies of Many DNA Informational Proteins." *Mol Microbiol* 33: 457–465.

Forterre, P. (2002) "A Hot Story from Comparative Genomics: Reverse Gyrase Is the Only Hyperthermophile-Specific Protein." *Trends Genet* 18: 236–237.

Forterre, P. (2006) "Three RNA Cells for Ribosomal Lineages and Three DNA Viruses to Replicate Their Genomes: A Hypothesis for the Origin of Cellular Domain." *Proc Natl Acad Sci USA* 103: 3,669–3,674.

Forterre, P., and D. Prangishvili. (2009) "The Great Billion-Year War Between Ribosome- and Capsid-Encoding Organisms (Cells and Viruses) As the Major Source of Evolutionary Novelties." *Ann N Y Acad Sci* 1,178: 65–77.

Foster, P. L. (2000) "Adaptive Mutation: Implications for Evolution." *Bioessays* 22: 1,067–1,074.

Foster, P. L. (2007) "Stress-Induced Mutagenesis in Bacteria." *Crit Rev Biochem Mol Biol* 42: 373–397.

Fowles, J. (1969). *The French Lieutenant's Woman*. Boston: Little & Brown.

Fox, S. W. (1976) "The Evolutionary Significance of Phase-Separated Microsystems." *Orig Life* 7: 49–68.

Frank, A. C., H. Amiri, and S. G. Andersson. (2002) "Genome Deterioration: Loss of Repeated Sequences and Accumulation of Junk DNA." *Genetica* 115: 1–12.

Fraser, C. M., J. D. Gocayne, O. White, M. D. Adams, R. A. Clayton, R. D. Fleischmann, C. J. Bult, A. R. Kerlavage, G. Sutton, J. M. Kelley, et al. (1995) "The Minimal Gene Complement of Mycoplasma Genitalium." *Science* 270: 397–403.

Fredrick, K., and H. F. Noller. (2002) "Accurate Translocation of mRNA by the Ribosome Requires a Peptidyl Group or Its Analog on the tRNA Moving into the 30s P-Site." *Mol Cell* 9: 1,125–1,131.

Freeland, S. J., and L. D. Hurst. (1998) "The Genetic Code Is One in a Million." *J Mol Evol* 47: 238–248.

French, S. L., T. J. Santangelo, A. L. Beyer, and J. N. Reeve. (2007) "Transcription and translation are coupled in Archaea." *Mol Biol Evol* 24: 893–895.

Friedberg, E. C., G. C. Walker, W. Siede, R. D. Wood, R. A. Schulz, and T. Ellenberger. (2005) *DNA Repair and Mutagenesis*. Washington, D.C.: ASM Press.

Fritz-Laylin, L. K., S. E. Prochnik, M. L. Ginger, J. B. Dacks, M. L. Carpenter, M. C. Field, A. Kuo, A. Paredez, J. Chapman, J. Pham, S. Shu, R. Neupane, M. Cipriano, J. Mancuso, H. Tu, A. Salamov, E. Lindquist, H. Shapiro, S. Lucas, I. V. Grigoriev, W. Z. Cande, C. Fulton, D. S. Rokhsar, and S. C. Dawson. (2010) "The Genome of Naegleria Gruberi Illuminates Early Eukaryotic Versatility." *Cell* 140: 631–642.

Frost, L. S., and G. Koraimann. (2010) "Regulation of Bacterial Conjugation: Balancing Opportunity with Adversity." *Future Microbiol* 5: 1,057–1,071.

Frost, L. S., R. Leplae, A. O. Summers, and A. Toussaint. (2005) "Mobile Genetic Elements: The Agents of Open Source Evolution." *Nat Rev Microbiol* 3: 722–732.

Fuerst, J. A. (2005) "Intracellular Compartmentation in Planctomycetes." *Annu Rev Microbiol* 59: 299–328.

Futuyma, D. (2005). *Evolution*. Sunderland, MA: Sinauer Associates.

Gago, S., S. F. Elena, R. Flores, and R. Sanjuan. (2009) "Extremely High Mutation Rate of a Hammerhead Viroid." *Science* 323: 1,308.

Galhardo, R. S., P. J. Hastings, and S. M. Rosenberg. (2007) "Mutation As a Stress Response and the Regulation of Evolvability." *Crit Rev Biochem Mol Biol* 42: 399–435.

Galperin, M. Y. (2005) "A Census of Membrane-Bound and Intracellular Signal Transduction Proteins in Bacteria: Bacterial Iq, Extroverts and Introverts." *BMC Microbiol* 5: 35.

Garriga, J., and A. Vilenkin. (2001) "Many Worlds in One." *Phys. Rev. D* 64: 043,511.

Garriss, G., M. K. Waldor, and V. Burrus. (2009) "Mobile Antibiotic Resistance Encoding Elements Promote Their Own Diversity." *PLoS Genet* 5: e1000775.

Gavrilets, S. (2004) *Fitness Landscapes and the Origin of Species.* Princeton, NJ: Princeton University Press.

Ge, F., L. S. Wang, and J. Kim. (2005) "The Cobweb of Life Revealed by Genome-Scale Estimates of Horizontal Gene Transfer." *PLoS Biol* 3: e316.

Gell-Mann, M. (1995) *The Quark and the Jaguar: Adventures in the Simple and the Complex* New York: St. Martin's Griffin.

Gibson, C. M., and M. S. Hunter. (2010) "Extraordinarily Widespread and Fantastically Complex: Comparative Biology of Endosymbiotic Bacterial and Fungal Mutualists of Insects." *Ecol Lett* 13: 223–234.

Gilbert, W. (1978) "Why Genes in Pieces?" *Nature* 271: 501.

Gillespie, J. H. (2000) "The Neutral Theory in an Infinite Population." *Gene* 261: 11–18.

Giovannoni, S. J., H. J. Tripp, S. Givan, M. Podar, K. L. Vergin, D. Baptista, L. Bibbs, J. Eads, T. H. Richardson, M. Noordewier, M. S. Rappe, J. M. Short, J. C. Carrington, and E. J. Mathur. (2005) "Genome Streamlining in a Cosmopolitan Oceanic Bacterium." *Science* 309: 1,242–1,245.

Glansdorff, N., Y. Xu, and B. Labedan. (2008) "The Last Universal Common Ancestor: Emergence, Constitution and Genetic Legacy of an Elusive Forerunner." *Biol Direct* 3: 29.

Gliboff, S. (2005) "'Protoplasm...Is Soft Wax in Our Hands': Paul Kammerer and the Art of Biological Transformation." *Endeavour* 29: 162–167.

Gogarten, J. P., W. F. Doolittle, and J. G. Lawrence. (2002) "Prokaryotic Evolution in Light of Gene Transfer." *Mol Biol Evol* 19: 2,226–2,238.

Gogarten, J. P., H. Kibak, P. Dittrich, L. Taiz, E. J. Bowman, B. J. Bowman, M. F. Manolson, R. J. Poole, T. Date, T. Oshima, et al. (1989) "Evolution of the Vacuolar H+-Atpase: Implications for the Origin of Eukaryotes." *Proc Natl Acad Sci USA* 86: 6,661–6,665.

Gogarten, J. P., and J. P. Townsend. (2005) "Horizontal Gene Transfer, Genome Innovation and Evolution." *Nat Rev Microbiol* 3: 679–687.

Gorbalenya, A. E., V. M. Blinov, A. P. Donchenko, and E. V. Koonin. (1985) "An NTP-Binding Motif Is the Most Conserved Sequence in a Highly Diverged Monophyletic Group of Proteins Involved in Positive Strand RNA Viral Replication." *Molek Genetika, Microbiol Virusol* 11: 30–36.

Gorbalenya, A. E., L. Enjuanes, J. Ziebuhr, and E. J. Snijder. (2006) "Nidovirales: Evolving the Largest RNA Virus Genome." *Virus Res* 117: 17–37.

Gorbalenya, A. E., and E. V. Koonin. (1989) "Viral Proteins Containing the Purine NTP-Binding Sequence Pattern." *Nucleic Acids Res* 17: 8,413–8,440.

Gorbalenya, A. E., F. M. Pringle, J. L. Zeddam, B. T. Luke, C. E. Cameron, J. Kalmakoff, T. N. Hanzlik, K. H. Gordon, and V. K. Ward. (2002) "The Palm Subdomain-Based Active Site Is Internally Permuted in Viral RNA-Dependent Rna Polymerases of an Ancient Lineage." *J Mol Biol* 324: 47–62.

Gould, S. J. (1997) "The Exaptive Excellence of Spandrels as a Term and Proto-type." *Proc Natl Acad Sci USA* 94: 10,750–10,755.

Gould, S. J. (1997). *Full House: The Spread of Excellence from Plato to Darwin.* San Francisco: Three Rivers Press.

Gould, S. J. (2002). *The Structure of Evolutionary Theory.* Cambridge, MA: Harvard University Press.

Gould, S. J., and R. C. Lewontin. (1979) "The Spandrels of San Marco and the Panglossian Paradigm: A Critique of the Adaptationist Programme." *Proc R Soc Lond B Biol Sci* 205: 581–598.

Graur, D., and W. Martin. (2004) "Reading the Entrails of Chickens: Molecular Timescales of Evolution and the Illusion of Precision." *Trends Genet* 20: 80–86.

Gray, M. W., J. Lukes, J. M. Archibald, P. J. Keeling, and W. F. Doolittle. (2010) "Cell Biology. Irremediable Complexity?" *Science* 330: 920–921.

Grilli, J., B. Bassetti, S. Maslov, and M. C. Lagomarsino. (2011) "Joint Scaling Laws in Functional and Evolutionary Categories in Prokaryotic Genomes." *arXiv:1101.5814v1 [q-bio.GN].*

Grishin, N. V., Y. I. Wolf, and E. V. Koonin. (2000) "From Complete Genomes to Measures of Substitution Rate Variability Within and Between Proteins." *Genome Res* 10: 991–1,000.

Groll, M., M. Bochtler, H. Brandstetter, T. Clausen, and R. Huber. (2005) "Molecu-lar Machines for Protein Degradation." *Chembiochem* 6: 222–256.

Guth, A. (1998) *The Inflationary Universe.* Boston: Basic Books.

Guth, A. H. (1998) *The Inflationary Universe: The Quest for a New Theory of Cos-mic Origins.* New York: Perseus Book Group.

Guth, A. H. (2001) "Eternal Inflation." *Ann N Y Acad Sci* 950: 66–82.

Guth, A. H., and D. I. Kaiser. (2005) "Inflationary Cosmology: Exploring the Uni-verse from the Smallest to the Largest Scales." *Science* 307: 884–890.

Haeckel, E. (1997). *The Wonders of Life: A Popular Study of Biological Philosophy.* London: General Books, LLC.

Haldane, J. B. S. (1928) "The Origin of Life." *Rationalist Annual* 148: 3–10.

Hale, C. R., P. Zhao, S. Olson, M. O. Duff, B. R. Graveley, L. Wells, R. M. Terns, and M. P. Terns. (2009) "RNA-Guided RNA Cleavage by a CRISPR RNA-Cas Protein Complex." *Cell* 139: 945–956.

Halfmann, R., and S. Lindquist. (2010) "Epigenetics in the Extreme: Prions and the Inheritance of Environmentally Acquired Traits." *Science* 330: 629–632.

Hampl, V., L. Hug, J. W. Leigh, J. B. Dacks, B. F. Lang, A. G. Simpson, and A. J. Roger. (2009) "Phylogenomic Analyses Support the Monophyly of Excavata and Resolve Relationships Among Eukaryotic 'Supergroups.'" *Proc Natl Acad Sci USA* 106: 3,859–3,864.

Harrison, P. M., and M. Gerstein. (2002) "Studying Genomes Through the Aeons: Protein Families, Pseudogenes, and Proteome Evolution." *J Mol Biol* 318: 1,155–1,174.

Hartl, D. L., and A. G. Clark. (2006) *Principles of Population Genetics*. Sunderland, MA: Sinauer Associates.

Hartung, S., and K. P. Hopfner. (2009) "Lessons from Structural and Biochemical Studies on the Archaeal Exosome." *Biochem Soc Trans* 37: 83–87.

Hawking, S. W. (1988) *A Brief History of Time: From the Big Bang to Black Holes*. London: Bantam.

Hawking, S. W., and L. Mlodinow. (2010) *The Grand Design*. London: Bantam.

Hazkani-Covo, E., R. M. Zeller, and W. Martin. (2010) "Molecular Poltergeists: Mitochondrial DNA Copies (Numts) in Sequenced Nuclear Genomes." *PLoS Genet* 6: e1000834.

Hilario, E., and J. P. Gogarten. (1993) "Horizontal Transfer of Atpase Genes—the Tree of Life Becomes a Net of Life." *Biosystems* 31: 111–119.

Hochstrasser, M. (2009) "Origin and Function of Ubiquitin-Like Proteins." *Nature* 458: 422–429.

Hollick, J. B. (2010) "Paramutation and Development." *Annu Rev Cell Dev Biol* 26: 557–579.

Holmes, E. C. (2009) *The Evolution and Emergence of RNA Viruses*. Oxford: Oxford University Press.

Horie, M., T. Honda, Y. Suzuki, Y. Kobayashi, T. Daito, T. Oshida, K. Ikuta, P. Jern, T. Gojobori, J. M. Coffin, and K. Tomonaga. (2010) "Endogenous Non-Retroviral RNA Virus Elements in Mammalian Genomes." *Nature* 463: 84–87.

Huber, C., W. Eisenreich, S. Hecht, and G. Wachtershauser. (2003) "A Possible Primordial Peptide Cycle." *Science* 301: 938–940.

Hudder, A., L. Nathanson, and M. P. Deutscher. (2003) "Organization of Mammalian Cytoplasm." *Mol Cell Biol* 23: 9,318–93,26.

Hulen, C., and J. Legault-Demare. (1975) "In Vitro Synthesis of Large Peptide Molecules Using Glucosylated Single-Stranded Bacteriophage T4d DNA Template." *Nucleic Acids Res* 2: 2,037–2,048.

Humbard, M. A., H. V. Miranda, J. M. Lim, D. J. Krause, J. R. Pritz, G. Zhou, S. Chen, L. Wells, and J. A. Maupin-Furlow. (2010) "Ubiquitin-Like Small Archaeal Modifier Proteins (Samps) in Haloferax Volcanii." *Nature* 463: 54–60.

Hurst, L. D. (2002) "The Ka/Ks Ratio: Diagnosing the Form of Sequence Evolution." *Trends Genet* 18: 486.

Hurst, L. D., and N. G. Smith. (1999) "Do Essential Genes Evolve Slowly?" *Curr Biol* 9: 747–750.

Hussain, T., V. Kamarthapu, S. P. Kruparani, M. V. Deshmukh, and R. Sankaranarayanan. (2010) "Mechanistic Insights into Cognate Substrate Discrimination During Proofreading in Translation." *Proc Natl Acad Sci USA* 107: 22,117–22,121.

Huxley, J. (2010) *Evolution: The Modern Synthesis: The Definitive Edition.* Cambridge: MIT Press.

Huxley, T. H. (1860) "Darwin on the Origin of Species." *Westminster Review:* 541–570.

Irimia, M., D. Penny, and S. W. Roy. (2007) "Coevolution of Genomic Intron Number and Splice Sites." *Trends Genet* 23: 321–325.

Irimia, M., S. W. Roy, D. E. Neafsey, J. F. Abril, J. Garcia-Fernandez, and E. V. Koonin. (2009) "Complex Selection on 5' Splice Sites in Intron-Rich Organisms." *Genome Res* 19: 2,021–2,027.

Iwabe, N., K. Kuma, M. Hasegawa, S. Osawa, and T. Miyata. (1989) "Evolutionary Relationship of Archaebacteria, Eubacteria, and Eukaryotes Inferred from Phylogenetic Trees of Duplicated Genes." *Proc Natl Acad Sci USA* 86: 9,355–9,359.

Iyer, L. M., A. M. Burroughs, and L. Aravind. (2006) "The Prokaryotic Antecedents of the Ubiquitin-Signaling System and the Early Evolution of Ubiquitin-Like Beta-Grasp Domains." *Genome Biol* 7: R60.

Iyer, L. M., E. V. Koonin, and L. Aravind. (2003) "Evolutionary Connection Between the Catalytic Subunits of DNA-Dependent RNA Polymerases and Eukaryotic RNA-Dependent RNA Polymerases and the Origin of RNA Polymerases." *BMC Struct Biol* 3: 1.

Iyer, L. M., E. V. Koonin, and L. Aravind. (2004a) "Evolution of Bacterial RNA Polymerase: Implications for Large-Scale Bacterial Phylogeny, Domain Accretion, and Horizontal Gene Transfer." *Gene* 335: 73–88.

Iyer, L. M., E. V. Koonin, D. D. Leipe, and L. Aravind. (2005) "Origin and Evolution of the Archaeo-Eukaryotic Primase Superfamily and Related Palm-Domain Proteins: Structural Insights and New Members." *Nucleic Acids Res* 33: 3,875–3,896.

Iyer, L. M., K. S. Makarova, E. V. Koonin, and L. Aravind. (2004b) "Comparative Genomics of the Ftsk-Hera Superfamily of Pumping Atpases: Implications for the Origins of Chromosome Segregation, Cell Division and Viral Capsid Packaging." *Nucleic Acids Res* 32: 5,260–5,279.

Jacob, F. (1977) "Evolution and Tinkering." *Science* 196: 1,161–1,166.

Jacob, F. (1993). *The Logic of Life.* Princeton: Princeton University Press.

Jacob, F., and J. Monod. (1961) "Genetic Regulatory Mechanisms in the Synthesis of Proteins." *J Mol Biol* 3: 318–356.

Jain, R., M. C. Rivera, and J. A. Lake. (1999) "Horizontal Gene Transfer Among Genomes: The Complexity Hypothesis." *Proc Natl Acad Sci USA* 96: 3,801–3,806.

Jeffares, D. C., T. Mourier, and D. Penny. (2006) "The Biology of Intron Gain and Loss." *Trends Genet* 22: 16–22.

Jensen, L. J., P. Julien, M. Kuhn, C. von Mering, J. Muller, T. Doerks, and P. Bork. (2008) "Eggnog: Automated Construction and Annotation of Orthologous Groups of Genes." *Nucleic Acids Res* 36: D250–254.

Johansson, M., M. Lovmar, and M. Ehrenberg. (2008) "Rate and Accuracy of Bacterial Protein Synthesis Revisited." *Curr Opin Microbiol* 11: 141–147.

Johnson, L. J., and P. J. Tricker. (2010) "Epigenomic Plasticity Within Populations: Its Evolutionary Significance and Potential." *Heredity* 105: 113–121.

Johnson, N. (2009) *Simply Complexity: A Clear Guide to Complexity Theory.* New York: Oneworld Publications.

Johnston, W. K., P. J. Unrau, M. S. Lawrence, M. E. Glasner, and D. P. Bartel. (2001) "RNA-Catalyzed Rna Polymerization: Accurate and General RNA-Templated Primer Extension." *Science* 292: 1,319–1,325.

Jordan, I. K., K. S. Makarova, J. L. Spouge, Y. I. Wolf, and E. V. Koonin. (2001) "Lineage-Specific Gene Expansions in Bacterial and Archaeal Genomes." *Genome Res* 11: 555–565.

Jordan, I. K., L. Marino-Ramirez, Y. I. Wolf, and E. V. Koonin. (2004) "Conservation and Coevolution in the Scale-Free Human Gene Coexpression Network." *Mol Biol Evol* 21: 2,058–2,070.

Jordan, I. K., I. B. Rogozin, Y. I. Wolf, and E. V. Koonin. (2002) "Essential Genes Are More Evolutionarily Conserved Than Are Nonessential Genes in Bacteria." *Genome Res* 12: 962–968.

Juhas, M., J. R. van der Meer, M. Gaillard, R. M. Harding, D. W. Hood, and D. W. Crook. (2009) "Genomic Islands: Tools of Bacterial Horizontal Gene Transfer and Evolution." *FEMS Microbiol Rev* 33: 376–393.

Kaneko, K. (2007) "Evolution of Robustness to Noise and Mutation in Gene Expression Dynamics." *PLoS One* 2: e434.

Karev, G. P., Y. I. Wolf, A. Y. Rzhetsky, F. S. Berezovskaya, and E. V. Koonin. (2002) "Birth and Death of Protein Domains: A Simple Model of Evolution Explains Power Law Behavior." *BMC Evol Biol* 2: 18.

Karginov, F. V., and G. J. Hannon. (2010) "The CRISPR System: Small RNA-Guided Defense in Bacteria and Archaea." *Mol Cell* 37: 7–19.

Kasting, J. F., and S. Ono. (2006) "Palaeoclimates: The First Two Billion Years." *Philos Trans R Soc Lond B Biol Sci* 361: 917–929.

Kauffman, S. (1996) *At Home in the Universe: The Search for the Laws of Self-Organization and Complexity.* Oxford: Oxford University Press.

Keating, K. S., N. Toor, P. S. Perlman, and A. M. Pyle. (2010) "A Structural Analysis of the Group II Intron Active Site and Implications for the Spliceosome." *RNA* 16: 1–9.

Keeling, P. J. (2007) "Genomics. Deep Questions in the Tree of Life." *Science* 317: 1,875–1,876.

Keeling, P. J., G. Burger, D. G. Durnford, B. F. Lang, R. W. Lee, R. E. Pearlman, A. J. Roger, and M. W. Gray. (2005) "The Tree of Eukaryotes." *Trends Ecol Evol* 20: 670–676.

Keeling, P. J., and J. D. Palmer. (2008) "Horizontal Gene Transfer in Eukaryotic Evolution." *Nat Rev Genet* 9: 605–618.

Keiler, K. C. (2008) "Biology of Trans-Translation." *Annu Rev Microbiol* 62: 133–151.

Kelley, D. S., J. A. Karson, G. L. Fruh-Green, D. R. Yoerger, T. M. Shank, D. A. Butterfield, J. M. Hayes, M. O. Schrenk, E. J. Olson, G. Proskurowski, M. Jakuba, A. Bradley, B. Larson, K. Ludwig, D. Glickson, K. Buckman, A. S. Bradley, W. J. Brazelton, K. Roe, M. J. Elend, A. Delacour, S. M. Bernasconi, M. D. Lilley, J. A. Baross, R. E. Summons, and S. P. Sylva. (2005) "A Serpentinite-Hosted Ecosystem: The Lost City Hydrothermal Field." *Science* 307: 1,428–1,434.

Kelly, S., B. Wickstead, and K. Gull. (2010) "Archaeal Phylogenomics Provides Evidence in Support of a Methanogenic Origin of the Archaea and a Thaumarchaeal Origin for the Eukaryotes." *Proc Biol Sci* 278: 1009–1018.

Khesin, R. B. (1984) *Inconstancy of the Genome*. Moscow: Nauka.

Kim, V. N., J. Han, and M. C. Siomi. (2009) "Biogenesis of Small RNAs in Animals." *Nat Rev Mol Cell Biol* 10: 126–139.

Kimura, M. (1983) *The Neutral Theory of Molecular Evolution*. Cambridge: Cambridge University Press.

Kimura, M. (1991) "Recent Development of the Neutral Theory Viewed from the Wrightian Tradition of Theoretical Population Genetics." *Proc Natl Acad Sci USA* 88: 5,969–5,973.

Kipling, R. (2009) *Just So Stories: For Little Children*. Oxford: Oxford University Press.

Kirschner, M., and J. Gerhart. (1998) "Evolvability." *Proc Natl Acad Sci USA* 95: 8,420–8,427.

Knight, R. D., and L. F. Landweber. (2000) "The Early Evolution of the Genetic Code." *Cell* 101: 569–572.

Knoll, A. H., E. J. Javaux, D. Hewitt, and P. Cohen. (2006) "Eukaryotic Organisms in Proterozoic Oceans." *Philos Trans R Soc Lond B Biol Sci* 361: 1,023–1,038.

Kobayashi, I. (1998) "Selfishness and Death: Raison D'etre of Restriction, Recombination, and Mitochondria." *Trends Genet* 14: 368–374.

Kobayashi, I. (2001) "Behavior of Restriction-Modification Systems As Selfish Mobile Elements and Their Impact on Genome Evolution." *Nucleic Acids Res* 29: 3,742–3,756.

Kogenaru, M., M. G. de Vos, and S. J. Tans. (2009) "Revealing Evolutionary Pathways by Fitness Landscape Reconstruction." *Crit Rev Biochem Mol Biol* 44: 169–174.

Kondrashov, F. A., I. B. Rogozin, Y. I. Wolf, and E. V. Koonin. (2002) "Selection in the Evolution of Gene Duplications." *Genome Biol* 3: RESEARCH0008.

Koonin, E. V. (1991) "Genome Replication/Expression Strategies of Positive-Strand RNA Viruses: A Simple Version of a Combinatorial Classification and Prediction of New Strategies." *Virus Genes* 5: 273–281.

Koonin, E. V. (2003) "Comparative Genomics, Minimal Gene-Sets and the Last Universal Common Ancestor." *Nature Rev Microbiol* 1: 127–136.

Koonin, E. V. (2004) "A Non-Adaptationist Perspective on Evolution of Genomic Complexity or the Continued Dethroning of Man." *Cell Cycle* 3: 280–285.

Koonin, E. V. (2005) "Orthologs, Paralogs and Evolutionary Genomics." *Annu Rev Genet* 39: 309–338.

Koonin, E. V. (2006) "The Origin of Introns and Their Role in Eukaryogenesis: A Compromise Solution to the Introns-Early Versus Introns-Late Debate?" *Biol Direct* 1: 22.

Koonin, E. V. (2007a) "The Biological Big Bang Model for the Major Transitions in Evolution." *Biol Direct* 2: 21.

Koonin, E. V. (2007b) "The Cosmological Model of Eternal Inflation and the Transition from Chance to Biological Evolution in the History of Life." *Biol Direct* 2: 15.

Koonin, E. V. (2007c) "An RNA-Making Reactor for the Origin of Life." *Proc Natl Acad Sci USA* 104: 9,105–9,106.

Koonin, E. V. (2009a) "Evolution of Genome Architecture." *Int J Biochem Cell Biol* 41: 298–306.

Koonin, E. V. (2009b) "Intron-Dominated Genomes of Early Ancestors of Eukaryotes." *J Hered* 100: 618–623.

Koonin, E. V. (2009c) "On the Origin of Cells and Viruses: Primordial Virus World Scenario." *Ann N Y Acad Sci* 1,178: 47–64.

Koonin, E. V. (2010a) "The Origin and Early Evolution of Eukaryotes in the Light of Phylogenomics." *Genome Biol* 11: 209.

Koonin, E. V. (2010b) "Preview. The Incredible Expanding Ancestor of Eukaryotes." *Cell* 140: 606–608.

Koonin, E. V. (2010c) "Taming of the Shrewd: Novel Eukaryotic Genes from RNA Viruses." *BMC Biol* 8: 2.

Koonin, E. V. (2010d) "The Two Empires and the Three Domains of Life in the Postgenomic Age." *Nature Education* 3: 27.

Koonin, E. V., and L. Aravind. (2002) "Origin and Evolution of Eukaryotic Apoptosis: The Bacterial Connection." *Cell Death Differ* 9: 394–404.

Koonin, E. V., L. Aravind, and A. S. Kondrashov. (2000a) "The Impact of Comparative Genomics on Our Understanding of Evolution." *Cell* 101: 573–576.

Koonin, E. V., and V. V. Dolja. (1993) "Evolution and Taxonomy of Positive-Strand RNA Viruses: Implications of Comparative Analysis of Amino Acid Sequences." *Crit Rev Biochem Mol Biol* 28: 375–430.

Koonin, E. V., N. D. Fedorova, J. D. Jackson, A. R. Jacobs, D. M. Krylov, K. S. Makarova, R. Mazumder, S. L. Mekhedov, A. N. Nikolskaya, B. S. Rao, I. B. Rogozin, S. Smirnov, A. V. Sorokin, A. V. Sverdlov, S. Vasudevan, Y. I. Wolf, J. J. Yin, and D. A. Natale. (2004) "A Comprehensive Evolutionary Classification of Proteins Encoded in Complete Eukaryotic Genomes." *Genome Biol* 5: R7.

Koonin, E. V., A. E. Gorbalenya, and K. M. Chumakov. (1989) "Tentative Identification of RNA-Dependent RNA Polymerases of dsRNA Viruses and Their Relationship to Positive Strand RNA Viral Polymerases." *FEBS Lett* 252: 42–46.

Koonin, E. V., and K. S. Makarova. (2009) "CRISPR-Cas: An Adaptive Immunity System in Prokaryotes." *F1000 Biol Rep* 1: 95.

Koonin, E. V., K. S. Makarova, and L. Aravind. (2001a) "Horizontal Gene Transfer in Prokaryotes: Quantification and Classification." *Annu Rev Microbiol* 55: 709–742.

Koonin, E. V., and W. Martin. (2005) "On the Origin of Genomes and Cells within Inorganic Compartments." *Trends Genet* 21: 647–654.

Koonin, E. V., and A. R. Mushegian. (1996) "Complete Genome Sequences of Cellular Life Forms: Glimpses of Theoretical Evolutionary Genomics." *Curr Opin Genet Dev* 6: 757–762.

Koonin, E. V., A. R. Mushegian, and K. E. Rudd. (1996) "Sequencing and Analysis of Bacterial Genomes." *Curr Biol* 6: 404–416.

Koonin, E. V., and A. S. Novozhilov. (2009) "Origin and Evolution of the Genetic Code: The Universal Enigma." *IUBMB Life* 61: 99–111.

Koonin, E. V., T. G. Senkevich, and V. V. Dolja. (2006) "The Ancient Virus World and Evolution of Cells." *Biol Direct* 1: 29.

Koonin, E. V., and Y. I. Wolf. (2006) "Evolutionary Systems Biology: Links Between Gene Evolution and Function." *Curr Opin Biotechnol* 17: 481–487.

Koonin, E. V., and Y. I. Wolf. (2008a). "Evolutionary Systems Biology." *Evolutionary Genomics and Proteomics*. Ed. by M. Pagel and A. Pomiankowski. Sunderland, MA: Sinauer Associates, Inc.: 11–25.

Koonin, E. V., and Y. I. Wolf. (2008b) "Genomics of Bacteria and Archaea: The Emerging Dynamic View of the Prokaryotic World." *Nucleic Acids Res* 36: 6,688–6,719.

Koonin, E. V., and Y. I. Wolf. (2009a) "The Fundamental Units, Processes and Patterns of Evolution, and the Tree of Life Conundrum." *Biol Direct* 4: 33.

Koonin, E. V., and Y. I. Wolf. (2009b) "Is Evolution Darwinian or/and Lamarckian?" *Biol Direct* 4: 42.

Koonin, E. V., and Y. I. Wolf. (2010a) "The Common Ancestry of Life." *Biol Direct* 5: 64.

Koonin, E. V., and Y. I. Wolf. (2010b) "Constraints and Plasticity in Genome and Molecular-Phenome Evolution." *Nat Rev Genet* 11: 487–498.

Koonin, E. V., Y. I. Wolf, and L. Aravind. (2000b) "Protein Fold Recognition Using Sequence Profiles and Its Application in Structural Genomics." *Adv Protein Chem* 54: 245–275.

Koonin, E. V., Y. I. Wolf, and L. Aravind. (2001b) "Prediction of the Archaeal Exo-some and Its Connections with the Proteasome and the Translation and Tran-scription Machineries by a Comparative-Genomic Approach." *Genome Res* 11: 240–252.

Koonin, E. V., Y. I. Wolf, and G. P. Karev. (2002) "The Structure of the Protein Uni-verse and Genome Evolution." *Nature* 420: 218–223.

Koonin, E. V., Y. I. Wolf, K. Nagasaki, and V. V. Dolja. (2008) "The Big Bang of Picorna-Like Virus Evolution Antedates the Radiation of Eukaryotic Super-groups." *Nat Rev Microbiol* 6: 925–939.

Koonin, E. V., and N. Yutin. (2010) "Origin and Evolution of Eukaryotic Large Nucleo-Cytoplasmic DNA Viruses." *Intervirology* 53: 284–292.

Kristensen, D. M., A. R. Mushegian, V. V. Dolja, and E. V. Koonin. (2010) "New Dimensions of the Virus World Discovered Through Metagenomics." *Trends Microbiol* 18: 11–19.

Krylov, D. M., Y. I. Wolf, I. B. Rogozin, and E. V. Koonin. (2003) "Gene Loss, Protein Sequence Divergence, Gene Dispensability, Expression Level, and Interactivity Are Correlated in Eukaryotic Evolution." *Genome Res* 13: 2,229–2,235.

Kuhn, T. S. (1962) *The Structure of Scientific Revolutions*. Chicago: University of Chicago Press.

Kun, A., M. Santos, and E. Szathmary. (2005) "Real Ribozymes Suggest a Relaxed Error Threshold." *Nat Genet* 37: 1,008–1,011.

Kunin, V., L. Goldovsky, N. Darzentas, and C. A. Ouzounis. (2005) "The Net of Life: Reconstructing the Microbial Phylogenetic Network." *Genome Res* 15: 954–959.

Kunin, V., and C. A. Ouzounis. (2003) "The Balance of Driving Forces During Genome Evolution in Prokaryotes." *Genome Res* 13: 1,589–1,594.

Kurland, C. G. (2005) "What Tangled Web: Barriers to Rampant Horizontal Gene Transfer." *Bioessays* 27: 741–747.

Kurland, C. G., B. Canback, and O. G. Berg. (2003) "Horizontal Gene Transfer: A Critical View." *Proc Natl Acad Sci USA* 100: 9,658–9,662.

Kurland, C. G., L. J. Collins, and D. Penny. (2006) "Genomics and the Irreducible Nature of Eukaryote Cells." *Science* 312: 1,011–1,014.

Lamarck, J.-B. (1809) *Philosophie Zoologique, Ou Exposition Des Considérations Relatives À L'histoire Naturelle Des Animaux*. Paris: Dentu.

Lambowitz, A. M., and S. Zimmerly. (2004) "Mobile Group Ii Introns." *Annu Rev Genet* 38: 1–35.

Lane, C. E., and J. M. Archibald. (2008) "The Eukaryotic Tree of Life: Endosym-biosis Takes Its Toll." *Trends Ecol Evol* 23: 268–275.

Lane, N., and W. Martin. (2010) "The Energetics of Genome Complexity." *Nature* 467: 929–934.

Lang, A. S., and J. T. Beatty. (2007) "Importance of Widespread Gene Transfer Agent Genes in Alpha-Proteobacteria." *Trends Microbiol* 15: 54–62.

Lang, B. F., M. W. Gray, and G. Burger. (1999) "Mitochondrial Genome Evolution and the Origin of Eukaryotes." *Annu Rev Genet* 33: 351–397.

Lapierre, P., and J. P. Gogarten. (2009) "Estimating the Size of the Bacterial Pan-Genome." *Trends Genet* 25: 107–110.

Lathe, W. C. III, B. Snel, and P. Bork. (2000) "Gene Context Conservation of a Higher Order Than Operons." *Trends Biochem Sci* 25: 474–479.

Lawrence, J. (1999) "Selfish Operons: The Evolutionary Impact of Gene Clustering in Prokaryotes and Eukaryotes." *Curr Opin Genet Dev* 9: 642–648.

Lawrence, J. G., R. W. Hendrix, and S. Casjens. (2001) "Where Are the Pseudo-genes in Bacterial Genomes?" *Trends Microbiol* 9: 535–540.

Le Hir, H., A. Nott, and M. J. Moore. (2003) "How Introns Influence and Enhance Eukaryotic Gene Expression." *Trends Biochem Sci* 28: 215–220.

Leipe, D. D., L. Aravind, and E. V. Koonin. (1999) "Did DNA Replication Evolve Twice Independently?" *Nucleic Acids Res* 27: 3,389–3,401.

Leipe, D. D., E. V. Koonin, and L. Aravind. (2004) "Stand, a Class of P-Loop NTPases Including Animal and Plant Regulators of Programmed Cell Death: Multiple, Complex Domain Architectures, Unusual Phyletic Patterns, and Evolution by Horizontal Gene Transfer." *J Mol Biol* 343: 1–28.

Leipe, D. D., Y. I. Wolf, E. V. Koonin, and L. Aravind. (2002) "Classification and Evolution of P-Loop GTPases and Related Atpases." *J Mol Biol* 317: 41–72.

Lenski, R. E., C. Ofria, R. T. Pennock, and C. Adami. (2003) "The Evolutionary Origin of Complex Features." *Nature* 423: 139–144.

Lespinet, O., Y. I. Wolf, E. V. Koonin, and L. Aravind. (2002) "The Role of Lineage-Specific Gene Family Expansion in the Evolution of Eukaryotes." *Genome Res* 12: 1,048–1,059.

Levy, S. F., and M. L. Siegal. (2008) "Network Hubs Buffer Environmental Variation in Saccharomyces Cerevisiae." *PLoS Biol* 6: e264.

Li, W. H. (1997). *Molecular Evolution*. Sunderland, MA: Sinauer.

Lincoln, T. A., and G. F. Joyce. (2009) "Self-Sustained Replication of an Rna Enzyme." *Science* 323: 1,229–1,232.

Lindberg, J., and J. Lundeberg. (2010) "The Plasticity of the Mammalian Transcriptome." *Genomics* 95: 1–6.

Linde, A. D. (1986) "Eternally Existing Self-Reproducing Chaotic Inflationary Universe." *Phys Lett B* 175: 395.

Lindell, D., J. D. Jaffe, Z. I. Johnson, G. M. Church, and S. W. Chisholm. (2005) "Photosynthesis Genes in Marine Viruses Yield Proteins During Host Infection." *Nature* 438: 86–89.

Ling, J., H. Roy, and M. Ibba. (2007) "Mechanism of Trna-Dependent Editing in Translational Quality Control." *Proc Natl Acad Sci USA* 104: 72–77.

Ling, J., and D. Soll. (2010) "Severe Oxidative Stress Induces Protein Mistranslation Through Impairment of an Aminoacyl-Trna Synthetase Editing Site." *Proc Natl Acad Sci USA* 107: 4,028–4,033.

Liu, Y., J. Zhou, M. V. Omelchenko, A. S. Beliaev, A. Venkateswaran, J. Stair, L. Wu, D. K. Thompson, D. Xu, I. B. Rogozin, E. K. Gaidamakova, M. Zhai, K. S. Makarova, E. V. Koonin, and M. J. Daly. (2003) "Transcriptome Dynamics of Deinococcus Radiodurans Recovering from Ionizing Radiation." *Proc Natl Acad Sci USA* 100: 4,191–4,196.

Livio, M., and M. J. Rees. (2005) "Cosmology. Anthropic Reasoning." *Science* 309: 1,022–1,023.

Lobkovsky, A. E., Y. I. Wolf, and E. V. Koonin. (2010) "Universal Distribution of Protein Evolution Rates As a Consequence of Protein Folding Physics." *Proc Natl Acad Sci USA* 107: 2,983–2,988.

Loewe, L. (2009) "A framework for evolutionary systems biology." *BMC Syst Biol* 3: 27.

Lopez-Garcia, P., and D. Moreira. (2006) "Selective Forces for the Origin of the Eukaryotic Nucleus." *Bioessays* 28: 525–533.

Lozada-Chavez, I., S. C. Janga, and J. Collado-Vides. (2006) "Bacterial Regulatory Networks Are Extremely Flexible in Evolution." *Nucleic Acids Res* 34: 3,434–3,445.

Luria, S. E., and J. Darnell. (1967). *General Virology*. New York: John Wiley.

Luscombe, N. M., J. Qian, Z. Zhang, T. Johnson, and M. Gerstein. (2002) "The Dominance of the Population by a Selected Few: Power-Law Behaviour Applies to a Wide Variety of Genomic Properties." *Genome Biol* 3: RESEARCH0040.

Lynch, M. (2006) "The Origins of Eukaryotic Gene Structure." *Mol Biol Evol* 23: 450–468.

Lynch, M. (2007a) "The Evolution of Genetic Networks by Non-Adaptive Processes." *Nat Rev Genet* 8: 803–813.

Lynch, M. (2007b) "The Frailty of Adaptive Hypotheses for the Origins of Organismal Complexity." *Proc Natl Acad Sci USA* 104 Suppl 1: 8,597–8,604.

Lynch, M. (2007c) *The Origins of Genome Archiecture*. Sunderland, MA: Sinauer Associates.

Lynch, M. (2010) "Evolution of the Mutation Rate." *Trends Genet* 26: 345–352.

Lynch, M., R. Burger, D. Butcher, and W. Gabriel. (1993) "The Mutational Meltdown in Asexual Populations." *J Hered* 84: 339–344.

Lynch, M., and J. S. Conery. (2000) "The Evolutionary Fate and Consequences of Duplicate Genes." *Science* 290: 1,151–1,155.

Lynch, M., and J. S. Conery. (2003) "The Origins of Genome Complexity." *Science* 302: 1,401–1,404.

Lynch, M., and V. Katju. (2004) "The Altered Evolutionary Trajectories of Gene Duplicates." *Trends Genet* 20: 544–549.

Lyotard, J. F. (1979) *La Condition Postmoderne: Rapport Sur Le Savoir*. Paris: Minuit.

Makarova, K. S., L. Aravind, N. V. Grishin, I. B. Rogozin, and E. V. Koonin. (2002) "A DNA Repair System Specific for Thermophilic Archaea and Bacteria Predicted by Genomic Context Analysis." *Nucleic Acids Res* 30: 482–496.

Makarova, K. S., L. Aravind, Y. I. Wolf, R. L. Tatusov, K. W. Minton, E. V. Koonin, and M. J. Daly. (2001a) "Genome of the Extremely Radiation-Resistant Bacterium Deinococcus Radiodurans Viewed from the Perspective of Comparative Genomics." *Microbiol Mol Biol Rev* 65: 44–79.

Makarova, K. S., N. V. Grishin, S. A. Shabalina, Y. I. Wolf, and E. V. Koonin. (2006) "A Putative RNA-Interference-Based Immune System in Prokaryotes: Computational Analysis of the Predicted Enzymatic Machinery, Functional Analogies with Eukaryotic Rnai, and Hypothetical Mechanisms of Action." *Biol Direct* 1: 7.

Makarova, K. S., M. V. Omelchenko, E. K. Gaidamakova, V. Y. Matrosova, A. Vasilenko, M. Zhai, A. Lapidus, A. Copeland, E. Kim, M. Land, K. Mavrommatis, S. Pitluck, P. M. Richardson, C. Detter, T. Brettin, E. Saunders, B. Lai, B. Ravel, K. M. Kemner, Y. I. Wolf, A. Sorokin, A. V. Gerasimova, M. S. Gelfand, J. K. Fredrickson, E. V. Koonin, and M. J. Daly. (2007a) "Deinococcus Geothermalis: The Pool of Extreme Radiation Resistance Genes Shrinks." *PLoS ONE* 2: e955.

Makarova, K. S., V. A. Ponomarev, and E. V. Koonin. (2001b) "Two C or Not Two C: Recurrent Disruption of Zn-Ribbons, Gene Duplication, Lineage-Specific Gene Loss, and Horizontal Gene Transfer in Evolution of Bacterial Ribosomal Proteins." *Genome Biol* 2: RESEARCH 0033.

Makarova, K. S., A. V. Sorokin, P. S. Novichkov, Y. I. Wolf, and E. V. Koonin. (2007b) "Clusters of Orthologous Genes for 41 Archaeal Genomes and Implications for Evolutionary Genomics of Archaea." *Biol Direct* 2: 33.

Makarova, K. S., Y. I. Wolf, and E. V. Koonin. (2003) "Potential Genomic Determinants of Hyperthermophily." *Trends Genet* 19: 172–176.

Makarova, K. S., Y. I. Wolf, and E. V. Koonin. (2009a) "Comprehensive Comparative-Genomic Analysis of Type 2 Toxin-Antitoxin Systems and Related Mobile Stress Response Systems in Prokaryotes." *Biol Direct* 4: 19.

Makarova, K. S., Y. I. Wolf, S. L. Mekhedov, B. G. Mirkin, and E. V. Koonin. (2005) "Ancestral Paralogs and Pseudoparalogs and Their Role in the Emergence of the Eukaryotic Cell." *Nucleic Acids Res* 33: 4,626–4,638.

Makarova, K. S., Y. I. Wolf, J. van der Oost, and E. V. Koonin. (2009b) "Prokaryotic Homologs of Argonaute Proteins Are Predicted to Function As Key Components of a Novel System of Defense Against Mobile Genetic Elements." *Biol Direct* 4: 29.

Makarova, K. S., N. Yutin, S. D. Bell, and E. V. Koonin. (2010) "Evolution of Diverse Cell Division and Vesicle Formation Systems in Archaea." *Nat Rev Microbiol* 8: 731–741.

Mans, B. J., V. Anantharaman, L. Aravind, and E. V. Koonin. (2004) "Comparative Genomics, Evolution, and Origins of the Nuclear Envelope and Nuclear Pore Complex." *Cell Cycle* 3: 1,612–1,637.

Mansy, S. S., J. P. Schrum, M. Krishnamurthy, S. Tobe, D. A. Treco, and J. W. Szostak. (2008) "Template-Directed Synthesis of a Genetic Polymer in a Model Protocell." *Nature* 454: 122–125.

Margulis, L. (2009) "Genome Acquisition in Horizontal Gene Transfer: Symbiogenesis and Macromolecular Sequence Analysis." *Methods Mol Biol* 532: 181–191.

Margulis, L., M. Chapman, R. Guerrero, and J. Hall. (2006) "The Last Eukaryotic Common Ancestor (LECA): Acquisition of Cytoskeletal Motility from Aerotolerant Spirochetes in the Proterozoic Eon." *Proc Natl Acad Sci USA* 103: 13,080–13,085.

Margulis, L., and D. Sagan. (2003). *Acquiring Genomes: The Theory of the Origin of Species.* New York: Basic Books.

Marinsek, N., E. R. Barry, K. S. Makarova, I. Dionne, E. V. Koonin, and S. D. Bell. (2006) "GINS, a Central Nexus in the Archaeal DNA Replication Fork." *EMBO Rep* 7: 539–545.

Marraffini, L. A., and E. J. Sontheimer. (2008) "CRISPR Interference Limits Horizontal Gene Transfer in Staphylococci by Targeting DNA." *Science* 322: 1,843–1,845.

Martin, W. (1999) "Mosaic Bacterial Chromosomes: A Challenge En Route to a Tree of Genomes." *Bioessays* 21: 99–104.

Martin, W., J. Baross, D. Kelley, and M. J. Russell. (2008) "Hydrothermal Vents and the Origin of Life." *Nat Rev Microbiol* 6: 805–814.

Martin, W., and E. V. Koonin. (2006a) "Introns and the Origin of Nucleus-Cytosol Compartmentalization." *Nature* 440: 41–45.

Martin, W., and E. V. Koonin. (2006b) "A Positive Definition of Prokaryotes." *Nature* 442: 868.

Martin, W., and K. V. Kowallik. (1999) "Annotated English Translation of Mereschkowsky's 1905 Paper 'über Natur Und Ursprung Der Chromatophoren Im Pflanzenreiche.'" *Eur J Phycol* 34: 287–296.

Martin, W., and M. Muller. (1998) "The Hydrogen Hypothesis for the First Eukaryote." *Nature* 392: 37–41.

Martin, W., and M. J. Russell. (2003) "On the Origins of Cells: A Hypothesis for the Evolutionary Transitions from Abiotic Geochemistry to Chemoautotrophic Prokaryotes, and from Prokaryotes to Nucleated Cells." *Philos Trans R Soc Lond B Biol Sci* 358: 59–83; discussion 83–55.

Martin, W., and M. J. Russell. (2007) "On the Origin of Biochemistry at an Alkaline Hydrothermal Vent." *Philos Trans R Soc Lond B Biol Sci* 362: 1,887–1,925.

Martinez, J. L. (2008) "Antibiotics and Antibiotic Resistance Genes in Natural Environments." *Science* 321: 365–367.

Masel, J., and A. Bergman. (2003) "The Evolution of the Evolvability Properties of the Yeast Prion [Psi+]." *Evolution* 57: 1,498–1,512.

Masel, J., and M. L. Siegal. (2009) "Robustness: Mechanisms and Consequences." *Trends Genet* 25: 395–403.

Masel, J., and M. V. Trotter. (2010) "Robustness and Evolvability." *Trends Genet* 26: 406–414.

Maslov, S., S. Krishna, T. Y. Pang, and K. Sneppen. (2009) "Toolbox Model of Evolution of Prokaryotic Metabolic Networks and Their Regulation." *Proc Natl Acad Sci USA* 106: 9,743–9,748.

Mayr, E. (1963). *Animal Species and Evolution*. Cambridge: Harvard University Press.

McCarthy, B. J., and J. J. Holland. (1965) "Denatured DNA as a Direct Template for in Vitro Protein Synthesis." *Proc Natl Acad Sci USA* 54: 880–886.

McClain, W. H., L. B. Lai, and V. Gopalan. (2010) "Trials, Travails, and Triumphs: An Account of RNA Catalysis in RNase P." *J Mol Biol* 397: 627–646.

McClintock, B. (1984) "The Significance of Responses of the Genome to Challenge." *Science* 226: 792–801.

McCutcheon, J. P., B. R. McDonald, and N. A. Moran. (2009) "Convergent Evolution of Metabolic Roles in Bacterial Co-Symbionts of Insects." *Proc Natl Acad Sci USA* 106: 15,394–15,399.

McDaniel, L. D., E. Young, J. Delaney, F. Ruhnau, K. B. Ritchie, and J. H. Paul. (2010) "High Frequency of Horizontal Gene Transfer in the Oceans." *Science* 330: 50.

McDonald, J. H., and M. Kreitman. (1991) "Adaptive Protein Evolution at the Adh Locus in Drosophila." *Nature* 351: 652–654.

McGeoch, A. T., and S. D. Bell. (2008) "Extra-Chromosomal Elements and the Evolution of Cellular DNA Replication Machineries." *Nat Rev Mol Cell Biol* 9: 569–574.

Medvedev, Z. A. (1969) *The Rise and Fall of T. D. Lysenko*. New York: Columbia University Press.

Mendell, J. E., K. D. Clements, J. H. Choat, and E. R. Angert. (2008) "Extreme Polyploidy in a Large Bacterium." *Proc Natl Acad Sci USA* 105: 6,730–6,734.

Mendes Soares, L. M., and J. Valcarcel. (2006) "The Expanding Transcriptome: The Genome as the 'Book of Sand.'" *Embo J* 25: 923–931.

Merkl, R. (2006) "A Comparative Categorization of Protein Function Encoded in Bacterial or Archeal Genomic Islands." *J Mol Evol* 62: 1–14.

Mielke, R. E., M. J. Russell, P. R. Wilson, S. E. McGlynn, M. Coleman, R. Kidd, and I. Kanik. (2010) "Design, Fabrication, and Test of a Hydrothermal Reactor for Origin-of-Life Experiments." *Astrobiology* 10: 799–810.

Mills, D. R., F. R. Kramer, and S. Spiegelman. (1973) "Complete Nucleotide Sequence of a Replicating RNA Molecule." *Science* 180: 916–927.

Mira, A., A. B. Martin-Cuadrado, G. D'Auria, and F. Rodriguez-Valera. (2010) "The Bacterial Pan-Genome: A New Paradigm in Microbiology." *Int Microbiol* 13: 45–57.

Mira, A., H. Ochman, and N. A. Moran. (2001) "Deletional Bias and the Evolution of Bacterial Genomes." *Trends Genet* 17: 589–596.

Mirkin, B. G., T. I. Fenner, M. Y. Galperin, and E. V. Koonin. (2003) "Algorithms for Computing Parsimonious Evolutionary Scenarios for Genome Evolution, the Last Universal Common Ancestor, and Dominance of Horizontal Gene Transfer in the Evolution of Prokaryotes." *BMC Evol Biol* 3: 2.

Mojica, F. J., C. Diez-Villasenor, J. Garcia-Martinez, and E. Soria. (2005) "Intervening Sequences of Regularly Spaced Prokaryotic Repeats Derive from Foreign Genetic Elements." *J Mol Evol* 60: 174–182.

Molina, N., and E. van Nimwegen. (2009) "Scaling Laws in Functional Genome Content Across Prokaryotic Clades and Lifestyles." *Trends Genet* 25: 243–247.

Monier, A., J. M. Claverie, and H. Ogata. (2008) "Taxonomic Distribution of Large DNA Viruses in the Sea." *Genome Biol* 9: R106.

Monod, J. (1972). *Chance and Necessity: An Essay on the Natural Philosophy of Modern Biology*. New York: Vintage.

Moreira, D., and P. Lopez-Garcia. (2009) "Ten Reasons to Exclude Viruses from the Tree of Life." *Nat Rev Microbiol* 7: 306–311.

Mulkidjanian, A. Y. (2009) "On the Origin of Life in the Zinc World: 1. Photosynthesizing, Porous Edifices Built of Hydrothermally Precipitated Zinc Sulfide as Cradles of Life on Earth." *Biol Direct* 4: 26.

Mulkidjanian, A. Y., M. Y. Galperin, K. S. Makarova, Y. I. Wolf, and E. V. Koonin. (2008) "Evolutionary Primacy of Sodium Bioenergetics." *Biol Direct* 3: 13.

Mulkidjanian, A. Y., E. V. Koonin, K. S. Makarova, S. L. Mekhedov, A. Sorokin, Y. I. Wolf, A. Dufresne, F. Partensky, H. Burd, D. Kaznadzey, R. Haselkorn, and M. Y. Galperin. (2006) "The Cyanobacterial Genome Core and the Origin of Photosynthesis." *Proc Natl Acad Sci USA* 103: 13,126–13,131.

Mulkidjanian, A. Y., K. S. Makarova, M. Y. Galperin, and E. V. Koonin. (2007) "Inventing the Dynamo Machine: The Evolution of the F-Type and V-Type Atpases." *Nat Rev Microbiol* 5: 892–899.

Müller, F., T. Brissac, N. Le Bris, H. Felbeck, and O. Gros. (2010) "First Description of Giant Archaea (Thaumarchaeota) Associated with Putative Bacterial Ectosymbionts in a Sulfidic Marine Habitat." *Env Microbiol* 12: 2,371–2,383.

Müller, H. J. (1964) "The Relation of Recombination to Mutational Advance." *Mutat Res* 106: 2–9.

Mushegian, A. R., and E. V. Koonin. (1996a) "Gene Order Is Not Conserved in Bacterial Evolution." *Trends Genet* 12: 289–290.

Mushegian, A. R., and E. V. Koonin. (1996b) "A Minimal Gene Set for Cellular Life Derived by Comparison of Complete Bacterial Genomes." *Proc Natl Acad Sci USA* 93: 10,268–10,273.

Nelson, K. E., R. A. Clayton, S. R. Gill, M. L. Gwinn, R. J. Dodson, D. H. Haft, E. K. Hickey, J. D. Peterson, W. C. Nelson, K. A. Ketchum, L. McDonald, T. R. Utterback, J. A. Malek, K. D. Linher, M. M. Garrett, A. M. Stewart, M. D. Cotton, M. S. Pratt, C. A. Phillips, D. Richardson, J. Heidelberg, G. G. Sutton, R. D. Fleischmann, J. A. Eisen, O. White, S. L. Salzberg, H. O. Smith, J. C. Venter, and C. M. Fraser. (1999) "Evidence for Lateral Gene Transfer

Between Archaea and Bacteria from Genome Sequence of Thermotoga Maritima." *Nature* 399: 323–329.

Netzer, N., J. M. Goodenbour, A. David, K. A. Dittmar, R. B. Jones, J. R. Schneider, D. Boone, E. M. Eves, M. R. Rosner, J. S. Gibbs, A. Embry, B. Dolan, S. Das, H. D. Hickman, P. Berglund, J. R. Bennink, J. W. Yewdell, and T. Pan. (2009) "Innate Immune and Chemically Triggered Oxidative Stress Modifies Translational Fidelity." *Nature* 462: 522–526.

Neupert, W., and J. M. Herrmann. (2007) "Translocation of Proteins into Mitochondria." *Annu Rev Biochem* 76: 723–749.

Novichkov, P. S., M. V. Omelchenko, M. S. Gelfand, A. A. Mironov, Y. I. Wolf, and E. V. Koonin. (2004) "Genome-Wide Molecular Clock and Horizontal Gene Transfer in Bacterial Evolution." *J Bacteriol,* in press.

Novichkov, P. S., Y. I. Wolf, I. Dubchak, and E. V. Koonin. (2009) "Trends in Prokaryotic Evolution Revealed by Comparison of Closely Related Bacterial and Archaeal Genomes." *J Bacteriol* 191: 65–73.

Novozhilov, A. S., and E. V. Koonin. (2009) "Exceptional Error Minimization in Putative Primordial Genetic Codes." *Biol Direct* 4: 44.

Nunoura, T., Y. Takaki, J. Kakuta, S. Nishi, J. Sugahara, H. Kazama, G. J. Chee, M. Hattori, A. Kanai, H. Atomi, K. Takai, and H. Takami. (2010) "Insights into the Evolution of Archaea and Eukaryotic Protein Modifier Systems Revealed by the Genome of a Novel Archaeal Group." *Nucleic Acids Res* Dec. 15 [Epub ahead of print].

O'Maille, P. E., A. Malone, N. Dellas, B. Andes Hess Jr., L. Smentek, I. Sheehan, B. T. Greenhagen, J. Chappell, G. Manning, and J. P. Noel. (2008) "Quantitative Exploration of the Catalytic Landscape Separating Divergent Plant Sesquiterpene Synthases." *Nat Chem Biol* 4: 617–623.

O'Malley, M. A. (2009) "What Did Darwin Say About Microbes, and How Did Microbiology Respond?" *Trends Microbiol* 17: 341–347.

O'Malley, M. A., ed. (2010) *Special Issue: The Tree of Life. Biology and Philosophy.*

O'Malley, M. A., and Y. Boucher. (2005) "Paradigm Change in Evolutionary Microbiology." *Stud Hist Philos Biol Biomed Sci* 36: 183–208.

Ochman, H. (2002) "Distinguishing the Orfs from the Elfs: Short Bacterial Genes and the Annotation of Genomes." *Trends Genet* 18: 335–337.

Ohno, S. (1970). *Evolution by Gene Duplication.* New York: Springer-Verlag.

Ohta, T. (2002) "Near-Neutrality in Evolution of Genes and Gene Regulation." *Proc Natl Acad Sci USA* 99: 16,134–16,137.

Omelchenko, M. V., M. Y. Galperin, Y. I. Wolf, and E. V. Koonin. (2010) "Non-Homologous Isofunctional Enzymes: A Systematic Analysis of Alternative Solutions in Enzyme Evolution." *Biol Direct* 5: 31.

Omelchenko, M. V., Y. I. Wolf, E. K. Gaidamakova, V. Y. Matrosova, A. Vasilenko, M. Zhai, M. J. Daly, E. V. Koonin, and K. S. Makarova. (2005) "Comparative Genomics of Thermus Thermophilus and Deinococcus Radiodurans: Divergent Routes of Adaptation to Thermophily and Radiation Resistance." *BMC Evol Biol* 5: 57.

Oparin, A. I. (1924) *The Origin of Life*. Moscow: Moscow Worker.

Oparin, A. I., and V. V. Fesenkov. (1956). *Life in the Universe*. Moscow: USSR Academy of Sciences Publisher.

Orgel, L. E. (1968) "Evolution of the Genetic Apparatus." *J Mol Biol* 38: 381–393.

Orgel, L. E., and F. H. Crick. (1980) "Selfish DNA: The Ultimate Parasite." *Nature* 284: 604–607.

Ostrowski, E. A., R. J. Woods, and R. E. Lenski. (2008) "The Genetic Basis of Parallel and Divergent Phenotypic Responses in Evolving Populations of Escherichia Coli." *Proc Biol Sci* 275: 277–284.

Ouzounis, C. A., V. Kunin, N. Darzentas, and L. Goldovsky. (2006) "A Minimal Estimate for the Gene Content of the Last Universal Common Ancestor—Exobiology from a Terrestrial Perspective." *Res Microbiol* 157: 57–68.

Pace, N. R. (1997) "A Molecular View of Microbial Diversity and the Biosphere." *Science* 276: 734–740.

Pace, N. R. (2006) "Time for a Change." *Nature* 441: 289.

Pace, N. R. (2009a) "Mapping the Tree of Life: Progress and Prospects." *Microbiol Mol Biol Rev* 73: 565–576.

Pace, N. R. (2009b) "Problems with 'Procaryote.'" *J Bacteriol* 191: 2,008–2,010; discussion 2,011.

Pais, A. (1994) *Niels Bohr's Times: In Physics, Philosophy and Polity*. Oxford: Oxford University Press.

Pal, C., B. Papp, and L. D. Hurst. (2001) "Highly Expressed Genes in Yeast Evolve Slowly." *Genetics* 158: 927–931.

Pal, C., B. Papp, and M. J. Lercher. (2005) "Adaptive Evolution of Bacterial Metabolic Networks by Horizontal Gene Transfer." *Nat Genet* 37: 1,372–1,375.

Pande, V. S., A. Grosberg, T. Tanaka, and D. S. Rokhsar. (1998) "Pathways for Protein Folding: Is a New View Needed?" *Curr Opin Struct Biol* 8: 68–79.

Patterson, D. J. (1999) "The Diversity of Eukaryotes." *Am Nat* 154: S96–124.

Paul, J. H. (2008) "Prophages in Marine Bacteria: Dangerous Molecular Time Bombs or the Key to Survival in the Seas?" *ISME J* 2: 579–589.

Pennisi, E. (1999) "Is It Time to Uproot the Tree of Life?" *Science* 284: 1,305–1,307.

Pennisi, E. (2009) "History of Science. The Case of the Midwife Toad: Fraud or Epigenetics?" *Science* 325: 1,194–1,195.

Penny, D. (2005) "An Interpretative Review of the Origin of Life Research." *Biol Philos*, in press.

Pereto, J., P. Lopez-Garcia, and D. Moreira. (2004) "Ancestral Lipid Biosynthesis and Early Membrane Evolution." *Trends Biochem Sci* 29: 469–477.

Perna, N. T., G. Plunkett III, V. Burland, B. Mau, J. D. Glasner, D. J. Rose, G. F. Mayhew, P. S. Evans, J. Gregor, H. A. Kirkpatrick, G. Posfai, J. Hackett, S. Klink, A. Boutin, Y. Shao, L. Miller, E. J. Grotbeck, N. W. Davis, A. Lim, E. T.

Dimalanta, K. D. Potamousis, J. Apodaca, T. S. Anantharaman, J. Lin, G. Yen, D. C. Schwartz, R. A. Welch, and F. R. Blattner. (2001) "Genome Sequence of Enterohaemorrhagic Escherichia Coli O157:H7." *Nature* 409: 529–533.

Pisani, D., J. A. Cotton, and J. O. McInerney. (2007) "Supertrees Disentangle the Chimerical Origin of Eukaryotic Genomes." *Mol Biol Evol* 24: 1,752–1,760.

Poole, A. M. (2009) "Horizontal Gene Transfer and the Earliest Stages of the Evolution of Life." *Res Microbiol* 160: 473–480.

Poole, A. M., D. C. Jeffares, and D. Penny. (1998) "The Path from the Rna World." *J Mol Evol* 46: 1–17.

Poole, A. M., and D. Penny. (2007) "Evaluating Hypotheses for the Origin of Eukaryotes." *Bioessays* 29: 74–84.

Popper, K. R. (1982) *Unended Quest: An Intellectual Autobiography*. London: Open Court Publishing Company.

Prangishvili, D., P. Forterre, and R. A. Garrett. (2006a) "Viruses of the Archaea: A Unifying View." *Nat Rev Microbiol* 4: 837–848.

Prangishvili, D., R. A. Garrett, and E. V. Koonin. (2006b) "Evolutionary Genomics of Archaeal Viruses: Unique Viral Genomes in the Third Domain of Life." *Virus Res* 117: 52–67.

Price, M.N., P. S. Dehal, and A. P. Arkin. (2010) "FastTree 2—approximately maximum-likelihood trees for large alignments." *PLoS One* 5: e9490.

Prusiner, S. B. (1998) "Prions." *Proc Natl Acad Sci USA* 95: 13,363–13,383.

Puigbo, P., Y. I. Wolf, and E. V. Koonin. (2009) "Search for a 'Tree of Life' in the Thicket of the Phylogenetic Forest." *J Biol* 8: 59.

Puigbo, P., Y. I. Wolf, and E. V. Koonin. (2010) "The Tree and Net Components of Prokaryote Evolution." *Genome Biol Evol* 2: 745–756.

Putnam, N. H., M. Srivastava, U. Hellsten, B. Dirks, J. Chapman, A. Salamov, A. Terry, H. Shapiro, E. Lindquist, V. V. Kapitonov, J. Jurka, G. Genikhovich, I. V. Grigoriev, S. M. Lucas, R. E. Steele, J. R. Finnerty, U. Technau, M. Q. Martindale, and D. S. Rokhsar. (2007) "Sea Anemone Genome Reveals Ancestral Eumetazoan Gene Repertoire and Genomic Organization." *Science* 317: 86–94.

Radman, M. (1975) "SOS Repair Hypothesis: Phenomenology of an Inducible DNA Repair Which Is Accompanied by Mutagenesis." *Basic Life Sci* 5A: 355–367.

Radman, M., I. Matic, and F. Taddei. (1999) "Evolution of Evolvability." *Ann N Y Acad Sci* 870: 146–155.

Ragan, M. A. (2001) "Detection of Lateral Gene Transfer among Microbial Genomes." *Curr Opin Genet Dev* 11: 620–626.

Ragan, M. A., J. O. McInerney, and J. A. Lake, eds. (2009) *Theme Issue: The Network of Life: Genome Beginnings and Evolution. Phil Trans R Soc B.*

Rao, V. B., and M. Feiss. (2008) "The Bacteriophage DNA Packaging Motor." *Annu Rev Genet* 42: 647–681.

Raoult, D. (2010) "The Post-Darwinist Rhizome of Life." *Lancet* 375: 104–105.

Raoult, D., S. Audic, C. Robert, C. Abergel, P. Renesto, H. Ogata, B. La Scola, M. Suzan, and J. M. Claverie. (2004) "The 1.2-Megabase Genome Sequence of Mimivirus." *Science* 306: 1,344–1,350.

Raoult, D., and P. Forterre. (2008) "Redefining Viruses: Lessons from Mimivirus." *Nat Rev Microbiol* 6: 315–319.

Ravasi, T., H. Suzuki, C. V. Cannistraci, S. Katayama, V. B. Bajic, K. Tan, A. Akalin, S. Schmeier, M. Kanamori-Katayama, N. Bertin, P. Carninci, C. O. Daub, A. R. Forrest, J. Gough, S. Grimmond, J. H. Han, T. Hashimoto, W. Hide, O. Hofmann, A. Kamburov, M. Kaur, H. Kawaji, A. Kubosaki, T. Lassmann, E. van Nimwegen, C. R. MacPherson, C. Ogawa, A. Radovanovic, A. Schwartz, R. D. Teasdale, J. Tegner, B. Lenhard, S. A. Teichmann, T. Arakawa, N. Ninomiya, K. Murakami, M. Tagami, S. Fukuda, K. Imamura, C. Kai, R. Ishihara, Y. Kitazume, J. Kawai, D. A. Hume, T. Ideker, and Y. Hayashizaki. (2010) "An Atlas of Combinatorial Transcriptional Regulation in Mouse and Man." *Cell* 140: 744–752.

Raymond, J., O. Zhaxybayeva, J. P. Gogarten, S. Y. Gerdes, and R. E. Blankenship. (2002) "Whole-Genome Analysis of Photosynthetic Prokaryotes." *Science* 298: 1,616–1,620.

Rees, M. (2001) *Our Cosmic Habitat*. Princeton: Princeton University Press.

Richards, E. J. (2006) "Inherited Epigenetic Variation—Revisiting Soft Inheritance." *Nat Rev Genet* 7: 395–401.

Richards, T. A., and T. Cavalier-Smith. (2005) "Myosin Domain Evolution and the Primary Divergence of Eukaryotes." *Nature* 436: 1,113–1,118.

Ridley, M. (2006) *Genome*. New York: Harper Perennial.

Rivera, M. C., and J. A. Lake. (2004) "The Ring of Life Provides Evidence for a Genome Fusion Origin of Eukaryotes." *Nature* 431: 152–155.

Robertson, M. P., S. M. Knudsen, and A. D. Ellington. (2004) "In Vitro Selection of Ribozymes Dependent on Peptides for Activity." *RNA* 10: 114–127.

Roger, A. J. (1999) "Reconstructing Early Events in Eukaryotic Evolution." *Am Nat* 154: S146–163.

Rogozin, I. B., M. K. Basu, M. Csuros, and E. V. Koonin. (2009) "Analysis of Rare Genomic Changes Does Not Support the Unikont-Bikont Phylogeny and Suggests Cyanobacterial Symbiosis As the Point of Primary Radiation of Eukaryotes." *Genome Biol Evol* 1: 99–113.

Rogozin, I. B., K. S. Makarova, J. Murvai, E. Czabarka, Y. I. Wolf, R. L. Tatusov, L. A. Szekely, and E. V. Koonin. (2002) "Connected Gene Neighborhoods in Prokaryotic Genomes." *Nucleic Acids Res* 30: 2,212–2,223.

Rogozin, I. B., Y. I. Wolf, A. V. Sorokin, B. G. Mirkin, and E. V. Koonin. (2003) "Remarkable Interkingdom Conservation of Intron Positions and Massive, Lineage-Specific Intron Loss and Gain in Eukaryotic Evolution." *Curr Biol* 13: 1,512–1,517.

Rokas, A., and S. B. Carroll. (2006) "Bushes in the Tree of Life." *PLoS Biol* 4: e352.

Rosenberg, S. M. (2001) "Evolving Responsively: Adaptive Mutation." *Nat Rev Genet* 2: 504–515.

Russell, M. J. (2007) "The Alkaline Solution to the Emergence of Life: Energy, Entropy, and Early Evolution." *Acta Biotheor* 55: 133–179.

Russell, M. J., and A. J. Hall. (1997) "The Emergence of Life from Iron Monosulphide Bubbles at a Submarine Hydrothermal Redox and Ph Front." *J Geol Soc London* 154: 377–402.

Russell, R. (2008) "RNA Misfolding and the Action of Chaperones." *Front Biosci* 13: 1–20.

Rutherford, S., Y. Hirate, and B. J. Swalla. (2007) "The Hsp90 Capacitor, Developmental Remodeling, and Evolution: The Robustness of Gene Networks and the Curious Evolvability of Metamorphosis." *Crit Rev Biochem Mol Biol* 42: 355–372.

Sagan, L. (1967) "On the Origin of Mitosing Cells." *J Theor Biol* 14: 255–274.

Salgado, H., G. Moreno-Hagelsieb, T. F. Smith, and J. Collado-Vides. (2000) "Operons in Escherichia Coli: Genomic Analyses and Predictions." *Proc Natl Acad Sci USA* 97: 6,652–6,657.

Sapp, J. (2009) *The New Foundations of Evolution: On the Tree of Life*. Oxford: Oxford University Press.

Saxinger, C., and C. Ponnamperuma. (1974) "Interactions Between Amino Acids and Nucleotides in the Prebiotic Milieu." *Orig Life* 5: 189–200.

Schrimpf, S. P., M. Weiss, L. Reiter, C. H. Ahrens, M. Jovanovic, J. Malmstrom, E. Brunner, S. Mohanty, M. J. Lercher, P. E. Hunziker, R. Aebersold, C. von Mering, and M. O. Hengartner. (2009) "Comparative Functional Analysis of the Caenorhabditis Elegans and Drosophila Melanogaster Proteomes." *PLoS Biol* 7: e48.

Schroedinger, E. *What Is Life?: With "Mind and Matter" and "Autobiographical Sketches."* Cambridge: Cambridge University Press.

Sella, G., and A. E. Hirsh. (2005) "The Application of Statistical Physics to Evolutionary Biology." *Proc Natl Acad Sci USA* 102: 9,541–9,546.

Seshasayee, A. S., G. M. Fraser, and N. M. Luscombe. (2010) "Comparative Genomics of Cyclic-Di-Gmp Signalling in Bacteria: Post-Translational Regulation and Catalytic Activity." *Nucleic Acids Res* 38: 5,970–5,981.

Shabalina, S. A., and E. V. Koonin. (2008) "Origins and Evolution of Eukaryotic RNA Interference." *Trends Ecol Evol* 23: 578–587.

Shiflett, A. M., and P. J. Johnson. (2010) "Mitochondrion-Related Organelles in Eukaryotic Protists." *Annu Rev Microbiol* 64: 409–429.

Shnol, S. E. (2001) *Heroes, Villains, Conformists of Russian Science*. Moscow: Kron-Press.

Sicheritz-Ponten, T., and S. G. Andersson. (2001) "A Phylogenomic Approach to Microbial Evolution." *Nucleic Acids Res* 29: 545–552.

Simpson, G. G. (1983) *Tempo and Mode in Evolution*. New York: Columbia University Press.

Smith, M. W., D. F. Feng, and R. F. Doolittle. (1992) "Evolution by Acquisition: The Case for Horizontal Gene Transfers." *Trends Biochem Sci* 17: 489–493.

Smolin, L. (1999) *The Life of the Cosmos*. Oxford: Oxford University Press.

Sniegowski, P. D., P. J. Gerrish, and R. E. Lenski. (1997) "Evolution of High Mutation Rates in Experimental Populations of E. coli." *Nature* 387: 703–705.

Snel, B., P. Bork, and M. A. Huynen. (2002) "Genomes in Flux: The Evolution of Archaeal and Proteobacterial Gene Content." *Genome Res* 12: 17–25.

Soyfer, V. N. (1994) *Lysenko and the Tragedy of Soviet Science*. New Brunswick, NJ: Rutgers University Press.

Soyfer, V. N. (2001) "The Consequences of Political Dictatorship for Russian Science." *Nat Rev Genet* 2: 723–729.

Spiegelman, S. (1971) "An Approach to the Experimental Analysis of Precellular Evolution." *Q Rev Biophys* 4: 213–253.

Srivastava, M., E. Begovic, J. Chapman, N. H. Putnam, U. Hellsten, T. Kawashima, A. Kuo, T. Mitros, A. Salamov, M. L. Carpenter, A. Y. Signorovitch, M. A. Moreno, K. Kamm, J. Grimwood, J. Schmutz, H. Shapiro, I. V. Grigoriev, L. W. Buss, B. Schierwater, S. L. Dellaporta, and D. S. Rokhsar. (2008) "The Trichoplax Genome and the Nature of Placozoans." *Nature* 454: 955–960.

Srivastava, M., O. Simakov, J. Chapman, B. Fahey, M. E. Gauthier, T. Mitros, G. S. Richards, C. Conaco, M. Dacre, U. Hellsten, C. Larroux, N. H. Putnam, M. Stanke, M. Adamska, A. Darling, S. M. Degnan, T. H. Oakley, D. C. Plachetzki, Y. Zhai, M. Adamski, A. Calcino, S. F. Cummins, D. M. Goodstein, C. Harris, D. J. Jackson, S. P. Leys, S. Shu, B. J. Woodcroft, M. Vervoort, K. S. Kosik, G. Manning, B. M. Degnan, and D. S. Rokhsar. (2010) "The Amphimedon Queenslandica Genome and the Evolution of Animal Complexity." *Nature* 466: 720–726.

Stalder, L., and O. Muhlemann. (2008) "The Meaning of Nonsense." *Trends Cell Biol* 18: 315–321.

Stanier, R. Y., and C. B. Van Niel. (1962) "The Concept of a Bacterium." *Arch Mikrobiol* 42: 17–35.

Stechmann, A., and T. Cavalier-Smith. (2003) "The Root of the Eukaryote Tree Pinpointed." *Curr Biol* 13: R665–666.

Stern, A., L. Keren, O. Wurtzel, G. Amitai, and R. Sorek. (2010) "Self-Targeting by CRISPR: Gene Regulation or Autoimmunity?" *Trends Genet* 26: 335–340.

Stoltzfus, A. (1999) "On the Possibility of Constructive Neutral Evolution." *J Mol Evol* 49: 169–181.

Stover, C. K., X. Q. Pham, A. L. Erwin, S. D. Mizoguchi, P. Warrener, M. J. Hickey, F. S. Brinkman, W. O. Hufnagle, D. J. Kowalik, M. Lagrou, R. L. Garber, L. Goltry, E. Tolentino, S. Westbrock-Wadman, Y. Yuan, L. L. Brody, S. N. Coulter, K. R. Folger, A. Kas, K. Larbig, R. Lim, K. Smith, D. Spencer, G. K. Wong, Z. Wu, I. T. Paulsen, J. Reizer, M. H. Saier, R. E. Hancock, S. Lory, and M. V. Olson. (2000) "Complete Genome Sequence of Pseudomonas Aeruginosa Pao1, an Opportunistic Pathogen." *Nature* 406: 959–964.

Sullivan, M. B., D. Lindell, J. A. Lee, L. R. Thompson, J. P. Bielawski, and S. W. Chisholm. (2006) "Prevalence and Evolution of Core Photosystem II Genes in Marine Cyanobacterial Viruses and Their Hosts." *PLoS Biol* 4: e234.

Susskind, L. (2003) "The Anthropic Landscape of String Theory." *arXiv: hep-th/0302219.*

Susskind, L. (2006a) *The Cosmic Landscape. String Theory and the Illusion of Intelligent Design.* New York-Boston: Little, Brown and Company.

Susskind, L. (2006b) *The Cosmic Landscape: String Theory and the Illusion of Intelligent Design.* San Francisco: Back Bay Books.

Suttle, C. A. (2005) "Viruses in the Sea." *Nature* 437: 356–361.

Suttle, C. A. (2007) "Marine Viruses—Major Players in the Global Ecosystem." *Nat Rev Microbiol* 5: 801–812.

Sydow, J. F., and P. Cramer. (2009) "RNA Polymerase Fidelity and Transcriptional Proofreading." *Curr Opin Struct Biol* 19: 732–739.

Syvanen, M. (1994) "Horizontal Gene Transfer: Evidence and Possible Consequences." *Annu Rev Genet* 28: 237–261.

Szathmary, E. (1993) "Coding Coenzyme Handles: A Hypothesis for the Origin of the Genetic Code." *Proc Natl Acad Sci USA* 90: 9,916–9,920.

Szathmary, E. (1999) "The Origin of the Genetic Code: Amino Acids as Cofactors in an RNA World." *Trends Genet* 15: 223–229.

Szathmary, E. (2000) "The Evolution of Replicators." *Philos Trans R Soc Lond B Biol Sci* 355: 1,669–1,676.

Szathmary, E., and L. Demeter. (1987) "Group Selection of Early Replicators and the Origin of Life." *J Theor Biol* 128: 463–486.

Szathmary, E., and J. Maynard Smith. (1997) "From Replicators to Reproducers: The First Major Transitions Leading to Life." *J Theor Biol* 187: 555–571.

Takeuchi, N., and P. Hogeweg. (2008) "Evolution of Complexity in RNA-Like Replicator Systems." *Biol Direct* 3: 11.

Takeuchi, N., P. Hogeweg, and E. V. Koonin. (2011) "On the Origin of DNA Genomes: Evolution of the Division of Labor Between Template and Catalyst in Model Replicator Systems." *PLoS Comput Biol,* in press.

Tatusov, R. L., N. D. Fedorova, J. D. Jackson, A. R. Jacobs, B. Kiryutin, E. V. Koonin, D. M. Krylov, R. Mazumder, S. L. Mekhedov, A. N. Nikolskaya, B. S. Rao, S. Smirnov, A. V. Sverdlov, S. Vasudevan, Y. I. Wolf, J. J. Yin, and D. A. Natale. (2003) "The Cog Database: An Updated Version Includes Eukaryotes." *BMC Bioinformatics* 4: 41.

Tatusov, R. L., E. V. Koonin, and D. J. Lipman. (1997) "A Genomic Perspective on Protein Families." *Science* 278: 631–637.

Tejero, H., A. Marin, and F. Montero. (2011) "The Relationship Between Error Catastrophe, Survival of the Flattest, and Natural Selection." *BMC Evol Biol* 11: 2.

Theobald, D. L. (2010) "A Formal Test of the Theory of Universal Common Ancestry." *Nature* 465: 219–222.

Toor, N., K. S. Keating, S. D. Taylor, and A. M. Pyle. (2008) "Crystal Structure of a Self-Spliced Group Ii Intron." *Science* 320: 77–82.

Treangen, T. J., and E. P. Rocha. (2011) "Horizontal Transfer, Not Duplication, Drives the Expansion of Protein Families in Prokaryotes." *PLoS Genet* 7: e1001284.

Trifonov, E. N. (2004) "The Triplet Code from First Principles." *J Biomol Struct Dyn* 22: 1–11.

Trifonov, E. N., I. Gabdank, D. Barash, and Y. Sobolevsky. (2006) "Primordia Vita. Deconvolution from Modern Sequences." *Orig Life Evol Biosph* 36: 559–565.

Tuite, M. F., and T. R. Serio. (2010) "The Prion Hypothesis: From Biological Anomaly to Basic Regulatory Mechanism." *Nat Rev Mol Cell Biol* 11: 823–833.

Turk, R. M., N. V. Chumachenko, and M. Yarus. (2010) "Multiple Translational Products from a Five-Nucleotide Ribozyme." *Proc Natl Acad Sci USA* 107: 4,585–4,589.

Tyedmers, J., M. L. Madariaga, and S. Lindquist. (2008) "Prion Switching in Response to Environmental Stress." *PLoS Biol* 6: e294.

Ulrich, L. E., E. V. Koonin, and I. B. Zhulin. (2005) "One-Component Systems Dominate Signal Transduction in Prokaryotes." *Trends Microbiol* 13: 52–56.

Vabulas, R. M., S. Raychaudhuri, M. Hayer-Hartl, and F. U. Hartl. (2010) "Protein Folding in the Cytoplasm and the Heat Shock Response." *Cold Spring Harb Perspect Biol* 2: a004390.

van den Born, E., M. V. Omelchenko, A. Bekkelund, V. Leihne, E. V. Koonin, V. V. Dolja, and P. O. Falnes. (2008) "Viral Alkb Proteins Repair RNA Damage by Oxidative Demethylation." *Nucleic Acids Res* 36: 5,451–5,461.

van der Giezen, M. (2009) "Hydrogenosomes and Mitosomes: Conservation and Evolution of Functions." *J Eukaryot Microbiol* 56: 221–231.

van der Oost, J., M. M. Jore, E. R. Westra, M. Lundgren, and S. J. Brouns. (2009) "CRISPR-Based Adaptive and Heritable Immunity in Prokaryotes." *Trends Biochem Sci* 34: 401–407.

Van Etten, J. L., L. C. Lane, and D. D. Dunigan. (2010) "DNA Viruses: The Really Big Ones (Giruses)." *Annu Rev Microbiol* 64: 83–99.

Van Melderen, L. (2010) "Toxin–Antitoxin Systems: Why So Many, What For?" *Curr Opin Microbiol* 13: 781–785.

Van Melderen, L., and M. Saavedra De Bast. (2009) "Bacterial Toxin–Antitoxin Systems: More Than Selfish Entities?" *PLoS Genet* 5: e1000437.

Van Niel, C. B. (1955) "Natural Selection in the Microbial World." *J Gen Microbiol* 13: 201–217.

van Nimwegen, E. (2003) "Scaling Laws in the Functional Content of Genomes." *Trends Genet* 19: 479–484.

Van Valen, L. (1973) "A New Evolutionary Law." *Evol. Tehory* 1: 1–30.

Vargas, A. O. (2009) "Did Paul Kammerer Discover Epigenetic Inheritance? A Modern Look at the Controversial Midwife Toad Experiments." *J Exp Zool B Mol Dev Evol* 312: 667–678.

Venn, J. (1866) *The Logic of Chance: An Essay on the Foundations and Province of the Theory of Probability, with Especial Reference to its Application to Moral and Socila Science*. London and Cambridge: MacMillan and Co.

Venters, B. J., and B. F. Pugh. (2009) "How Eukaryotic Genes Are Transcribed." *Crit Rev Biochem Mol Biol* 44: 117–141.

Veretnik, S., C. Wills, P. Youkharibache, R. E. Valas, and P. E. Bourne. (2009) "Sm/Lsm Genes Provide a Glimpse into the Early Evolution of the Spliceosome." *PLoS Comput Biol* 5: e1000315.

Vetsigian, K., C. Woese, and N. Goldenfeld. (2006) "Collective Evolution and the Genetic Code." *Proc Natl Acad Sci USA* 103: 10,696–10,701.

Vilenkin, A. (1983) "The Birth of Inflationary Universes." *Phys. Rev. D* 27: 2,848.

Vilenkin, A. (2007) *Many Worlds in One: The Search for Other Universes*. Boston: Hill and Wang.

Vogel, G. (1997) "Prusiner Recognized for Once-Heretical Prion Theory." *Science* 278: 214.

Vogel, G., and E. Pennisi. (2009) "Physiology Nobel. U.S. Researchers Recognized for Work on Telomeres." *Science* 326: 212–213.

Volker, C., and A. N. Lupas. (2002) "Molecular Evolution of Proteasomes." *Curr Top Microbiol Immunol* 268: 1–22.

von Dohlen, C. D., S. Kohler, S. T. Alsop, and W. R. McManus. (2001) "Mealybug Beta-Proteobacterial Endosymbionts Contain Gamma-Proteobacterial Symbionts." *Nature* 412: 433–436.

Von Neumann, J. (1966) *Theory of Self-Reproducing Automata*. Urbana, IL: University of Illinois Press.

Wachtershauser, G. (1997) "The Origin of Life and Its Methodological Challenge." *J Theor Biol* 187: 483–494.

Waddington, C. H., and E. Robertson. (1966) "Selection for Developmental Canalisation." *Genet Res* 7: 303–312.

Wagner, A. (2008a) "Neutralism and Selectionism: A Network-Based Reconciliation." *Nat Rev Genet* 9: 965–974.

Wagner, A. (2008b) "Robustness and Evolvability: A Paradox Resolved." *Proc Biol Sci* 275: 91–100.

Walker, J. E., M. Saraste, M. J. Runswick, and N. J. Gay. (1982) "Distantly Related Sequences in the Alpha- and Beta-Subunits of Atp Synthase, Myosin, Kinases, and Other Atp-Requiring Enzymes and a Common Nucleotide Binding Fold." *Embo J* 1: 945–951.

Wang, E. T., R. Sandberg, S. Luo, I. Khrebtukova, L. Zhang, C. Mayr, S. F. Kingsmore, G. P. Schroth, and C. B. Burge. (2008) "Alternative Isoform Regulation in Human Tissue Transcriptomes." *Nature* 456: 470–476.

Wang, X., Y. Kim, Q. Ma, S. H. Hong, K. Pokusaeva, J. M. Sturino, and T. K. Wood. (2010) "Cryptic Prophages Help Bacteria Cope with Adverse Environments." *Nat Commun* 1: 147.

Wang, Z., and J. Zhang. (2009) "Why Is the Correlation Between Gene Importance and Gene Evolutionary Rate So Weak?" *PLoS Genet* 5: e1000329.

Watson, J. D., and F. H. Crick. (1953a) "Genetical Implications of the Structure of Deoxyribonucleic Acid." *Nature* 171: 964–967.

Watson, J. D., and F. H. Crick. (1953b) "Molecular Structure of Nucleic Acids; a Structure for Deoxyribose Nucleic Acid." *Nature* 171: 737–738.

Watts, D. J. (2004) *Six Degrees: The Science of a Connected Age*. New York: W. W. Norton & Co.

Weinberg, S. (1994) *Dreams of a Final Theory: The Scientist's Search for the Ultimate Laws of Nature*. New York: Vintage.

Weinreich, D. M., N. F. Delaney, M. A. Depristo, and D. L. Hartl. (2006) "Darwinian Evolution Can Follow Only Very Few Mutational Paths to Fitter Proteins." *Science* 312: 111–114.

Weissmann, A. (1893) *The Germ-Plasm. A Theory of Heredity.* London: Charles Scribner's Sons.

Whitehead, D. J., C. O. Wilke, D. Vernazobres, and E. Bornberg-Bauer. (2008) "The Look-Ahead Effect of Phenotypic Mutations." *Biol Direct* 3: 18.

Wilke, C. O., J. L. Wang, C. Ofria, R. E. Lenski, and C. Adami. (2001) "Evolution of Digital Organisms at High Mutation Rates Leads to Survival of the Flattest." *Nature* 412: 331–333.

Wilson, A. C., S. S. Carlson, and T. J. White. (1977) "Biochemical Evolution." *Annu Rev Biochem* 46: 573–639.

Wochner, A., J. Attwater, A. Coulson, and P. Holliger. (2011) "Ribozyme-Catalyzed Transcription of an Active Ribozyme." *Science* 332: 209–212.

Woese, C. (1967) *The Genetic Code*. New York: Harper & Row.

Woese, C. (1998) "The Universal Ancestor." *Proc Natl Acad Sci U S A* 95: 6,854–6,859.

Woese, C. R. (1987) "Bacterial Evolution." *Microbiol Rev* 51: 221–271.

Woese, C. R. (2002) "On the Evolution of Cells." *Proc Natl Acad Sci USA* 99: 8,742–8,747.

Woese, C. R., and G. E. Fox. (1977) "The Concept of Cellular Evolution." *J Mol Evol* 10: 1–6.

Woese, C. R., and N. Goldenfeld. (2009) "How the Microbial World Saved Evolution from the Scylla of Molecular Biology and the Charybdis of the Modern Synthesis." *Microbiol Mol Biol Rev* 73: 14–21.

Woese, C. R., O. Kandler, and M. L. Wheelis. (1990) "Towards a Natural System of Organisms: Proposal for the Domains Archaea, Bacteria, and Eucarya." *Proc Natl Acad Sci USA* 87: 4,576–4,579.

Wohlgemuth, I., M. Beringer, and M. V. Rodnina. (2006) "Rapid Peptide Bond Formation on Isolated 50s Ribosomal Subunits." *EMBO Rep* 7: 699–703.

Wolf, Y. I., L. Aravind, N. V. Grishin, and E. V. Koonin. (1999a) "Evolution of Aminoacyl-tRNA Synthetases—Analysis of Unique Domain Architectures and Phylogenetic Trees Reveals a Complex History of Horizontal Gene Transfer Events." *Genome Res* 9: 689–710.

Wolf, Y. I., S. E. Brenner, P. A. Bash, and E. V. Koonin. (1999b) "Distribution of Protein Folds in the Three Superkingdoms of Life." *Genome Res* 9: 17–26.

Wolf, Y. I., L. Carmel, and E. V. Koonin. (2006) "Unifying Measures of Gene Function and Evolution." *Proc Biol Sci* 273: 1,507–1,515.

Wolf, Y. I., I. V. Gopich, D. J. Lipman, and E. V. Koonin. (2010) "Relative Contributions of Intrinsic Structural-Functional Constraints and Translation Rate to the Evolution of Protein-Coding Genes." *Genome Biol Evol* 2: 190–199.

Wolf, Y. I., and E. V. Koonin. (2007) "On the Origin of the Translation System and the Genetic Code in the RNA World by Means of Natural Selection, Exaptation, and Subfunctionalization." *Biol Direct* 2: 14.

Wolf, Y. I., P. S. Novichkov, G. P. Karev, E. V. Koonin, and D. J. Lipman. (2009) "The Universal Distribution of Evolutionary Rates of Genes and Distinct Characteristics of Eukaryotic Genes of Different Apparent Ages." *Proc Natl Acad Sci USA* 106: 7,273–7,280.

Wolf, Y. I., I. B. Rogozin, N. V. Grishin, and E. V. Koonin. (2002) "Genome Trees and the Tree of Life." *Trends Genet* 18: 472–479.

Wolf, Y. I., I. B. Rogozin, A. S. Kondrashov, and E. V. Koonin. (2001) "Genome Alignment, Evolution of Prokaryotic Genome Organization, and Prediction of Gene Function Using Genomic Context." *Genome Res* 11: 356–372.

Woods, R. J., J. E. Barrick, T. F. Cooper, U. Shrestha, M. R. Kauth, and R. E. Lenski. (2011) "Second-Order Selection for Evolvability in a Large Escherichia Coli Population." *Science* 331: 1,433–1,436.

Woodson, S. A. (2010) "Taming Free Energy Landscapes with RNA Chaperones." *RNA Biol* 7: 38–47.

Wozniak, R. A., and M. K. Waldor. (2010) "Integrative and Conjugative Elements: Mosaic Mobile Genetic Elements Enabling Dynamic Lateral Gene Flow." *Nat Rev Microbiol* 8: 552–563.

Wright, G. D. (2007) "The Antibiotic Resistome: The Nexus of Chemical and Genetic Diversity." *Nat Rev Microbiol* 5: 175–186.

Wu, D., S. C. Daugherty, S. E. Van Aken, G. H. Pai, K. L. Watkins, H. Khouri, L. J. Tallon, J. M. Zaborsky, H. E. Dunbar, P. L. Tran, N. A. Moran, and J. A. Eisen. (2006) "Metabolic Complementarity and Genomics of the Dual Bacterial Symbiosis of Sharpshooters." *PLoS Biol* 4: e188.

Wunderlich, Z., and L. A. Mirny. (2009) "Different Gene Regulation Strategies Revealed by Analysis of Binding Motifs." *Trends Genet* 25: 434–440.

Yang, D., Y. Oyaizu, H. Oyaizu, G. J. Olsen, and C. R. Woese. (1985) "Mitochondrial Origins." *Proc Natl Acad Sci USA* 82: 4,443–4,447.

Yang, Z. (2007) "Paml 4: Phylogenetic Analysis by Maximum Likelihood." *Mol Biol Evol* 24: 1,586–1,591.

Yarus, M. (1998) "Amino Acids as RNA Ligands: A Direct-RNA-Template Theory for the Code's Origin." *J Mol Evol* 47: 109–117.

Yarus, M., J. G. Caporaso, and R. Knight. (2005) "Origins of the Genetic Code: The Escaped Triplet Theory." *Annu Rev Biochem* 74: 179–198.

Yarus, M., J. J. Widmann, and R. Knight. (2009) "RNA–Amino Acid Binding: A Stereochemical Era for the Genetic Code." *J Mol Evol* 69: 406–429.

Yooseph, S., G. Sutton, D. B. Rusch, A. L. Halpern, S. J. Williamson, K. Remington, J. A. Eisen, K. B. Heidelberg, G. Manning, W. Li, L. Jaroszewski, P. Cieplak, C. S. Miller, H. Li, S. T. Mashiyama, M. P. Joachimiak, C. van Belle, J. M. Chandonia, D. A. Soergel, Y. Zhai, K. Natarajan, S. Lee, B. J. Raphael, V. Bafna, R. Friedman, S. E. Brenner, A. Godzik, D. Eisenberg, J. E. Dixon, S. S. Taylor, R. L. Strausberg, M. Frazier, and J. C. Venter. (2007) "The Sorcerer Ii Global Ocean Sampling Expedition: Expanding the Universe of Protein Families." *PLoS Biol* 5: e16.

Yutin, N., K. S. Makarova, S. L. Mekhedov, Y. I. Wolf, and E. V. Koonin. (2008) "The Deep Archaeal Roots of Eukaryotes." *Mol Biol Evol* 25: 1,619–1,630.

Yutin, N., M. Y. Wolf, Y. I. Wolf, and E. V. Koonin. (2009) "The Origins of Phagocytosis and Eukaryogenesis." *Biol Direct* 4: 9.

Zhang, Y., C. Laing, M. Steele, K. Ziebell, R. Johnson, A. K. Benson, E. Taboada, and V. P. Gannon. (2007) "Genome Evolution in Major Escherichia Coli O157:H7 Lineages." *BMC Genomics* 8: 121.

Zuckerkandl, E., and L. Pauling. (1965). "Evolutionary Divergence and Convergence of Proteins" in Bryson, V., and H. J. Vogel (eds). *Evolving Genes and Proteins*. New York: Academic Press, pp. 97–166.

Endnotes

Chapter 1

1   In a rather remarkable feat of serendipity, Lamarck's *magnum opus, La Philosophie Zoologique,* appeared in print the year Darwin was born.

2   A wonderfully insightful account of the immediate impact and public perception of Darwin's book can be found in John Fowles's novel *The French Lieutenant's Woman* (Fowles, J. 1969. *The French lieutenant's woman.* Boston: Little & Brown).

3   The term *Darwinism* obviously could not have been used by Darwin himself, and it is a rather unfortunate, even if well-intended, neologism of Darwin's prime follower and defender, Thomas Henry Huxley, in a review of *Origin* (Huxley, T. H. 1860. "Darwin on the origin of Species." *Westminster Review:* 541-570). The word seems to have a distinct dogmatic, even pseudo-scientific connotation, by association with other well-known "isms" such as Marxism or Freudianism or even Lysenkoism (see Chapter 9 on that one). Indeed, no one has ever spoken of Newtonism or Einsteinism, whereas Mendelism (typically, as part of a tandem Mendelism-Weissmanism or Mendelism-Morganism) was only used in the dismal context of the Lysenkoist antiscience in the Soviet Union. Having said this, Huxley's term has firmly stuck, and it does have the advantage of brevity. I use it in this book exclusively to describe the "original synthesis" of evolutionary biology that was primarily accomplished by Darwin in *Origin* but was solidified and refined in the subsequent work of Huxley, Wallace, Weissmann, Haeckel, and other early Darwin followers.

4   The phrase "irreducible complexity" was coined by Michael Behe, one of the chief advocates of the antievolutionary intelligent design (ID) concept, in his (in)famous *Darwin's Black Box* book (Behe, M. J. 2006. *Darwin's Black Box: The Biochemical Challenge to Evolution.* New York: Free Press). To Behe and other ID advocates, the "irreducibility" of complex biological structures is evidence

(even proof) of the inevitability of ID. Of course, ID is malicious nonsense, but the term "irreducible complexity" is quite evocative; however, evolutionary biologists might prefer to speak of "apparent" or "purported" irreducibility of complex structures.

5   Sir Ronald Fisher was a true genius. (Fisher Box, J. 1978. *R.A. Fisher: The Life of a Scientist*. New York: Wiley.) He effectively founded not only population genetics, but, in many ways, modern statistics and introduced a mathematical definition of information long before Claude Shannon. We include more examples of his remarkable scientific prescience in this book. Sir Ronald also dedicated much of his career to the cause of eugenics, an endeavor that nowadays is viewed as pseudo-scientific and bordering on criminal. We should be careful not to judge great minds of even the relatively recent past by today's standards.

6   This is done in a variety of textbooks and monographs on all levels, both introductory and highly technical. For a balanced, moderately technical presentation, see D. L. Hartl and A. G. Clark (2006), *Principles of Population Genetics*, Sunderland, MA: Sinauer Associates.

7   In principle, if a fitness landscape is constructed for a gene, its dimensionality would equal the number of nucleotide sites. Interactions between sites (epistasis) decrease the landscape dimensionality.

8   Technically, Fisher's theorem does not prohibit all downward movement because it applies only to the fraction of the fitness change attributable to selection. In practice, however, Fisher thought that most, if not all, populations were far too large for the phenomenon that Wright denoted as drift to be of any importance. This was the subject of a bitter debate between Fisher and Wright. The ultimate winner was certainly Wright.

9   This famous phrase is the title of Dobzhansky's essay published in the *American Biology Teacher* magazine (Dobzhansky, T. 1973. Nothing in biology makes sense except in the light of evolution. The American Biology Teacher 35: 125–129.). Taken in its entirety, the essay is rather astonishing. Although much of the text explains evolutionary concepts with remarkable lucidity, the concluding paragraphs are dedicated to eloquent propaganda of the compatibility of evolution and Christian faith, and it is difficult to avoid the impression that this was the author's principal purpose. According to Dobzhansky, who was a devout Russian Orthodox Christian, God implemented His plan of creation as unraveling the grand scenario of the evolution of life. Moreover, Dobzhansky ingeniously brands denial of evolution a blasphemy because this position implies that God is a cheater who deliberately misleads humankind by presenting them with plentiful evidence of evolution. I suspect that not everyone who quotes Dobzhansky's motto in discussions on teaching evolution has actually read the essay.

Chapter 2

1 By themselves, these principles are elementary textbook knowledge, but the information-theoretical approach to evolution developed here is not as trivial, so I felt it necessary to explicitly restate them.

2 Chargaff did not appreciate the crucial importance of his own finding until it was too late, and the fact that two arrogant youths who knew no chemistry succeeded in discovering the secret of life that he, the expert chemist, failed to grasp embittered Chargaff for the rest of his long life, inspiring his poignant, even if excessively caustic, books. (E. Chargaff. *Heraclitean Fire: Sketches from a Life Before Nature*, New York: Rockefeller University Press, 1978.)

3 It is not entirely clear who was the first to formulate the principle that I here denote as EPR. An account tells of eminent Russian geneticists Nikolai Kol'tzov and Nikolai Timofeev-Resovski expressing this idea in the 1930s (S. E. Shnol, *Heroes, Villains, Conformists of Russian Science*, Moscow: Kron-Press, 2001), but I am unaware of a formal publication. Erwin Schroedinger comes close to the idea in the famous book *What Is Life?: with Mind and Matter and Autobiographical Sketches* (Cambridge University Press, 1992) but does not quite get there. In the English-language literature, Richard Dawkins clearly formulates the idea in the 1976 classic *The Selfish Gene* (R. Dawkins. *The Selfish Gene: 30th Anniversary Edition—with a new Introduction by the Author* Oxford: Oxford University Press, 2006). From a different, abstract standpoint, a completely substrate-free theory of self-reproducing, evolving automata was developed by the great mathematician John von Neumann (*Theory of Self-Reproducing Automata*, Urbana, IL: University of Illinois Press, 1966).

4 This seems to be a good place to say a few words about metaphors in biology, especially as the "selfish gene" has been singled out as a metaphor particularly prone to mislead (Ball, 2011). Beyond doubt, much caution is due in the use and especially interpretation of metaphors, and any shade of anthropomorphic perception of the "selfishness" of genes should be avoided at all costs. Nevertheless, I think that metaphors are necessary for the advancement of science, and as long as science is communicated through natural language (not only mathematical expressions), metaphors are unavoidable. Moreover, a really good metaphor—that is, one that is both brief and catchy, and that captures an important general trend in a broad field of observation—has the potential to greatly stimulate new thinking and research. For example, I believe that "selfish gene," "junk DNA," and "fitness landscape" are excellent metaphors.

5 In the days before complete genome sequences, the vision of dynamic genomes was probably best captured in the comprehensive monograph by the eminent Russian geneticist Roman Khesin (*Inconstancy of the Genome.* Moscow: Nauka, 1984). This prescient book was published shortly before Khesin's untimely death and became rather legendary among Russian biologists. Unfortunately, it did not seem to have much impact outside Russia.

6 Having remarked on Sir Ronald Fisher's involvement with eugenics, it would be unfair not to mention that another founder of population genetics, J. B. S. Haldane, was a long-term member of the British Communist Party and, apparently out of the characteristic party loyalty, for years supported the Lysenkoist

pseudoscience (at least by giving it the benefit of the doubt). Haldane was a scientist of no lesser dimensions than Fisher, one of the last great polymaths in the history of science (K. R. Dronamraju, *Haldane and Modern Biology,* Johns Hopkins University Press, 1968). To his credit are not only numerous important results in mathematical genetics (including the theory of genetic load), but also major contributions to the field of kinetics of enzymatic reactions and, perhaps most important, the plethora of incredibly prescient ideas on a huge variety of subjects in his books and articles (we return to some of these in Chapter 10). Haldane also wrote hundreds of brilliant popular essays on all aspects of science many of which he published in the communist newspaper *Daily Worker.* In 1950, Haldane quit the Communist Party after realizing the degree of devastation that Lysenko and his gang had wrought upon Soviet genetics and geneticists. Haldane's may be a prime example that even the greatest scientists cannot be viewed in isolation from the historical context.

7   Emile Zuckerkandl has pointed out to me that in his early molecular evolution papers with Linus Pauling, the distinction between homologs evolved by vertical descent and those evolved by duplication is drawn clearly, even though no special terms were used to denote these distinct classes of homologs. Every discovery or conceptual breakthrough has its predecessors.

8   *Just So Stories* is a delectable collection of children tales by Rudyard Kipling (R. Kipling. *Just So Stories: for Little Children.* Oxford: Oxford Univ Press, 2009). In it, the origin of some striking features of animal morphology, such as the elephant proboscis and the armadillo shell, is tracked back to various peculiar accidents. Kipling seemingly already realized the fallacy of pan-adaptationism, although his conclusions were not necessarily the same as the conclusions of Gould and Lewontin.

9   I readily confess my partiality toward viruses. In my sophomore year at Moscow State University, I chose the Department of Virology for my major. In part, the choice was dictated by extraneous considerations, such as the apparent interest in real science and liberal atmosphere in this department, which was quite unlike some of the other departments. This was important at the time, and was no mistake. But the more fundamental incentive was my fascination with the diversity of genetic mechanisms and genome organization among viruses, leading to the idea that viruses might be directly relevant for understanding the earliest stages in the evolution of life. I still think this idea is right on the mark, as discussed in Chapters 10 and 11. All experimental work I ever did myself was in virology; it might have been inconsequential by itself, but it was enormously instructive for all subsequent research in computational biology. Perhaps most important, my first forays into comparative genomics, which happened to coincide with the onset of this whole field of inquiry, had to do with virus genomes. These small genomes were the ideal learning ground: Even with the primitive computational means of the time (but with all the eagerness of a novice, of course), one could pretty much study the evolution of every amino acid in viral proteins.

10   Much worse than Fisher or Haldane, Mereschkowsky went public with extremely abhorrent views, with a clearly fascistic slant. Nevertheless, his papers on endosymbiosis strike one as exemplary scholarship to this day. (W. Martin and K. V. Kowallik, "Annotated English Translation of Mereschkowsky's 1905 Paper

'Über Natur und Ursprung der Chromatophoren im Pflanzenreiche,'" *European Journal of Psychology* 34 [1999]: 287–296.)

11 Note the similarity to Wright's fitness landscape—to my knowledge, these are two convergent ideas.

Chapter 3

1 Apparently, the term was first used by German botanist Hans Winkler in 1920 (M. Ridley, *Genome*, New York: Harper Perennial, 2006).

2 The *Oxford Dictionary of the English Language* defines *cog* as follows: "*a wheel or bar with a series of projections on its edge, which transfers motion by engaging with projections on another wheel or bar.*" It defines the phrase "a cog in a machine (or wheel)" as "*a small or insignificant member of a larger organization or system.*"

3 Finding that there is so little gene order conservation between the first sequenced bacterial genomes was so striking that Arcady Mushegian and I entitled a brief article describing this observation "Gene Order Is Not Conserved in Bacterial Evolution" (A. R. Mushegian and E. V. Koonin *Trends in Genetics* 12 (1996a): 289–290). Nothing seems to be wrong with the actual statements in the article, but if I were to publish it again, I would strive for a more nuanced title. The original one captures our astonishment over the dissociation between the conservation of gene sequences and variability of gene order.

4 That is, of course, as long as the biosphere survives.

5 The article describing the genome of *M. genitalium* is entitled "The Minimal Gene Complement of *Mycoplasma genitalium*" (C. M. Fraser, J. D. Gocayne, O. White, M. D. Adams, R. A. Clayton, R. D. Fleischmann, C. J. Bult, A. R. Kerlavage, G. Sutton, J. M. Kelley, et al., *Science* 270 (1995): 397–403). However, despite having a small number of genes, this organism is quite specialized and cannot be "really" minimal.

Chapter 4

1 To make a long story short, the simplest models of protein folding caricature the process by fitting the toy "sequence" onto an orthogonal lattice. Somewhat more realistic (or, rather, less unrealistic) models give up the lattice to allow free-folding of the modeled polymer chain. This technique better approximates the real protein-folding process but is much more expensive computationally. A number of considerations suggested that the lattice approximation would be too crude for the task discussed here, so we employed an off-lattice model.

2 Most readers will recall Kevin Bacon's six degrees of separation, and even more are familiar with in-flight magazines that are freely provided by most airlines. Next time you are on a plane, take a moment to look at the inevitable map of airline connections at the back of the magazine; that is an excellent example of a scale-free network, with airline hubs such as Atlanta, Chicago, or Denver. And,

of course, the Internet is a scale-free network as well. Fascinating—and technically accurate—discussions of networks in all spheres of life can be found in popular books by Albert-Laszlo Barabasi, one of the pioneers of network biology (*Linked: The New Science of Networks,* New York: Perseus Press, 2002) and Duncan Watts (*Six Degrees: The Science of a Connected Age,* New York: W.W. Norton & Co., 2004).

3   Every important discovery has its predecessors—one only has to search with some care. I believe this to be one of the "universal laws" of history of science. It seems that "Van Nimwegen law" was first described in the article on the genome of the bacterium *Pseudomonas aeruginosa,* without much ado or much analysis (C. K. Stover, X. Q. Pham, A. L. Erwin, S. D. Mizoguchi, P. Warrener, M. J. Hickey, F. S. Brinkman, W. O. Hufnagle, D. J. Kowalik, M. Lagrou, R. L. Garber, L. Goltry, E. Tolentino, S. Westbrock-Wadman, Y. Yuan, L. L. Brody, S. N. Coulter, K. R. Folger, A. Kas, K. Larbig, R. Lim, K. Smith, D. Spencer, G. K. Wong, Z. Wu, I. T. Paulsen, J. Reizer, M. H. Saier, R. E. Hancock, S. Lory, and M. V. Olson, "Complete Genome Sequence of Pseudomonas Aeruginosa PAO1, an Opportunistic Pathogen," *Nature* 406 (2000): 959–964).

Chapter 5

1   Actually, although Darwin did not discuss microbes in print, some of his letters show considerable interest and insights in this subject (M. A. O'Malley, "What Did Darwin Say About Microbes, and How Did Microbiology Respond?" *Trends in Microbiology* 17 [2009]: 341–347).

2   In all likelihood, this is a biased perspective, but to me, the excitement about these first genomes was next to none.

3   Finding an obvious bimodal distribution of any quantity is rare anywhere in nature—this really hints that "something is going on."

4   An alternative reading of *F* in ELFs should be obvious.

5   The first publication on the genome sequence of a hyperthermophilic bacterium, *Aquifex aeolicus,* failed to report the excess of "archaeal" genes and actually explicitly claimed the lack of such an excess (G. Deckert, P. V. Warren, T. Gaasterland, W. G. Young, A. L. Lenox, D. E. Graham, R. Overbeek, M. A. Snead, M. Keller, M. Aujay, R. Huber, R. A. Feldman, J. M. Short, G. J. Olsen, and R. V. Swanson, "The Complete Genome of the Hyperthermophilic Bacterium Aquifex Aeolicus," *Nature* 392 [1998]: 353–358). The idea of possible gene exchange between archaeal and bacterial hyperthermophiles certainly crossed the authors' mind. The only reason it was not detected was that the genome of *Aquifex* is quite small, but the authors neglected to normalize the count of "archaeal" genes by the total number of genes (or genome size). Once done, such normalization immediately revealed the striking prevalence of "archaeal" genes compared to genomes of mesophilic bacteria (L. Aravind, R. L. Tatusov, Y. I. Wolf, D. R. Walker, and E. V. Koonin, "Evidence for Massive Gene Exchange Between Archaeal and Bacterial Hyperthermophiles," *Trends in Genetics* 14 [1998]: 442–444).

Chapter 6

1 This chapter is deliberately brief and cuts to the chase by describing several recent studies that, in my view, reflect the current state of the TOL problem. An interesting historical overview can be found in the book by Jan Sapp that also includes a discussion of the modern status of tree thinking (J. Sapp, *The New Foundations of Evolution: On the Tree of Life* Oxford: Oxford University Press, 2009). An even more detailed panorama of the current view on the TOL is given in two series of articles written by both biologists and philosophers (O'Malley, M. A., ed. (2010) *Special Issue: The Tree of Life. Biology and Philosophy*; Ragan, M. A., J. O. McInerney, and J. A. Lake, eds. (2009) *Theme Issue: The Network of Life: Genome Beginnings and Evolution. Phil Trans R Soc B*).

2 The analogy with the Big Bang models of the beginning of our universe in cosmology is obvious. In Chapter 12 and Appendix B, I touch upon the interpretation of the Big Bang in modern cosmological models.

Chapter 7

1 Species abbreviations: *Aureococcus anophagefferens* (Aano), *Aedes aegypti* (Aaeg), *Agaricus bisporus* (Abis), *Anopheles gambiae* (Agam), *Allomyces macrogynus ATCC 38327* (Amac), *Apis mellifera* (Amel), *Aspergillus nidulans FGSC A4* (Anid), *Acyrthosiphon pisum* (Apis), *Arabidopsis thaliana* (Atha), *Babesia bovis* (Bbov), *Batrachochytrium dendrobatidis* (Bden), *Branchiostoma floridae* (Bflo), *Botryotinia fuckeliana B05.10* (Bfuc), *Brugia malayi* (Bmal), *Bombyx mori* (Bmor), *Coccomyxa sp. C-169* (C169), *Chlorella sp. NC64a* (C64a), *Caenorhabditis briggsae* (Cbri), *Caenorhabditis elegans* (Cele), *Coprinopsis cinerea okayama7#130* (Ccin), *Cochliobolus heterostrophus C5* (Chet), *Coccidioides immitis RS* (Cimm), *Ciona intestinalis* (Cint), *Cryptococcus neoformansvar. neoformans* (Cneo), *Chlamydomonas reinhardtii* (Crei), *Capitella teleta* (Ctel), *Capsaspora owczarzaki ATCC 30864* (Cowc), *Dictyostelium discoideum* (Ddis), *Dictyostelium purpureum* (Dpur), *Drosophila melanogaster* (Dmel), *Drosophila mojavenis* (Dmoj), *Daphnia pulex* (Dpul), *Danio rerio* (Drer), *Entamoeba dispar* (Edis), *Entamoeba histolytica* (Ehis), *Emiliania huxleyi* (Ehux), *Fragilariopsiscylindrus* (Fcyl), *Phanerochaete chrysosporium* (Fchr), *Phaeodactylum tricornutum* (Ftri), *Gallus gallus* (Ggal), *Gibberella zeae PH-1* (Gzea), *Hydra magnipapillata* (Hmag), *Helobdella robusta* (Hrob), *Homo sapiens* (Hsap), *Ixodes scapularis* (Isca), *Laccaria bicolor* (Lbic), *Lottia gigantea* (Lgig), *Micromonas sp. RCC299* (M299), *Monosiga brevicollis* (Mbre), *Mucor circinelloides* (Mcir), *Mycosphaerella fijiensis* (Mfij), *Mycosphaerella graminicola* (Mgra), *Magnaporthe grisea 70-15* (Mgri), *Melampsora laricis-populina* (Mlar), *Micromonas pusilla CCMP1545* (Mpus), *Neurospora crassa OR74A* (Ncra), *Nematostella vectensis* (Nvec), *Nasonia vitripennis* (Nvit), *Ostreococcus sp. RCC809* (O809), *Ostreococcus lucimarinus* (Oluc), *Oryza sativa japonica* (Osat), *Ostreococcus tauri* (Otau), *Phytophthora capsici* (Pcap), *Plasmodium falciparum* (Pfal), *Puccinia graminis* (Pgra), *Pediculus humanus* (Phum), *Phaeosphaeria nodorum SN15* (Pnod), *Physcomitrella patens subsp. patens* (Ppat), *Phytophthora ramorum* (Pram),

Pyrenophora tritici-repentis Pt-1C-BFP (Prep), *Proterospongia sp. ATCC 50818,* (Prsp), *Phytophthora sojae* (Psoj), *Paramecium tetraurelia* (Ptet), *Plasmodium vivax* (Pviv), *Plasmodium yoelii yoelii* (Pyoe), *Rhizopus oryzae* (Rory), *Sorghum bicolor* (Sbic), *Saccharomyces cerevisiae* (Scer), *Schizosaccharomyces japonicas yFS275* (Sjap), *Schistosoma mansoni* (Sman), *Selaginella moellendorffii* (Smoe), *Schizosaccharomyces pombe* (Spom), *Spizellomyces punctatus DAOM BR1173* (Spun), *Strongylocentrotus purpuratus* (Spur), *Sporobolomyces roseus* (Sros), *Sclerotinia sclerotiorum 1980 UF-70* (Sscl), *Trichoplax adhaerens* (Tadh), *Theileria annulata* (Tann), *Tribolium castaneum* (Tcas), *Toxoplasma gondii* (Tgon), *Taenopygia guttata* (Tgut), *Theileria parvum* (Tpar), *Thalassiosira pseudonana* (Tpse), *Tetrahymena thermophila* (Tthe), *Ustilago maydis 521* (Umay), *Uncinocarpus reesii 1704* (Uree), *Volvox carteri* (Vcar), and *Vitis vinifera* (Vvin).

Chapter 8

1   An unwieldy number of (semi) popular books on various aspects of complexity have been published, including one by the great physicist Murray Gell-Mann, the author of the quark theory (M. Gell-Mann, *The Quark and the Jaguar: Adventures in the Simple and the Complex,* New York: St. Martin's Griffin, 1995). The more technical book by Stuart Kauffman presents many original ideas on the evolution of complexity (S. Kauffman, *At Home in the Universe: The Search for the Laws of Self-Organization and Complexity,* Oxford: Oxford University Press, 1996). Another up-to-date, concise introductory text is (N. Johnson, *Simply Complexity: A Clear Guide to Complexity Theory,* New York: Oneworld Publications, 2009).

2   Information is equivalent to Kolmogorov complexity only for strictly random sequences with defined frequencies of the symbols. The genomic sequences are generally not like that—they embody various dependences between nucleotides in different positions. Despite the intuitive appeal of the Kolmogorov complexity concept, there is no general formula to calculate it.

3   These probabilities are not simply frequencies but in theory should come from unbiased statistical models for individual sites that, although never known precisely, can be approximated by various mathematical models and are also approached with the increase of the number of sequences in an alignment.

4   The renowned eighteenth-century natural theologist William Paley saw the problem with perfect clarity when he reasonably submitted that a watch found during a walk in the fields implied a watchmaker. This sensible train of thought led both to the modern ID movement and to the famous retort of Richard Dawkins (R. Dawkins, *The Blind Watchmaker: Why the Evidence of Evolution Reveals a Universe without Design,* London: W.W.Norton & Co., 1996).

5   The metaphor is widely used in mathematical studies of stochastic processes.

6   I do not imply any ridiculous anthropomorphic perception of progress—just the simple intuition that equates progress with gradually increasing complexity.

Chapter 9

1   In the 1880s, the famous German biologist August Weismann, in the context of his theory of germ plasm and germline-soma barrier, set out to directly falsify the inheritance of acquired characters in a series of experiments that became as famous as Lamarck's giraffe (A. Weissmann, *The Germ-Plasm. A Theory of Heredity*, London: Charles Scribner's Sons, 1893). Almost needless to say, cutting tails off Weismann's experimental rats not just failed to produce any tail-less pups, but did not result in any shortening of the tail of the progeny whatsoever. Weismann's experiments delivered a serious blow to the public perception of the inheritance of acquired characters, although, technically, they may be considered irrelevant to Lamarck's concept that, as already mentioned, insisted on the inheritance of beneficial changes primarily caused by intense use of organs, not senseless mutilation (which was generally known to have no effect on progeny long before Weismann, for instance, in the case of human circumcision, although claims to the contrary were common enough in Weismann's day and apparently were the direct incentive for his experiments).

2   Inspired by ideas of progress in biological evolution, the flamboyant Viennese researcher and popularizer of science Paul Kammerer embarked at the beginning of the twentieth century on a two-decade-long quest to demonstrate inheritance of acquired characters (S. Gliboff, "'Protoplasm ... Is Soft Wax in Our Hands': Paul Kammerer and the Art of Biological Transformation," *Endeavour* 29 [2005]: 162–167; E. Pennisi, "History of Science. The Case of the Midwife Toad: Fraud or Epigenetics?" *Science* 325 [2009]: 1,194–1,195; A. O. Vargas, "Did Paul Kammerer Discover Epigenetic Inheritance? A Modern Look at the Controversial Midwife Toad Experiments," *Journal of Experimental Zoology Part B: Molecular and Developmental Evolution* 312 [2009]: 667–678). Kammerer's work included mostly experiments with amphibians that changed their color patterns and breeding habits, depending on environmental factors such as temperature and humidity. Strikingly, Kammerer insisted that the induced changes he observed were fully inheritable. Kammerer's experiments drew criticism due to his sloppy documentation and suspicious, apparently doctored drawings and photographs. Kammerer defended his conclusions energetically, but in 1923, his career came to a bitter end after the famous British geneticist William Bateson found that Kammerer's showcase midwife toad, which supposedly had acquired black mating pads, a trait that was passed to the progeny, actually had been injected with black ink. Kammerer killed himself within two years after this disgraceful revelation. Whether or not Kammerer was a fraud in the worst sense of the word remains unclear; it is thought that he might have used ink to "augment" a color change that he actually observed, a scientific practice that was not approved of even then, let alone now, but that was a far cry from flagrant cheating. Kammerer's findings might have their explanation in hidden variations among his animals that, unbeknownst to him, became subject to selection or, alternatively, in epigenetic inheritance. Under the most charitable of explanations, Kammerer ran a seriously sloppy operation, even if he unknowingly stumbled over important phenomena. Regardless of the specifics, the widely publicized "l'affaire Kammerer" hardly improved the reputation of Lamarckian inheritance. The worst for Lamarck was yet to come, though.

3   In a cruel irony, the Bolshevik leaders of the Soviet Union warmly welcomed
     Kammerer and nearly ended up moving his laboratory to that country (S.
     Gliboff, "'Protoplasm ... Is Soft Wax in Our Hands": Paul Kammerer and the Art
     of Biological Transformation," *Endeavour* 29 [2005]: 162–167). Despite the
     striking successes of Russian genetics in the 1920s (recall the names of Sergei
     Chetverikov and Nikolai Vavilov), the party leaders cherished the ideas of fast,
     planned, no-nonsense improvement of nature, including human nature. So
     when the general situation in the country gravitated toward mass terror and
     hunger around 1930, a suitable team was found, under the leadership of the
     agronomist Trofim Lysenko. Lysenko and his henchmen were not scientists at
     all, but utterly shameless criminals who exploited the abnormal situation in the
     country to amass extraordinary power over Soviet scientific establishment and
     beyond (V. N. Soyfer, "The Consequences of Political Dictatorship for Russian
     Science," *Nature Reviews Genetics* 2 [2001]: 723–729; V. N. Soyfer, *Lysenko and
     the Tragedy of Soviet Science*, New Brunswick, NJ: Rutgers University Press,
     1994). Lamarckian inheritance, which the Lysenkoists, not without a certain per-
     verse cleverness (to the modern reader, with a distinct Orwellian tint), touted as
     a "true Darwinian" mechanism of evolution, was the keystone of their "theory."
     They took Lamarck's idea to grotesque extremes by claiming, for instance, that
     cuckoos repeatedly emerged *de novo* from eggs of small birds as a particularly
     remarkable adaptation. In his later years, after he fell from power, Lysenko
     retained an experimental facility where he reportedly fed cows butter and choco-
     late, in an attempt to produce a breed that would stably give high-fat milk.
     Mostly, the Lysenkoist "science of true Darwinism" was not even fraudulent
     because its adepts often did not bother to fake any "experiments," but simply
     told their ideologically inspired tales. All this could have been comical, if not for
     the fact that many dissenters literally paid with their lives, whereas almost all
     research in biology in the Soviet Union was hampered for decades. There is no
     reason to discuss Lysenko any further here; detailed accounts have been pub-
     lished (Zh. A. Medvedev, *The Rise and Fall of T. D. Lysenko,* New York:
     Columbia University Press, 1969; V. N. Soyfer, "The Consequences of Political
     Dictatorship for Russian Science," *Nature Reviews Genetics* 2 [2001]: 723–729;
     V. N. Soyfer, *Lysenko and the Tragedy of Soviet Science,* New Brunswick, NJ:
     Rutgers University Press, 1994). The proceedings of the infamous 1948 session
     of the Soviet Agricultural Academy, where genetics was officially banished,
     remain a fascinating and harrowing read (*On the Situation in Biological Science,
     A Transcript of the Session of the V. I. Lenin All-Union Academy of Agricultural
     Sciences,* July 31–August 7, 1948, Moscow, USSR: The State Agricultural
     Literature Publishers, 1948).

4   The history of the discovery of the *cas* genes is interesting and instructive in its
     own right, even if tangential to the main subject of this book. In our 2002 study
     of overlapping gene arrays in the genomes of prokaryotes (see Chapter 5), this
     set of genes came across as the second-largest connected neighborhood, after
     the ribosomal superoperon (I. B. Rogozin, K. S. Makarova, J. Murvai, E.
     Czabarka, Y. I. Wolf, R. L. Tatusov, L. A. Szekely, and E. V. Koonin, "Connected
     Gene Neighborhoods in Prokaryotic Genomes," *Nucleic Acids Research* 30
     [2002]: 2,212–2,223). After a careful and painstaking analysis of the Cas protein
     sequences, we predicted that these proteins constituted a novel DNA repair

system (K. S. Makarova, L. Aravind, N. V. Grishin, I. B. Rogozin, and E. V. Koonin, "A DNA Repair System Specific for Thermophilic Archaea and Bacteria Predicted by Genomic Context Analysis," *Nucleic Acids Research* 30 [2002]: 482–496), a prediction that seemed to make a lot of sense, given the diverse roles of nucleases, helicases, and polymerases in repair. Unfortunately, we failed to examine the adjacent repeats. Only after the independent discovery of the phage-specific spacers (A. Bolotin, B. Quinquis, A. Sorokin, and S. D. Ehrlich, "Clustered Regularly Interspaced Short Palindrome Repeats (CRISPRs) Have Spacers of Extrachromosomal Origin," *Microbiology* 151 [2005]: 2,551–2,561; F. J. Mojica, C. Diez-Villasenor, J. Garcia-Martinez, and E. Soria, "Intervening Sequences of Regularly Spaced Prokaryotic Repeats Derive from Foreign Genetic Elements," Journal of Molecular Evolution 60 [2005]: 174–182) did everything come together, and the hypothesis on the mechanism of the antivirus immunity mediated by CRISPR was proposed (K. S. Makarova, N. V. Grishin, S. A. Shabalina, Y. I. Wolf, and E. V. Koonin, "A Putative RNA-Interference-Based Immune System in Prokaryotes: Computational Analysis of the Predicted Enzymatic Machinery, Functional Analogies with Eukaryotic RNAi, and Hypothetical Mechanisms of Action," *Biology Direct* 1 [2006]: 7). Subsequently, in its main aspects, it was validated by experiments (R. Barrangou, C. Fremaux, H. Deveau, M. Richards, P. Boyaval, S. Moineau, D. A. Romero, and P. Horvath, "CRISPR Provides Acquired Resistance Against Viruses in Prokaryotes," *Science* 315 [2007]: 1,709–1,712; F. V. Karginov and G. J. Hannon, "The CRISPR System: Small RNA-Guided Defense in Bacteria and Archaea," *Molecular Cell* 37 [2010]: 7–19). The important (and, in retrospect, obvious) lesson is that it really pays off to take into account as much evidence as you possibly can when interpreting your observations. A remarkable latest twist to the story is that at least one of the Cas proteins, Cas1, found in all CRISPR systems, does seem to contribute not only to the insertion of spacers into CRISPR cassettes, but also to several forms of repair (M. Babu, N. Beloglazova, R. Flick, C. Graham, T. Skarina, B. Nocek, A. Gagarinova, O. Pogoutse, G. Brown, A. Binkowski, S. Phanse, A. Joachimiak, E. V. Koonin, A. Savchenko, A. Emili, J. Greenblatt, A. M. Edwards, and A. F. Yakunin, "A Dual Function of the CRISPR-Cas System in Bacterial Antivirus Immunity and DNA Repair," *Molecular Microbiology* 79 [2011]: 484–502). After all, it seems like the original prediction did not completely miss the mark, even if the principal novelty was overlooked.

5   This is indeed what prions have become famous for, when it has been shown that mysterious "viruses" causing scrapie in sheep, mad cow disease in cattle, and several rare but devastating neurological disorders in humans are "infectious proteins" (M. F. Tuite and T. R. Serio, "The Prion Hypothesis: From Biological Anomaly to Basic Regulatory Mechanism," *Nature Reviews Molecular Cell Biology* 11 [2010]: 823–833). As it usually happens with startling discoveries in biology, it turned out that the mechanisms of prion replication underlying the infectivity involved autocatalytic propagation of protein aggregates (the 1997 Nobel Prize in Physiology or Medicine went to Stanley Prusiner; S. B. Prusiner, "Prions," *Proceedings of the National Academy of Sciences USA* 95 [1998]: 13,363–13,383; G. Vogel, "Prusiner Recognized for Once-Heretical Prion Theory," *Science* 278 [1997]: 214) and, although novel and unanticipated (except, in a sense, by Kurt Vonnegut in *Cat's Cradle*), did not violate the Central Dogma and the EPR principle.

Chapter 10

1   Another personal note is in order here: My own research in evolutionary genomics began with viruses in 1984. At that time, small genomes of viruses were the only complete genomic sequences available, and there were already 30 or so of these, from diverse hosts (animals, plants, bacteria), awaiting comparative analysis. It seems that, in those days, only a few researchers realized the crucial difference between a genome, the full complement of genetic information of a distinct agent, with its own evolutionary history (even one that heavily depends on its hosts like a virus), and fragmentary sequences that were available for genomes of cellular life forms. Back then, no clear distinction existed between functional and evolutionary genomics because most of the sequence of each genome was *terra incognita:* Attempts to predict the functions of viral proteins went hand in hand with efforts to reconstruct evolutionary relationships (not that such synergy does not exist now, but often it is less obvious). To a large extent, both directions were fruitful. This was an incredibly exciting time.

2   Much of this discussion is based on the original Virus World hypothesis article (E. V. Koonin, T. G. Senkevich, and V. V. Dolja, "The Ancient Virus World and Evolution of Cells," *Biology Direct* 1 (2006): 29), where numerous references can be found. In this chapter, I cite mostly publications on aspects of virus evolution that are not covered in the 2006 article, or the key references that have appeared since its publication.

3   In the case of selfish elements, identifying which form of agent-specific nucleic acid should be designated genome is not entirely obvious. For bona fide viruses, the genome traditionally—and sensibly—is defined as the form that is incapsidated (incorporated into the capsid and virion) and so serves as the transmitted genome. For capsidless genetic elements, the definition is more difficult but it still makes sense to suggest that the infectious form, whenever one exists, is the genome.

4   My general approach in this book is to relegate any personal recollections and reminiscences to these endnotes. Let this be the only exception—the subject is too important to me personally and also could be of some general interest.

5   The 2009 Nobel Prize in Physiology or Medicine was awarded to Elizabeth Blackburn, Carol Greider, and Jack Szostak for the discovery of telomerase (G. Vogel and E. Pennisi, "Physiology Nobel. U.S. Researchers Recognized for Work on Telomeres," *Science* 326 (2009): 212–213). It is certainly a remarkable discovery at any rate, but conceivably, some of the incentives for the award were the implications of the extension or shortening of the telomeres caused, respectively, by high or low levels of telomerase expression for cancer and aging. So it is particularly interesting that such an important and medically relevant enzyme comes right from the Virus World.

6   The metaphor of the Theseus ship that Antoine Danchin used extensively in discussions of genome evolution certainly comes to mind: As purportedly asked of the Oracle of Delphi, if every plank in the Hero's ship (for which grateful citizens lovingly tend) is eventually replaced by a new one, perhaps made of a different

material and even of somewhat different shape, but the overall contour remains recognizable, is it still the same ship? There can be no "correct" scientific definition of sameness (see Appendix A), but within the context and logic of our exploration of the Virus World, the sensible answer seems to be yes (even as many *buts* can be added as qualifications).

7   Leigh Van Valen proposed the hypothesis in an explicit form in 1973 (L. Van Valen, "A New Evolutionary Law," *Evol. Tehory* 1: 1–30). Its name comes from a famous scene in Lewis Carroll's *Through the Looking Glass and What Alice Found There* (L. Carroll, *Through the Looking Glass and What Alice Found There*, London: Macmillan, 1872). The Red Queen explains to Alice, "Now, *here*, you see, it takes all the running *you* can do to keep in the same place."

Chapter 11

1   Douglas Theobald published an ambitious article in which he claimed to have provided a formal demonstration of the existence of LUCA that was supposedly independent of the sequence similarity between the universal proteins in archaea, bacteria, and eukaryotes (D. L. Theobald, "A Formal Test of the Theory of Universal Common Ancestry," *Nature* 465 [2010]: 219–222). However, a more careful analysis of his approach indicates that the argument still contained a hidden assumption of sequence conservation (E. V. Koonin and Y. I. Wolf, "The Common Ancestry of Life," *Biology Direct* 5 [2010a]: 64).

2   The archaeal membrane phospholipids are isoprenoid ethers of glycerol 1-phosphate, whereas bacterial phospholipids are fatty acid esthers of glycerol 3-phosphate—that is, the lipids in the two domains differ not only in their chemical composition, but also in chirality (J. P. Pereto, P. Lopez-Garcia, and D. Moreira, "Ancestral Lipid Biosynthesis and Early Membrane Evolution," Trends in Biochemical Science 29 [2004]: 469–477).

3   Analogies with the history of human civilization (including, of course, science) are obvious and perhaps illuminating: The existence of a *lingua franca* greatly accelerates progress. Conversely, isolated communities are stalled in their development and doomed to eventual extinction.

Chapter 12

1   Those who have been forced to study "dialectical materialism", the strange concoction that was supposed to form the philosophical foundation of Marxism in the Soviet Union and other countries of the socialist camp, will never forget the definition Friedrich Engels gave: *Life is the mode of existence of protein bodies.* If one sets aside the disgust from the relentless drilling of this formula into our poor brains, along with other jewels of Marxist wisdom, it does not sound so bad now, even if trivial and largely beside the point.

2   A notable exception are the self-propagating prions that we briefly discussed in Chapter 9. Although the prions technically represent an ultimate form of strong

epigenetic inheritance (even producing infectious agents) based solely on proteins, the synthesis of the prion proteins still fully depends on the regular, nucleic acid–based information-transmission system of the cell.

3   John Walker (the future Nobel Prize winner for the structure of the membrane ATPase) and coworkers first described the P-loop in 1982 (J. E. Walker, M. Saraste, M. J. Runswick, and N. J. Gay, "Distantly Related Sequences in the Alpha- and Beta-Subunits of ATP Synthase, Myosin, Kinases, and other ATP-Requiring Enzymes and a Common Nucleotide Binding Fold," *EMBO Journal* 1 [1982]: 945–951) as a motif that is shared by two proton ATPase subunits and several other ATP-binding proteins that otherwise showed little or no sequence similarity to one another. The then-unsuspected conservation of this motif in a great variety of viral proteins was the subject of my very first paper in computational biology (A. E. Gorbalenya, V. M. Blinov, A. P. Donchenko, and E. V. Koonin, "An NTP-Binding Motif Is the Most Conserved Sequence in a Highly Diverged Monophyletic Group of Proteins Involved in Positive Strand RNA Viral Replication," Molekularnaya Genetika, Microbiologiya i Virusologiya 11 [1985]: 30–36). I suppose that hitting on the most conserved protein motif in that early work was a combination of chance (in this case, good luck) and "preferential attachment."

4   Oparin's "theory" was construed partly in response to the demands of the philosophy of "dialectical materialism" that was indoctrinated in the Soviet Union as part of the overall Marxist worldview and called for straightforward materialistic (in practice, often mechanistic and comically oversimplified) explanations for all natural phenomena. It is perhaps not by accident that Oparin's first short book was issued by a publishing house named Moscow Worker (A. I. Oparin, *The Origin of Life*, Moscow: Moscow Worker, 1924). Alexander Ivanovich Oparin himself was a rather odious character who successfully played to the Communist party tune throughout his long and highly successful career, during which he made it to the top of the Soviet scientific hierarchy. His behavior in the Lysenkoist era, and particularly in the aftermath of the 1948 pogrom of genetics, was as deplorable as it gets. Of course, all this was sheer striving for survival, as Oparin understood the importance of genetics quite well and made it clear as soon as Lysenko was out of power. I had an opportunity to meet Oparin in person in 1971, when I received from his hands a prize for second place in the biochemistry "olympics" for middle-schoolers. He gave the impression of a rather detached (in part, most likely due to a severe hearing impairment that was poorly compensated by the primitive hearing aid available at the time) but kindly old professor.

5   Of course, this is a formulation of the "weak" anthropic principle that is the only scientifically sensible rendering of anthropic reasoning. The so-called "strong" anthropic principle is the teleological notion that our (human) existence is, in some mysterious sense, the "goal" of the evolution of the universe. As such, this idea does not belong in the scientific domain (J. D. Barrow and F. J. Tipler, *The Anthropic Cosmological Principle*, Oxford: Oxford University Press, 1988).

Chapter 13

1   In using the term *rhizome*, I follow here the lead of the eminent microbiologist Didier Raoult (D. Raoult, "The Post-Darwinist Rhizome of Life," *Lancet* 375 [2010]: 104–105) and quote the famous postmodernist text of Deleuze and Guattari, for whom the rhizome was a metaphor of the complexity of the world in general (G. Deleuze and F. Guattari, *Thousand Plateaus: Capitalism and Schizophrenia*, Minneapolis, MN: University of Minnesota Press, 1987). The switch from one biological metaphor (tree) to another (rhizome) seems fitting.

2   We discuss here progress in research that definitely exists unlike progress in evolution.

3   In their latest popular book, *The Grand Design*, Stephen Hawking and Leonard Mlodinov write: "Regarding the laws that govern the universe, what we can say is this: There seems to be no single mathematical model or theory that can describe every aspect of the universe. Instead...there seems to be the network of theories that is called M-theory. Each theory in the M-theory network is good at describing phenomena within a certain range. Wherever their ranges overlap, the various theories in the network agree, so they can all be said to be parts of the same theory. But no single theory within the network can describe every aspect of the universe—all the forces of nature, the particles that feel these forces, and framework of space and time in which it all plays out. Though this situation does not fulfill the traditional physicists' dream of a single unified theory, it is acceptable within the framework of model-dependent realism" (S. W. Hawking and L. Mlodinov, *The Grand Design*, London: Bantam, 2010; p. 58). A sober assessment perhaps, and a far cry from the expression of hope that "we will know the mind of God," which ends Hawking's 1988 classic *A Brief history of Time.* (S. W. Hawking, *A Brief History of Time: From the Big Bang to Black Holes*, London: Bantam, 1988).

Appendix B

1   Adapted from E. V. Koonin, "The Cosmological Model of Eternal Inflation and the Transition from Chance to Biological Evolution in the History of Life," *Biology Direct* 2 (2007): 15.

Acknowledgments

The Acknowledgments section is an essential part of any book, as no scientific idea, hypothesis, or generalization worth its salt is born in an isolated brain. Science lives only in networks of scientists, and an author can say something worth reading only when he or she extensively taps into these networks. On one hand, an Acknowledgements section should be easy to write because it is relatively standard and formal; on the other hand, it becomes a difficult task because so many colleagues and friends have been important to the author's work and thinking in various ways, even if rather few have helped with the book directly. I try to be inclusive without going into too fine differentiation.

This book could not have been completed without the most generous help and support of Yuri Wolf, my collaborator of many years and coauthor of some of the key work that underlies the concepts developed here. Yuri not only read the entire text and made a variety of essential corrections and suggestions, but also created many of the figures. Three other close long-term collaborators, Valerian Dolja, Bill Martin and Tatiana Senkevich (also my wife), read the entire manuscript and made numerous helpful comments. I would like to specifically thank several colleagues and friends who, at different stages of my work in science, have been not only mentors or collaborators, but also sources of intellectual inspiration, and without whose support, wisdom, and criticism I would have had neither the incentive nor the ability to even start this book. These formative influences are Vadim Agol (my mentor in virology), Alex Gorbalenya (my senior partner in the first comparative genomic efforts), L. Aravind, Peer Bork, Jim Carrington, Konstantin Chumakov, Valerian Dolja, Patrick Forterre, Andrei Gudkov, Alexei Kondrashov, David Lipman, Bill Martin, (the late) Alexander Neyfakh, Pavel Pevzner, and Didier Raoult. Discussions with W. Ford Doolittle and Michael Lynch on genome evolution; Allan

Drummond and Claus Wilke on protein evolution; Sergei Maslov, Eric Van Nimwegen, and Michael Lässig on statistical physical approaches to genome evolution; and Alex Vilenkin on cosmology were indispensible to my entire way of thinking.

I am grateful to the former and current members of my research group for their essential contributions to the specific studies that are discussed in this book and many other projects that I did not have a chance to address: Vivek Anantharaman, Liran Carmel, Diana Chernikova, Michael Galperin, Nick Grishin, Laks Iyer, I. King Jordan, Georgy Karev, Fyodor Kondrashov, David Kristensen, Alex Lobkovsky, Kira Makarova, David Managadze, Ben Mans, Raja Mazumder, Sergei Mekhedov, Arcady Mushegian, Anastasiya Nikolskaya, Pavel Novichkov (also for help with Figure 5-2), Artem Novozhilov, Marina Omelchenko, Pere Puigbo (also for the generous help with the figures for Chapter 6), Alissa Resch, Igor Rogozin, Svetlana Shabalina, Alexei Spiridonov, Nikolai Spiridonov, Aleks Sverdlov, Nobuto Takeuchi, Roman Tatusov, Panayiotis Tsaparas, Sona Vasudevan, Roland Walker, Maxim Wolf, Itai Yanai, and Natalya Yutin. The various valuable contributions of many collaborators and visitors to the group are acknowledged with no lesser gratitude: Stephen Altschul, Fred Antson, Oded Beja, Stephen Bell, Stan Brouns, Miklos Csüros (also for help with Figure 7-8), Tal Dagan, Michael Daly, Ed De Long, Vishva Dixit, Michael Gelfand, Dan Hartl, Martijn Huynen, Detlef Leipe, Leonardo Mariño-Ramírez, Boris Mirkin, Andrey Mironov, Bernard Moss, Armen Mulkidjanian, Luca Pellegrini, Chris Ponting, David Prangishvili, Teresa Przytycka, Michael Roytberg, Ken Rudd, Alexei Savchenko, Shamil Sunyaev, Eva Szabarka, Laszlo Székely, John Van Der Oost, Alexander Yakunin, and Wei Yang.

I am deeply grateful to Dante for being the best friend ever.

Parts of the initial draft were written during the inspiring 2010 Winter Conference at the Aspen Center for Physics, and I wish to acknowledge the ACP. The final touches were made while attending the equally stimulating Program on Microbial and Viral Evolution at the Kavli Institute for Theoretical Physics in Santa Barbara in February 2011, so I am delighted to acknowledge KITP.

Last, but certainly not least, I thank my editor, Kirk Jensen, for the invitation to publish this book that came at the right moment and for the gentle prodding, and him and the entire team of editors at Pearson Education for the excellent, expedient editing.

About the author

Eugene V. Koonin is a Senior Investigator at the National Center for Biotechnology Information (National Library of Medicine, National Institutes of Health), as well as the Editor-in-Chief of the journal *Biology Direct*. Dr. Koonin's group performs research in many areas of evolutionary genomics, with a special emphasis on whole-genome approaches to the study of major transitions in life's evolution, such as the origin of eukaryotes, the evolution of eukaryotic gene structure, the origin and evolution of different classes of viruses, and evolutionary systems biology. Dr. Koonin is the author of more than 600 scientific articles and a previous book *Sequence—Evolution—Function: Computational Approaches in Comparative Genomics* (with Michael Galperin [2002] New York: Springer).

Index